The Changing Prairie

THE
CHANGING
PRAIRIE

North American Grasslands

Edited by
ANTHONY JOERN
KATHLEEN H. KEELER
School of Biological Sciences
University of Nebraska–Lincoln
Lincoln, Nebraska

New York   Oxford
OXFORD UNIVERSITY PRESS
1995

Oxford University Press

Oxford   New York
Athens   Auckland   Bangkok   Bombay
Calcutta   Cape Town   Dar es Salaam   Delhi
Florence   Hong Kong   Istanbul   Karachi
Kuala Lumpur   Madras   Madrid   Melbourne
Mexico City   Nairobi   Paris   Singapore
Taipei   Tokyo   Toronto

and associated companies in
Berlin   Ibadan

Copyright © 1995 by Oxford University Press, Inc.

Published by Oxford University Press, Inc.,
200 Madison Avenue, New York, New York 10016

Oxford is a registered trademark of Oxford University Press

All rights reserved. No part of this publication may be reproduced,
stored in a retrieval system, or transmitted, in any form or by any means,
electronic, mechanical, photocopying, recording, or otherwise,
without the prior permission of Oxford University Press.

Library of Congress Cataloging-in-Publication Data
The changing prairie : North American grasslands /
edited by Anthony Joern, Kathleen H. Keeler.
p.   cm.
Includes bibliographical references and index.
ISBN 0–19–507410–6
1. Prairies—North America.
2. Prairie ecology—North America.
3. Prairie conservation—North America.
I. Joern, Anthony.
II. Keeler, Kathleen H.
QK110.C47   1995   574.5'2643'097—dc20   94-6307

9 8 7 6 5 4 3 2 1

Printed in the United States of America
on acid-free paper

The time has come when scientific truth must cease to be the property of the few, when it must become woven into the common life of the world; for we have reached the point when the results of science touch the very problem of existence.

Louis Agassiz

Preface

North American grasslands have figured prominently in our North American heritage. Prairies first provided significant barriers to westward expansion, and then offered both economic and sociological opportunity, as well as heartache, for settlers. Many artists have gained significant inspiration from the beauty as well as the harshness of this region and its biota. And because of ideal climate and soil conditions, these grasslands have provided the agricultural foundation on which much of the economic growth and stability of the United States has historically depended.

Yet many see North American prairies as beautiful only when manipulated or exploited: Green croplands or manicured park lawns are attractive; native grasslands are "those ugly weeds." In the past, plowing virgin prairie could be easily defended on both economic and sociological grounds. And, historically, North American prairies must have seemed threatening in both their wildness and their endlessness.

The preservation of remaining North American prairies is now an urgent need. Many existing prairie types can be considered as threatened as or more threatened than tropical forests. No tallgrass prairie was saved in the sense of maintaining widely ranging species that link patches and regions (bison, elk, wolves); only plants remain as a reasonable legacy of this past system. Midgrass prairie has been almost completely plowed. More of western shortgrass prairie remains, but present human activity is exacting great stress on this ecosystem. The California grasslands, historically dominated by perennial bunchgrasses, were nearly fully invaded by annual grasses from the Mediterranean region; exotics nearly replaced native species in about half a century, leading to a significant impact on grassland dynamics. In sum, North American grasslands are a vanishing resource.

This book has three goals: (1) to provide noneconomic arguments for the value of prairies (economic arguments are available elsewhere); (2) to present a current synthesis of prairie ecology (dynamic processes regulating species distributions and abundances as well as nutrient cycling and landscape processes) to the literate public, including advocates and managers, in order to facilitate the best possible decision making; and (3) to introduce conservation and management issues relevant to prairies, draw attention of managers to the costs and benefits of alternative actions, inform conservationists of lessons already learned from grasslands, and highlight selected unsolved problems that we hope will stimulate future research to answer them.

The last decade and a half has been a renaissance period for research on North American prairies. Driven by as well as contributing to changing paradigms in modern evolutionary and ecosystem ecology, this research is generating a new vision of prairie structure and dynamics. Contrast the emphasis of David Costello's *The Prairie World*

(1969) with that of Jim Reichman's *Konza Prairie: A Tallgrass Natural History* (1987). Both describe prairies to a general audience. Costello approaches the topic in terms of seasonal changes and the types of plants and animals that one might encounter on a leisurely walk, whereas Reichman considers the dynamics of vegetation patches, at several scales, and emphasizes the *interactions* among plants and animals. This shifted emphasis more accurately captures the underlying driving forces that become the prairie that we see and provides more valuable insights for devising appropriate management. We have asked our contributors, active members of the research community generating new and exciting results, to approach their chapters with this in mind. A major goal of this book is to synthesize our present ecological understanding of the grassland ecosystem in sufficient but suitably nontechnical detail so that educated citizens can understand and appreciate key ecological principles that underlie the structure and function of grasslands.

Armed with these basic principles, development of an appropriate conservation ethic toward North American prairies will be possible. Within the context of shared grassland experiences of both the present and the past, this book will provide a basis for developing sound tactics to preserve North American prairies.

Too often, important technical research results are not readily accessible except to those performing research in a given specialized area. Even more important, individual results are often not sufficiently useful except when placed in a larger context, a process that requires familiarity with a whole field. As a result, it becomes increasingly difficult for intellectually competent but otherwise nonexpert individuals to exploit important results. Prairie-preservation activities often suffer from this problem. Those individuals most actively involved in prairie preservation and restoration often need more background to guide their efforts. Unfortunately, it is very difficult to locate adequate syntheses that meet the needs of serious prairie enthusiasts. Yet it is members of this interested and educated lay public who will be the prime movers behind any rational prairie-preservation movement. Our explicit goal is to produce a synthesis compatible with these needs in a way that strips away needless jargon and any excessive detail unnecessary for understanding basic principles. We hope that our readers will gain a full appreciation of the interactive processes controlling grassland function and acquire the ecological understanding necessary to successfully promote prairie-preservation and -restoration programs. We will further provide our readers with appropriate background to develop and justify logical arguments that they must construct for initiating preservation activities and to convincingly communicate and educate legislators, donors, and funding agencies for support of prairie conservation.

We wish to personally thank other authors contributing to this book in our most heartfelt manner; we have learned much from the exchange of ideas. In addition, we thank our many colleagues at the University of Nebraska who have been extremely supportive. The multidisciplinary Center for Great Plains Study at the University of Nebraska, in particular, has shown great interest in this project, for which we are grateful. Glenn Humphress and Cathy Parde have been extremely helpful in pulling the final stages of this project together, and we extend many thanks. Finally, we thank many past colleagues, students, and staff at Cedar Point Biological Station, located in a stunning grasslands ecosystem at the nexus of sandhills and shortgrass prairie in western Nebraska, who have provided a stimulating intellectual and social environment in supporting our teaching and research activities.

Lincoln, Neb.   A.J.
April 1994   K.H.K.

Contents

Contributors, xiii

Introduction, 3

I
Perceptions

1. Getting the Lay of the Land: Introducing North American Native Grasslands, 11

   Anthony Joern and Kathleen H. Keeler

2. Cultural Perception and Great Plains Grasslands, 25

   Paul A. Olson

3. Personal Journal, 42

   Keith Jacobshagen

II
Grassland Ecology

4. The Physical Environment of Great Plains Grasslands, 49

   Thomas B. Bragg

5. Population Processes, 82

   David C. Hartnett and Kathleen H. Keeler

6. The Entangled Bank: Species Interactions in the Structure and Functioning of Grasslands, 100

   Anthony Joern

7. Grassland Ecosystem and Landscape Dynamics, 128

   Scott L. Collins and Susan M. Glenn

8. Soil Systems and Nutrient Cycles of the North American Prairie, 157

   Timothy R. Seastedt

III
Conservation and Restoration

9. Grassland Management: Ecosystem Maintenance and Grazing, 177
   Linda L. Wallace and M. I. Dyer

10. The Challenges of Grassland Conservation, 199
    Jane H. Bock and Carl E. Bock

Scientific Names, 223

Glossary, 227

Subject Index, 235

Species Index, 241

Contributors

CARL E. BOCK
EPO Biology
University of Colorado
Boulder, Colo. 80309

JANE H. BOCK
EPO Biology
University of Colorado
Boulder, Colo. 80309

THOMAS B. BRAGG
Department of Biology
University of Nebraska
Omaha, Nebr. 68182

SCOTT L. COLLINS
Ecology Program
National Science Foundation
4201 Wilson Blvd.
Arlington, Va. 22230

M. I. DYER
Institute of Ecology
University of Georgia
Athens, Ga. 30602

SUSAN M. GLENN
Oklahoma Natural Heritage Inventory
Oklahoma Biological Survey
Norman, Okla. 73019

DAVID C. HARTNETT
Division of Biology
Kansas State University
Manhattan, Kans. 66506

KEITH JACOBSHAGEN
Department of Art
University of Nebraska
Lincoln, Nebr. 68588

ANTHONY JOERN
School of Biological Sciences
University of Nebraska
Lincoln, Nebr. 68588

KATHLEEN H. KEELER
School of Biological Sciences
University of Nebraska
Lincoln, Nebr. 68588

PAUL A. OLSON
Department of English
University of Nebraska
Lincoln, Nebr. 68588

TIMOTHY R. SEASTEDT
EPO Biology and Institute of Arctic &
Alpine Research
University of Colorado
Boulder, Colo. 80309

LINDA L. WALLACE
Department of Botany/Microbiology
University of Oklahoma
Norman, Okla. 73019

The Changing Prairie

Introduction

Preserving remaining North American grasslands requires a multiability approach. In this book, we investigate three aspects of an admittedly larger problem: (1) how we as humans perceive grasslands; (2) the ecology of grasslands, in order to define the framework within which conservation and preservation efforts must operate; and (3) conservation issues. Additional sociological, economic, philosophical, and cultural considerations will provide important additional insights to preserving and managing grasslands, but are not included here. By restricting our focus to only three issues, we feel that we can provide a basic, but appropriate, understanding of grassland ecosystems for the prairie enthusiast. This provides an essential framework required for what we perceive to be necessary quick action.

PART I: PERCEPTIONS

As humans, we are trapped by our own perceptions and experience of the world, and we act within these constraints. Our first concern in this book is to highlight this issue. We are not in a position to exploit such insights fully and then offer prescriptions for identifying the best approach to describe natural landscapes. However, we feel that it is important that both professionals and lay students think about this problem. These issues are not new. Language is now known to be more than just a vehicle for communication. Our choice of words and metaphors for describing things develops an atmosphere surrounding our subject that limits our intellectual opportunities to investigate the object (Lakoff and Johnson 1980; Hayakawa and Hayakawa 1990). Language often shapes our thinking modes as well as constrains the prescriptions that we offer to solve a problem. In the present context, for example, the phrase "balance of nature" provides a very different investigative framework for grassland study than does "struggle for existence" even though both are routinely used and accurate—even complementary in their own way.

Written impressions of grasslands obtained from well-crafted novels by Willa Cather and Mari Sandoz, or the children's writer Laura Ingalls Wilder, each written from a pioneer's perspective, suggest that pioneers and settlers brought order (often through agriculture) to an otherwise unruly natural enterprise. Of course, this is just another view of order. Similarly, others have argued that religious, political, and moral backgrounds shape the metaphors and models that we construct to describe natural systems and hence subsequent research. These cultural and psychological issues clearly direct our actions toward appreciating and conserving native grasslands.

Olson (Chapter 2) takes a historical perspective to examine culturally dependent, shifting perceptions held by European explorers and settlers toward North American prairie, especially in contrast to the European world that they knew best. This point becomes strikingly evident when European-based views are compared with those of an indigenous native culture, the Lakota. It becomes very clear that the native culture that arose within North American grasslands developed a very different perception of the North American grasslands compared with an overlay derived from a very different European background.

In a modern context, Jacobshagen (Chapter 3) an artist, documents his own visceral feelings toward prairie landscapes. Raised in Kansas, but well schooled in European painting traditions, Jacobshagen examines his own understanding of the dominant horizons and wide-open spaces in which he grew up and which now characterize his own paintings. His artistic impressions provide a clear sense of his creative visions dominated by landscape features truly representative of grassland. By his own admission, academic training within the European traditions channeled his early perceptions of grassland systems. However, it is clear that the plains environment has directed his paintings: Light and distance, sky and horizon are dominant themes. We include this chapter to emphasize that different aspects of the environment in the prairies catch the eye than would in forests, for example. This region shapes the reality of its inhabitants.

PART II: GRASSLAND ECOLOGY

At a general level, precipitation, soil characteristics, seasonality, and fire largely drive the emergence of typical grassland characteristics: an abundance of grasses with few or no woody plants. Yet this is an incomplete and unsatisfactory explanation of grassland structure and function. Factors other than just biogeographic considerations contribute to a functioning grassland. Examples arranged in a hierarchical fashion (MacMahon et al. 1978; O'Neill et al. 1986) include (1) the physiological responses of individual plant or animal species to local physical, abiotic conditions; (2) factors influencing population dynamics; (3) interactions among species that dictate population sizes and relative abilities to exist and persist in an area; (4) community-level, multispecies responses to environmental conditions such as changes in resource availability along gradients or frequency and strength of disturbances; and (5) ecosystem-level processes that direct nutrient and energy flow, ultimately contributing to attributes of biomass accumulation.

Traditionally, each of these issues is considered largely independent of the others. Certainly, all contribute importantly and simultaneously to grassland structure and function. This complexity presents a thorny problem to individuals interested in reconstructing the important biological events underlying grassland ecology as well as to those individuals responsible for directing rehabilitation efforts—at the correct level—in order to work within naturally occurring processes. Besides the problems inherent in directing efforts at the correct level in this ecological hierarchy, a similar problem is faced regarding the spatial or temporal scale of the study (O'Neill et al. 1986; Turner 1987; Turner and Gardner 1991). How large an area for what length of time must be considered in order to include all the component parts and processes that contribute to maintaining grassland structure and function? How big is big enough? How long is long enough?

This is the problem of scale (O'Neill et al. 1986; Turner and Gardner 1991). Scale is a problem that is not particularly well understood at this time in ecological studies, especially for grassland systems. Does understanding detailed processes at the local level allow us to scale up and predict responses of entire watersheds or landscapes, and vice versa? Scale, in an ecological sense, requires that we (1) understand how spatial and temporal boundaries contribute to target problems and (2) understand how ecological processes acting at each level in the hierarchy translate across hierarchical boundaries to affect processes taking place at other levels (O'Neill et al. 1986). This is the problem of scaling up or scaling down, problems still in need of detailed understanding.

In addition to considering scale, heterogeneity must be accommodated (Pickett and White 1985; Kolasa and Pickett 1991). This is particularly important, since real-world heterogeneity can be either stabilizing or destabilizing, depending on the situation or question.

Chapter topics represent each of these different approaches, ranging from basic population processes emphasizing individual, species-specific responses within localized areas to ecosystem-level processes that deal with the flux of matter and energy among whole functional groups. An introduction to each of these topics along with some independent thought and synthesis will provide an excellent scaffold for framing individual questions to our readers.

We asked each contributor to discuss a topic related to his or her research area, drawing generalizations and describing the world as each sees it. We cannot cover all topics, so we asked authors to emphasize new approaches and revised syntheses. As a result, detailed description of prairie vegetation and animals is not included. For the natural history of prairie species—which is fascinating reading—we recommend Roe (1951), Weaver (1954, 1960, 1965, 1968), Costello (1969), Risser et al. (1981), and Runkel and Roosa (1989). This book is not intended as an encyclopedia of prairie biology, but a taste of how our understanding of this dynamic ecosystem is changing.

The field is currently receiving rigorous study, although much remains to be done. Some inconsistencies exist between the present chapters and traditional views of prairie. We have made no attempt to reconcile these differences and, indeed, encouraged them. It is not yet clear whether the differences of opinion are attributable to differences in the ecosystems studied (tallgrass versus shortgrass prairie, Oklahoma versus Colorado) or in the eye or evaluation of the researcher. We do not understand the amount and importance of geographic variation in prairies, although we know that species change across the range of a prairie type (e.g., porcupine-grass dominates eastern sandhills prairie, whereas the closely related needle-and-thread grass dominates western sandhills prairie; compass plant is currently rare to absent in eastern Nebraska prairies; catsclaw sweetbriar is absent from prairies on loess soils). Implications surrounding the variety of responses has rarely been studied. Perhaps we will find that all the published views of prairie are valid, but only for particular prairies at particular times. We hope that inconsistencies will stimulate our readers, when looking at prairies, to evaluate the views for themselves.

Answers to specific problems at actual sites will require detailed understanding of the natural history and ecology of local species. However, important questions can be quickly formulated based on the principles raised in the accompanying chapters. Briefly, Bragg (Chapter 4) describes the prairie vegetation, with emphasis on the current understanding of the role of fire. Perceptions of fire have evolved from being an enemy of preservation to being a widely used tool. Currently, prairie ecologists are considering the complexity of fire: Fires can burn rapidly or slowly, patchily or very completely, relatively hot or warm; and the plants can be stimulated or injured, depending on time of year, soil types, and available water. Equally complex effects are seen on the animal community. Bragg's chapter considers this emerging picture.

Populations consist of groups of potentially interbreeding individuals, a process providing the year-to-year, generation-to-generation continuity. Hartnett and Keeler (Chapter 5) discuss characteristic patterns of population biology of prairie plants and animals. Individual plants and animals interact with dozens of other individuals of the same and different species. These interactions compose the webs of relationships that underlie communities. Joern (Chapter 6) describes representative interactions that develop into the complexity and diversity of the prairie community as we know it.

One of the most difficult problems in prairie ecology concerns understanding of stability and change. Collins and Glenn (Chapter 7) examine this issue head-on. They first discuss the history of the region, including the impacts still evident from the ice ages. They then describe results from their studies illustrating how changeable prairie plant communities are as normal features of natural prairies.

Nearly half the prairie plant occurs belowground as roots; it may be discomforting to some when they realize that we walk across the middle of the plants in our leisurely prairie hikes. Many of the animals—prairie dogs, gophers, mice, ants—live underground. But their impact results from linking significant aboveground and belowground processes, resulting in rich prairie soils. Portions of the Great Plains became the highly productive Corn Belt because of the outstanding quality of the soils. In Chapter 8, Seastedt discusses soil structure and function, drawing attention to the importance and dynamic nature of this often overlooked half of the ecosystem.

PART III: CONSERVATION AND RESTORATION

Certain North American grasslands are clearly under siege. The extreme eastern tallgrass prairie as an integrated ecosystem has ceased to exist. Especially along much of what was previously the Prairie Peninsula, only tiny pieces remain with no significant large vertebrates. Efforts to preserve the tiniest remaining remnants are important, but mostly for educational purposes or to preserve genotypes specifically associated with such environments. Midgrass prairie has largely become wheat fields. At the other extreme, large sections of shortgrass prairie remain, often in good shape or within reach of ready rehabilitation. Possibly because of the large remaining expanses, little threat is seen, although this view is premature (Bock and Bock, Chapter 10).

Few efforts presently exist that are aimed at conserving large tracts to preserve an intact, integrated prairie system that includes the long-range dispersing ungulate herbivores. Other systems, such as Nebraska sandhills grassland, retain important large-scale integrity, probably because the sandy soils are not readily tilled and ranchers have largely grazed intelligently. Recent innovations such as center-pivot irrigation have resulted in increasing tilled acreage in many grassland systems, including the sandhills grassland, and will bear close scrutiny for future impacts. Other grasslands illustrate intermediate levels of human impact, but have largely lost key features of an integrated grassland system, mostly because of fragmentation and the missing migrating herbivores.

All chapters bring up important issues dealing with future conservation and restoration needs. In particular, Wallace and Dyer (Chapter 9) and Bock and Bock (Chapter 10) focus on these issues by highlighting the impact of different grazing practices and their consequences for future grassland integrity. We intend this section to be only an opening discussion of the problems of grassland preservation, and many issues are left untouched here. Mostly, an educated and concerned public will provide the energy, interest, and resources to drive serious prairie preservation and rehabilitation. In the process, many new questions will arise, most begging for new insights at both the basic and applied levels. Our goal is to provide a basic foundation in this quest.

THE CHALLENGE AND THE FUTURE

The North American grasslands, especially the tallgrass and midgrass prairies, are among the most completely altered (i.e., destroyed) ecosystems in North America. Furthermore, interest in their preservation is only recently emerging, lagging long behind mountains and seashores.

Grasslands offer a terrific opportunity to develop methods for understanding how an ecosystem can chug along even after it was largely destroyed or to provide a model of ecosystem rehabilitation. The region will continue to be the major site of U.S. food production. In the face of this legitimate human pressure, preserving native species or intact prairie communities, either in preserves or in isolated patches interspersed among cornfields and housing developments, provides a challenge of major proportions. However, such conditions also provide a chance to lead the world in maintaining native species and ecosystems, just as we lead the world in agricultural productivity. All elements of this book are a response to this challenge: how we see the region, how the region functioned before the modern era, and how to integrate sound ecological principles into maintaining regional biodiversity. We must meet this challenge or lose our natural legacy!

LITERATURE CITED

Costello, D. F. 1969. *The Prairie World.* Crowell, New York.

Hayakawa, S. I., and A. R. Hayakawa. 1990. *Language in Thought and Action.* Harcourt Brace Jovanovich, San Diego.

Kolasa, J., and S.T.A. Pickett (eds.). 1991. *Ecological Heterogeneity.* Springer-Verlag, New York.

Lakoff, G., and M. Johnson. 1980. *Metaphors We Live By.* University of Chicago Press, Chicago.

MacMahon, J. A., D. L. Phillips, J. V. Robinson, and D. J. Schimpf. 1978. Levels of biological organization: a organism-centered approach. Bioscience 28:700–704.

O'Neill, R. V., D. L. DeAngelis, J. B. Waide, and T.F.H. Allen. 1986. *A Hierarchical Concept of Ecosystems.* Princeton University Press, Princeton, N.J.

Pickett, S.T.A., and P. S. White (eds.). 1985. *The*

Ecology of Natural Disturbance and Patch Dynamics. Academic Press, Orlando, Fla.

Risser, P. G., E. C. Birney, H. D. Blocker, S. W. May, W. J. Parton, and J. A. Wiens. 1981. *The True Prairie Ecosystem.* Hutchinson Ross, Stroudsburg, Pa.

Roe, F. G. 1951. *The North American Buffalo.* University of Toronto Press, Toronto.

Runkel, S. T., and D. M. Roosa, 1989. *Wildflowers of the Tallgrass Prairie.* University of Iowa Press, Ames.

Turner, M. G. (ed.). 1987. *Landscape Hetergeneity and Disturbance.* Springer-Verlag, New York.

Turner, M. G., and R. H. Gardner (eds.). 1991. *Quantitative Methods in Landscape Ecology: The Analysis and Interpretation of Landscape Heterogeneity.* Springer-Verlag, New York.

Weaver, J. E. 1954. *North American Prairie.* Johnson, Lincoln, Nebr.

Weaver, J. E. 1960. *Grasslands of the Great Plains.* Johnson, Lincoln, Nebr.

Weaver, J. E. 1965. *Native Vegetation of Nebraska.* University of Nebraska Press, Lincoln.

Weaver, J. E. 1968. *Prairie Plants and Their Environment: A Fifty-Year Study in the Midwest.* University of Nebraska Press, Lincoln.

I

PERCEPTIONS

1

Getting the Lay of the Land: Introducing North American Native Grasslands

ANTHONY JOERN
KATHLEEN H. KEELER

The expected catastrophic extinction of species (already under way in many places) will alter the planet's biological diversity so profoundly that, at the known rate of extinction, it will take millions of years to recover. Yet few ecologists study extinction. Indeed, very little ecology deals with any processes that last more than a few years, involve more than a handful of species, and cover an area of more than a few hectares. The temporal, spatial and organizational scales of most ecological studies are such that one can read entire issues of major journals and see no hint of impending catastrophe. The problems that ecologists face are so large; how do we contemplate processes that last longer than our research careers and that involve more species than we can count, over areas far too large for conventional experiments? The problems are also complex; understanding ecological processes at these large scales is far more of an intellectual challenge than is the stupefyingly tedious sequence of the human genome. The problems are also more important. With complete certainty, I predict that human genomes will be around in fifty years to sequence; with somewhat less certainty, I predict that there will be ten billion of them, dying from many causes each of which is orders of magnitude more important than the genetic causes the human genome sequencing will uncover. If we do not understand ecological processes better than at present, these ten billion humans will be destroying our planet more rapidly than we are now.

S. L. Pimm, *The Balance of Nature*

On July 6, 1892, Roscoe Pound and Jared Smith set out from Alliance, Nebraska, with a small pony named Moses, a road cart, and field supplies to initiate the Nebraska Sand Hills Botanical Expedition of 1892 (Hill, in press). During this trip, Pound and Smith systematically covered approximately 450 miles in Cherry County over some 30-odd days while carefully recording the vegetation of the region—a first. By the end of their journey, they had collected 286 plant species. Per Rydberg returned to the Sand Hills the following year to continue the quest begun by Pound and Smith (Rydberg 1885). Along with much larger reconnaissance ventures, including that of Lewis and Clark (Cutright 1989), collecting expeditions such as these throughout the Great Plains region typified early grassland study.

While seemingly parochial in nature, the efforts of Pound, Frederik Clements, and other young colleagues, in association with well-established botanists such as Charles Bessey, redirected the course of plant ecology in the United States (Tobey 1981). With Pound at the helm, the Botanical Survey of Nebraska culminated in far-reaching accomplishments (Tobey 1981), including Pound and Clements's *Phytogeography of Nebraska* (1900). Such pioneering efforts in plant ecology by this group continued, with many critical contributions including Clements's notions of communities (Clements 1916), the development of experimental plant ecology (Clements and Goldsmith 1924), and the lasting contributions of J. E. Weaver, who recorded long-term events critical to our understanding grassland dynamics (Weaver 1968). Grassland study continues to develop and mature. Modern efforts built on the foundations laid by these botanical pioneers now reveal significant new insights on grassland functions while challenging early propositions. Since these early contributions in grassland ecology, many have continued the quest, now offering new critical, significant insights about how grasslands function at a rapidly increasing rate (Coupland 1979; French 1979; Tobey 1981; Risser et al. 1981; Risser 1985, 1988). As a result, many early paradigms have been overturned (Collins and Glenn, Chapter 7; Seastedt, Chapter 8). Our present views represent an evolving synthesis of past and newly uncovered insights. Like the prairie itself, the study of grasslands is dynamic and interactive, with present views likely requiring further modifications in light of future discoveries.

About 100 years has elapsed since these early exploratory events transpired, almost no time at all considering that the generation length of many dominant prairie grasses easily exceeds this period. This short time span, however, has witnessed the virtual extinction of the eastern tallgrass prairie. This potential for rapid human impact was apparent, even in 1892, to the young Pound and Smith. A sense of urgency permeates this prospectus for their inaugural journey of the Survey of Nebraska: "The changes which are taking place in the flora of the state have already been noted. . . . The rapid settlement of the western portions of the state is undoubtedly accelerating these changes, and requires that those regions be examined at once" (Botanical Survey of Nebraska 1892, p. 5).

Compared with the present integrity of most North American grasslands, the setting in 1892 seems, perhaps nostalgically, pristine. Yet the seriousness of the damage caused by human encroachment into the prairie was a strong motivation for considerable botanical exploration in Nebraska even then, and undoubtedly elsewhere as well. As Hill (in press) clearly indicates, "One hundred years ago, Pound and his colleagues understood that, ecological pioneers that they might be, they were nonetheless engaged in a salvage operation to document the original plant geography of the sand hills before it was irretrievably lost." How serious was the human encroachment of 1892, the noticeable impact that required urgency in the actions of young botanists? Compared with now, the threat seems small—but correctly perceived when time remained for a comprehensive solution.

Today, individuals fortunate to drive through remote grasslands in parts of central and western North America may experience scenes such as those encountered by Pound and Smith. However, even on foot, one cannot travel for even the better part of a day without witnessing extensive, often invasive, human impact. Remnants of the prairie ecosystem are frequently all that one can find, particularly along the eastern edge of the prairie–forest border. In a favorite photographic series of an eastern tall-grass prairie remnant in Illinois, the golden arches of a well-known restaurant chain appear in all angles but one—the ultimate prairie remnant. Clearly, such patches are inadequate repositories for protecting both the functional and the aesthetic attributes of an important North American ecosystem.

That North American native grassland is a seriously threatened and poorly understood ecosystem is a major theme of this book. North American grasslands can no longer be taken for granted. Their worth should not be measured primarily in the number of cattle that can be grazed or in the crop-growing potential of the rich prairie soils that underlie much of the region.

Admittedly, some parts of what was only recently an extensive, unfragmented ecosystem have been so thoroughly altered that the best we can hope for is to preserve remnant traces. Native North American prairie, once tilled and converted to farming operations, is irreversibly altered (even extinguished). Equally important, pervasive fragmentation has broken a vast, continuous expanse into mere habitat islands, destroying critical large-scale links in the process. These small remnants no longer periodically host vast numbers of bison and other wide-ranging grazers such as elk and pronghorn.

Under normal situations, natural disturbance leads to local-scale extinction, then followed by reinvasion by seeds or colonizing individuals of species outside the disturbed area. Now, it is often the case that the ''outside'' is cropland or some other greatly disturbed system so that local extinction is no longer followed by natural replenishment. Since unplowed prairie units are now routinely interspersed among other types of land use, especially agricultural, many regenerating forces have been seriously curtailed. Examples include (1) small disturbances and patches in close proximity that harbored fugitive species, (2) extensive movement by large grazers in a landscape that fostered ready travel, which now is very difficult, and (3) fire, which often promoted natural regeneration of grassland but now is controlled to the point of extinction in many areas, severely disrupting many past processes.

Conservation biologists are rapidly recognizing that fragmentation is a major threat to most remaining natural ecosystems (Soule 1987), and North American grasslands are no exception. Collins and Glenn (Chapter 7) examine the importance of natural disturbance and spatially explicit processes as well as the consequences of fragmentation to the successful functioning of native grasslands in greater detail.

Native grasslands in this country are arguably a historical legacy as important as Independence Hall for preserving key elements of America's history. Without exposure to sufficiently large expanses of grassland, how can we expect to appropriately interpret the history, art, and literature arising from this region? Biologically, North American grasslands as an ecosystem are as important in their own right to the functioning of the hemisphere as the neotropical forests that are readily recognized as threatened. Our present hope is that remaining grasslands can be preserved, the worn-out tracts even restored (''rehabilitated,'' to borrow Seastedt's apt term) whenever possible. Considered use of our grassland heritage must include the scientific, educational, and aesthetic values of the land, often subordinating single-minded economic objectives in the process. Hence, we wrote this book.

The interested, educated citizen will have a much greater impact on prairie preservation and rehabilitation than will a small band of academic scholars. Here, we introduce what is known of processes responsible for maintaining present North American grasslands as understood today. Our intent is to reinforce emerging interests in grasslands as well as provide additional background for ongoing conservation and preservation efforts of native prairies. As interest and expertise increase among the citizenry, more detailed prairie-repair manuals will be required for hands-on action in managing and restoring native grassland. Hopefully, this need will be soon. Prairie-minded citizens are our best hope, in the best spirit of Roscoe Pound and his colleagues. Pound, by the way, became a rare interdisciplinary scholar in botany, sociology, and law, ultimately serving as dean of the Harvard law faculty as a theorist of jurisprudence. He retained a lifelong interest in grasslands, providing an admirable role model for citizens developing interest in native grasslands (Wigdor 1974; Hill 1989, in press).

HOW CAN ONE DELINEATE NORTH AMERICAN GRASSLANDS?

What are grasslands? How variable are North American grasslands? Are there obvious boundaries that delineate different grassland types in response to clearly identifiable environmental features? Given the large extent of grasslands on the North American continent, answers to each of these questions help us identify seeming discontinuities, such as (1) the total amount of plant material that grows each year, (2) the types of plant species that predominate (warm versus cool-season, tall versus short grass) and (3) the actual mixes of species (or groups of species) involved. Often, readily identifiable climatic and soil features associated with perceived breaks in vegetation type aid our efforts to classify prairie. A good understanding of most important patterns defining North American grasslands now exists. General descriptions of North American prairie, largely based on results of a large-scale grassland ecosystem project sponsored by the U.S. International Biological Program (IBP) in the 1970s include Risser et al. (1981), Innis (1978), and French (1979). Some additional recent general references on North American grasslands that we recommend for background include Weaver (1954, 1968), Costello (1969), Risser (1985, 1988), Capinera (1986), Reichman (1987), Detling (1988), Huenneke and Mooney (1989), and McNaughton (1991).

Variously called grassland, prairie, and rangeland, the historically grass-dominated regions in North America account for approximately 50% of the land surface area (400 million hectares) (Sims et al. 1978). While land-use practices in the last century greatly altered the landscape, vegetation in this region was historically composed of grasses and forbs with some woody shrubs; trees were not a major feature! Regional climate, often coupled with fire and sometimes soils, directly controls the limits of grasslands and even the transitions from one grassland type to another. Typically, we find grasslands in areas that receive insufficient rainfall to support fully developed forest, but enough rainfall to often support a closed perennial herbaceous layer with little bare soil visible (Coupland 1979). In other cases, sufficient precipitation exists so that woody plants routinely encroach and then take over a grassland tract unless periodic fires kill the invaders. Conversely, arid shortgrass systems seldom exhibit closed canopies, although they may display maximal productivity for the local conditions. Use of the term "prairie" emphasizes a special feature, grasslands maintained by naturally occurring forces representing years of interplay among countervailing pressures. Compare this with the term "grassland," which includes any grass-covered tract, from native prairie to recent plantings with non-native species. How important are these differences? At the very least, comparative study can provide important insights into grassland community functioning.

The unique environmental demands on grassland species resulted in the evolution of adaptations to persist in environments subjected to grazing, burning, frequent large temperature extremes, and the climatic and soil conditions associated with short- and long-term drought (Risser 1985). As a consequence, sufficient background for prairie enthusiasts to develop an adequate preservation plan will include an understanding and appreciation for (1) individual adaptations, many of them physiological, that prairie plants and animals exhibit in response to the prairie environment; (2) the physical environment that defines local and regional climates; (3) the factors that permit populations to persist at a locality over time or to colonize new areas and begin new populations; (4) the importance of interactions among species for determining patterns of coexistence and ultimately the factors that constrain the number of species that can coexist in an area; (5) the constraints on energy flow and nutrient cycling, especially regarding soil characteristics and development and the regional or landscape interactions; and (6) biogeographic features, such as the origin and maintenance of species pools available to fill grassland communities with species, and dispersal responses to newly emerging spatial configurations brought about by system fragmentation into small, discrete units. We believe that to successfully preserve or restore native grasslands as a functioning, enduring ecosystem, one must appreciate the links between the dynamic ecological and evolutionary processes, including chance events, and unique natural histories of the component species.

While it is unlikely that absolutely discrete and ideal grassland types exist, an acceptable division of grassland types in North America is depicted in Figure 1.1. A significant east–west climatic gradient exists such that grasslands along the eastern edge of this region receive much more rain (40–60 inches) than those in the intermountain region or those just east of the Rockies (10–15 inches). Important north–south gradients also affect species composition and seasonal variation in function, largely driven by differences in temperature, season length, and maximum day length. For example, there is a striking latitudinal gradient in the proportion of cool- versus warm-season plants at a site that highlights the importance of such a feature. Biogeographic patterns provide significant insight into key organizational processes in grasslands. Bragg (Chapter 4) describes these patterns and processes in detail.

For our purposes, we distinguish seven grassland types. While local variations of the following grassland categories are important, often representing important examples of shifts in ongoing ecological processes, we are striving for a broad overview at this point. In the Great Plains region, the *tallgrass prairie,* also called the *true prairie,* exists along the eastern, high-rainfall prairie–forest boundary and is dominated by tallgrass species typically 1–2 meters high (especially big bluestem, Indiangrass, and switchgrass). *Shortgrass steppe* (shortgrass prairie) dominates the western bound-

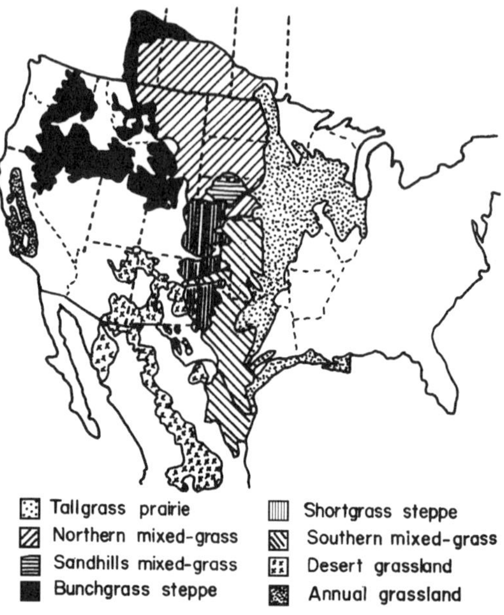

Figure 1.1. Major grassland types in North America. See text for explanation. (Modified from Risser et al. [1981], with permission of Hutchinson Ross)

ary of this region, a grassland type characteristic of low rainfall and dominated by shortgrass (less than .25 meter tall), including blue grama and buffalo grass. We recognize two grassland types located between the tallgrass and shortgrass regions: *mixed-grass prairie* and *sandhills prairie*. Mixed-grass prairie contains species from both tallgrass (confined to the moistest places) and shortgrass steppe, but is generally dominated by midgrass species (at about 1 meter height), such as needle-grasses, little bluestem, and wheatgrasses. Sandhills grassland is located primarily in Nebraska and Colorado (edging into adjacent states) and represents large-scale edaphic anomalies based on widespread sand-dune systems composing the soil substrate. While sandhills prairie often has been overlooked as a separate grassland type, we believe that the organizing principles of these grasslands are sufficiently different from the others, largely resulting from the sandy, nutrient-poor, unstable soil. Species characteristic of mixed-grasslands dominate sandhills prairie, but it also includes dominant sand specialists such as prairie sandreed (Weaver 1965).

Several important North American grasslands are not found in the Great Plains region. *Desert grasslands* are dominated by grama grasses and typically include woody dominants such as creosote bush and mesquite. *Annual grasslands* characteristics of the Central Valley of California consist largely of introduced annual bromegrass and wild oats. Finally, *bunchgrass steppe* dominates the arid intermountain regions of the Northwest, and is characterized by bluebunch wheatgrass and sagebrush.

Water availability, in association with temperature, clearly defines many key features of what we call grassland in its many guises. Precipitation ranges from about 100–200 millimeters to 1000 millimeters of rainfall in North American grasslands.

Grassland regions with greater water availability support increased total plant biomass and species mixes vary (Bragg, Chapter 4). The gradient tends to run east–west, so eastern tallgrass prairie produces considerably more plant material per annum than does the shortgrass prairie in the west. Because of species-specific attributes such as maximal potential height, water-use efficiency, ability to tolerate extreme temperatures, and past biogeographic influences, the actual taxa among grassland types vary considerably. In most grasslands, however, the forbs contribute most to species diversity, including often striking floral arrays, while grasses provide most of the biomass.

In the Great Plains region, about 75% of the rainfall falls from April to September; rainfall from October to March is more important in the mountain, northwestern bunchgrass or California annual grasslands (Sims et al. 1978). In semiarid areas of the shortgrass region, precipitation events of less than 10 millimeters accounted for 41% of the growing-season rainfall and 83% of the rainfall events (Sala and Lauenroth 1982). Water-use experiments performed in the field, for example, documented that blue grama sod responded to 5-millimeter precipitation events following drought conditions, responses that lasted several days. As a baseline, Lauenroth (1979) calculated that 60 millimeters of water is needed annually in a shortgrass prairie to just sustain the plant without growth—the threshold of growth. These figures are minimal estimates of water needed to support growth, levels that typically result in desert if they are sustained. Except for the interplay between such incredible ability to use small amounts of water and the favorable distribution of rainfall throughout the growing season, we would typically observe desert landscapes in the western Great Plains.

HOW RESILIENT ARE NATIVE GRASSLANDS?

Prairie is often distinguished from mere grasslands by a sense that the species associations are well suited for one another and better able to withstand disruption than other combinations of species. Early views provide an extreme comparison to illustrate this point. Grasslands ecologists such as J. E. Weaver, for example, argued that the natural stability and tough perseverance of prairie was central to its understanding. Grassland-dominated communities were so perfectly constructed, he argued, that after thousands of years of evolution, weeds could not penetrate, even when the land was settled and grazed (Tobey 1981). Current beliefs emphasize a more dynamic view, incorporating significantly more flux in the species assemblage in response to routine disturbances (Collins and Glenn, Chapter 7). Disturbance is now seen as part of a native prairie, and so required for appropriate maintenance of features we identify with prairie.

Likelihood of Invasion by Exotic Species

Grassland communities may successfully resist change in species composition in the face of external challenges from invaders. In other cases, often for unknown reasons, these communities may bend or even collapse when confronted with pervasive invasion or disturbance. The likelihood that prairie species will remain unaffected and coherent in response to severe perturbation and return to previous states is a measure of the resilience of the system (Pimm 1991).

Understanding the situations that lead to such invasions is central to present conservation and restoration efforts. How susceptible are native prairies or restored grasslands to disruption by external and/or atypical events? How long will a current grassland community retain its species composition, and what changes can be regarded as normal? If grasslands routinely vary on either short or long time scales, how seriously should we take invasion by non-native, weedy plants into a heavily grazed pasture? Will temperature changes predicted to accompany global climate change seriously impact the remaining native prairies in North America? If natural mechanisms can control invading species (e.g., Hartnett and Keeler, Chapter 5; Joern, Chapter 6; Collins and Glenn, Chapter 7), how well do present range management policies fit into natural processes (Wallace and Dyer, Chapter 9; Bock and Bock, Chapter 10)? A wealth of past evidence indicates that we must take related issues such as these seriously, even though we do not presently understand exactly how each species or ecosystem will respond.

Ongoing, documented invasion of grasslands by exotic plants is ubiquitous throughout North American grasslands (Mack 1986, 1989; Johnson and Mayeux 1992). In general, such invasions are localized and mostly nuisances, seldom altering the character of the ecosystem. This is not always the case, with the exceptions providing worrisome examples.

If you had taken a horseback ride through central California in the mid–nineteenth century, perennial bunchgrasses would have predominated, with flowering heads sometimes brushing against your reins. By the 1880s, "improved farmland" covered up to 75% of the area previously supporting native grassland (Huenneke 1989), and the remaining bunchgrass grassland persisted under siege. Exotic annual grasses from the Mediterranean region, probably introduced in the early part of the century, took over the region such that the character of the whole grassland changed drastically. While the exotic annual grasses were present in California for decades, the actual conversion from perennial bunchgrass prairie to introduced annual grassland seemingly occurred abruptly, over the course of 10 to 15 years in the late nineteenth century (Heady 1977; Jackson 1985). These events occurred approximately 100 years ago. We still do not understand what happened and why, except that the final result was a completely altered grassland (Huenneke and Mooney 1989).

The message is quite clear. In most cases, under low to moderate pressure, grasslands are remarkably resilient. Under concentrated attack, however, the original character of a grassland is lost, perhaps forever. Brown and Heske (1990) describe a revealing experiment at the arid grassland–desert transition zone in eastern Arizona. Experimental disruption of present-day interactions among several species of seed-eating rodents resulted in the decline in abundance of native species and the successful takeover by an exotic grass from South Africa. For the California and intermountain grasslands, a variety of poorly understood activities contributed to the shift in grassland from bunchgrass to annual grassland, possibly including drought, significant fragmentation of the region, increased grazing pressure from cattle, grasshopper outbreaks, and chance events (Mack 1986; Huenneke and Mooney 1989). In this case, the native grassland ecosystem seemingly bent and then broke under the weight of a rapidly changing environment. Is this situation typical, providing the key to understanding grassland resilience and persistence? For western and central Great Plains grasslands, the question remains unanswered.

Natural Variability in the Composition of Native Grassland

Present evidence indicates that most North American biotas, including grasslands, have experienced a history of flux rather than maintained a stable, strongly conserved community composition (Wells 1970a, 1970b; Tomanek and Hulett 1970; Davis 1986; Graham 1986; Huntley and Webb 1988; Johnson and Mayeux 1992; Collins and Glenn, Chapter 7). Data to assess long-term shifts are sketchy, based on relatively few studies using relatively short-term time sequences or spottily distributed and erratic fossil and pollen remains (Wells 1970a, 1970b; McAndrews 1988; Webb 1988). However, clear evidence of radical shifts in plant and animal distributions emerges.

Various lines of evidence document that North American grasslands exhibited significant past variation at several time scales. Over a short term (ca. 100 years or less), clear shifts in grassland composition have been repeatedly documented, often in association with changed precipitation (Weaver and Albertson 1943, 1944; Albertson et al. 1957; Tomanek and Hulett 1970), but also in response to variable fire (Daubenmire 1968; Collins and Wallace 1990; Bragg, Chapter 4) or grazing regimes (Coppock et al. 1983; Detling 1988; Wallace and Dyer, Chapter 9). As an example, Figure 1.2 illustrates largely out-of-phase responses by two dominant shortgrass species to periods of high versus low rainfall in Kansas. Variable climatic conditions can greatly

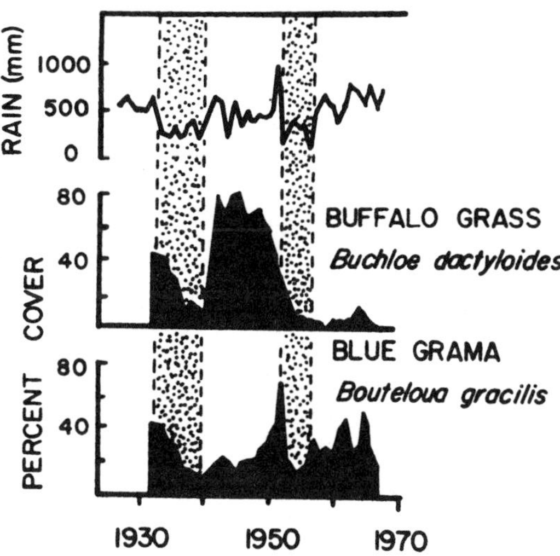

Figure 1.2. Change in biomass of two dominant shortgrass species in relation to precipitation. Percent cover of blue grama and buffalo grass in a shortgrass prairie community near Hays, Kansas, was followed for 40 years. As seen in the rainfall record, two droughts are evident (stippled): 1932–1939 and 1952–1957. The two species were equally abundant in 1932. (Modified from Tomanek and Hulett [1970] and Davis [1986], with permission of the Ecological Society of America)

shift plant species dominance over approximately a 10-year period (Albertson et al. 1957; Albertson and Tomanek 1965), especially when extreme, such as severe droughts, largely a result of different water-use capabilities of different plant species (Kemp and Williams 1980; Sala et al. 1981; Monson et al. 1983). The Great Drought of the 1930s, for example, resulted in significant vegetational changes in North American grasslands (Albertson and Weaver 1942, 1946; Weaver and Albertson 1943). Because of such responses, we obtain an inadequate picture of grassland structure and function without examining the full scope of natural conditions in what might be called ecological time. Ecological time marks the events of an already assembled set of species without accounting for significant evolutionary or biogeographic responses to longer term environmental shifts. Equally important, these responses point out that even small changes to influential driving variables in grasslands can have impact, often reversible. However, as some of the examples indicate, we do not know when small changes become irreversible.

The ecological and floristic history of North American native grasslands has also greatly fluctuated in response to long-term climatic changes, especially those associated with glacial advances and retreats (Bryson et al. 1970; Wells 1970a, 1970b; Huntley and Webb 1988), changes that can be measured over the course of several thousands of years rather than tens or hundreds of years. Reconstructions of ancient vegetation over larger areas can be inferred from the pollen record. Associated with periods of glacial ice advance and retreat, characteristic grassland species separated and then recongealed into species mixes typical of present-day grassland (Figure 1.3) (Gleason 1922; Axelrod 1985; Brown and Gersmehl 1985), but the individual species dispersed in an independent fashion.

In addition to dramatic changes in the vegetation of the present grasslands regions, fossil evidence from many sites indicates that most animals responded in a similar manner, including insects (Ross 1970), fish (Cross 1970), birds (Mengel 1970), and mammals (Hibbard 1970; Hoffman and Jones 1970). Several critical features typify these historical features of prairie structure. Distribution (range) limits of plant and animal species moved around, tracking climatic shifts. Range limits expanded and contracted, often over fairly large distances, before reassembling into our present grassland ecosystems. But different species did not move in a coordinated fashion (Huntley and Webb 1988). The community did not remain intact and shift as a single total-species unit. Instead, individual species appear to have dispersed in independent, species-specific ways. Since glacial advances and retreats occurred repeatedly, among other sources of significant climatic variation, it appears unlikely that the same grassland communities existed after each reshuffling.

The presence of species with a restricted, localized distribution often provides important information regarding the isolation of the biota in either time or space. Such plant or animal species are termed endemic to the region in recognition of their local evolutionary and biogeographic history. The presence of many endemics provides evidence for a long, stable biogeographic history. There are few grassland endemic plant (Wells 1970a) or insect species (Ross 1970) in North American plains grasslands; species found in present-day North American grasslands communities and ecosystems can typically be found in other North American biotas. Such data suggest that North American grasslands are a newly assembled group of species, each with

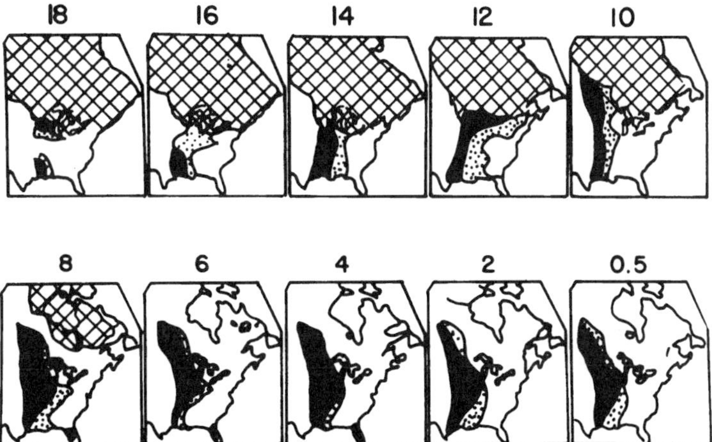

Figure 1.3. Generalized vegetation maps based on pollen analyses from sites throughout the region for prairie forb species to depict changes in the floristic history of the region for the past 18,000 years. Forb species included in the analysis include the sum of *Ambrosia, Artemisia,* other Compositae, Chenopodiaceae, and Amaranthaceae pollen. Stippled regions represent areas with 5% forb pollen in samples, and the black areas represent samples with 10% forb pollen. The cross-hatched areas indicate the location of the Laurentide ice sheet. Numbers above each map panel indicate the number of thousands of years before present. (Modified from Webb [1988], with permission of Kluwer Academic)

antecedents in other geographic regions. It appears that insufficient time and opportunity for speciation events forming species unique to that region existed.

From our perspective, an important message of such results is that North American grasslands, in their present state, are not monolithic species assemblies that have existed for millennia. The evidence does not support a generalization that prairie is an eternal, slowly evolved and necessarily finely tuned entity, an example of Clements's "super-organism" (Clements 1916, 1920). Rather, as entities, many present associations are comparatively young, following periods of significant independent shuffling, Certainly, sufficient time has passed since the last reassembly (probably less than 10,000 years in many cases) to try many natural experiments, time to shake out species recombinations that did not work well in one sense or another. These species assemblies that survived the "shakeouts" are the grassland ecosystems that we see today.

The Question of Planted Prairies

In many areas where prairies were plowed up, people have planted the native species together again. One concept that is particularly appropriate for immediate investigation concerns re-created ecosystems. When does a planted grassland become a prairie? What features are required for it to function so like an unplowed prairie that the differences are negligible? What activities can humans perform to speed up the convergence of planted prairie structure and function to native prairie structure and function?

A traditional line of thinking has been that humans can never restore a plowed ecosystem, and yet we believe that ecological succession will do the job, given 1000 years or so. Is that really the case? Several old planted prairies exist; some sites plowed in the 1930s are very similar to native prairies already; some are still clearly disturbed. Some areas of some prairie states have so thoroughly destroyed their native grasslands that if people want to experience walking in a prairie, one will have to be planted. Thus methods for restoration of the experience of prairies are needed, and then the relation of those prairies to "real" prairies considered. If there is hope of restoring ecosystems, then we can consider expanding native grasslands so they are once again abundant throughout their original range: An urban greenbelt in the central United States might rather be prairie than forest. If planted prairies will not be real prairies for 1000 years, we *must* expand our preservation efforts to protect the virgin remnants that remain.

Acknowledgments
   We thank R. D. Alward for very helpful comments on this chapter.

LITERATURE CITED

Albertson, F. W., and G. W. Tomanek. 1965. Vegetation changes during a 30-year period on grassland communities near Hays, Kansas. Ecology 46:714–720.

Albertson, F. W., and J. E. Weaver. 1942. History of the native vegetation of western Kansas during seven years of continuous drought. Ecological Monographs 12:23–51.

Albertson, F. W., and J. E. Weaver. 1946. Reduction of ungrazed mixed prairie to short grass as a result of drought and dust. Ecological Monographs 16:49–63.

Albertson, F. W., G. W. Tomanek, and A. Riegel. 1957. Ecology of drought cycles and grazing intensity on grasslands of Central Great Plains. Ecological Monographs 27:27–44.

Axelrod, D. I. 1985. Rise of the grassland biome. Botanical Review 51:163–201.

Brown, D. A., and P. J. Gersmehl. 1985. Migration models for grasses in the American midcontinent. Annals Association American Geographers 75:383–394.

Brown, J. H., and E. J. Heske. 1990. Control of a desert–grassland transition by a keystone rodent guild. Science 250:1705–1708.

Bryson, R. A., D. A. Baerreis, and W. M. Wendland. 1970. The character of late-glacial and postglacial climatic changes. Pp. 53–74 in W. Dort, Jr., and J. K. Jones, Jr. (eds.). *Pleistocene and Recent Environments of the Central Great Plains.* University Press of Kansas, Lawrence.

Capinera, J. L. (ed.). 1986. *Integrated Pest Management on Rangeland: A Shortgrass Prairie Perspective.* Westview Press, Boulder, Colo.

Clements, F. E. 1916. *Plant Succession: An Analysis of the Development of Vegetation.* Publication No. 242. Carnegie Institution, Washington, D.C.

Clements, F. E. 1920. *Plant Indicators.* Publication No. 290. Carnegie Institution, Washington, D.C.

Clements, F. E., and G. W. Goldsmith. 1924. *The Phytometer Method in Ecology: The Plant and Community as Instruments.* Carnegie Institution, Washington, D.C.

Clements, F. E., J. E. Weaver, and H. C. Hanson. 1929. *Plant Competition: Analysis of Community Functions.* Carnegie Institution, Washington, D.C.

Collins, S. L., and L. L. Wallace (eds.). 1990. *Fire in North American Tallgrass Prairie.* University of Oklahoma Press, Norman.

Coppock, D. L., J. E. Ellis, J. K. Detling, and M. I. Dyer. 1983. Plant–herbivore interactions in a North American mixed-grass prairie. II. Response of bison to modification of vegetation by prairie dogs. Oecologia 56:10–15.

Costello, D. F. 1969. *The Prairie World.* Crowell, New York.

Coupland, R. T. (ed.). 1979. *Grassland Ecosystems of the World: Analysis of Grasslands and Their Uses.* Cambridge University Press, Cambridge.

Cross, F. B. 1970. Fishes as indicators of Pleistocene and recent environments in the Central Plains. Pp. 241–257 in W. Dort, Jr., and J. K. Jones, Jr. (eds.). *Pleistocene and Recent Environments of the Central Great Plains.* University Press of Kansas, Lawrence.

Cutright, P. R. 1989. *Lewis & Clark: Pioneering Naturalists.* University of Nebraska Press, Lincoln.

Daubenmire, R. 1968. The ecology of fire in grasslands. Advances Ecological Research 5:209–266.

Davis, M. B. 1986. Climatic instability, time lags, and community disequilibrium. Pp. 269–284 in J. Diamond and T. J. Case (eds.). *Community Ecology.* Harper & Row, New York.

Detling, J. K. 1988. Grasslands and savannas: regulation of energy flow and nutrient cycling by herbivores. Pp. 131–148 in L. R. Pomeroy and J. J. Alberts (eds.). *Concepts of Ecosystem Ecology: A Comparative View.* Springer-Verlag, New York.

French, N. R. (ed.). 1979. *Perspectives in Grassland Ecology.* Springer-Verlag, New York.

Gleason, H. A. 1922. Vegetational history of the Middle West. Annals Association American Geographers 12:39–85.

Graham, R. W. 1986. Responses of mammalian communities to environmental changes during the late Quaternary. Pp. 300–313 in J. Diamond and T. J. Case (eds.). *Community Ecology.* Harper & Row, New York.

Heady, H. F. 1977. Valley grassland. Pp. 491–514 in M. G. Barbour and J. Major (eds.). *Terrestrial Vegetation of California.* Wiley Interscience, New York.

Hibbard, C. W. 1970. Pleistocene mammalian local faunas from the Great Plains and central lowland provinces of the United States. Pp. 395–413 in W. Dort, Jr., and J. K. Jones, Jr. (eds.). *Pleistocene and Recent Environments of the Central Great Plains.* University Press of Kansas, Lawrence.

Hill, M. R. 1989. Roscoe Pound and American sociology: a study in archival frame analysis, sociobiography, and sociological jurisprudence. Ph.D. Dissertation, Department of Sociology, University of Nebraska, Lincoln.

Hill, M. R. In press. Roscoe Pound and the Sand Hills Expedition of 1892. In *Exploring the Great Plains.* Center for Great Plains Studies, University of Nebraska, Lincoln.

Hoffman, R. S., and J.K.J. Jones. 1970. Influence of late-glacial and post-glacial events on the distribution of recent mammals on the northern Great Plains. Pp. 355–394 in W. Dort, Jr., and J. K. Jones, Jr. (eds.). *Pleistocene and Recent Environments of the Central Great Plains.* University Press of Kansas, Lawrence.

Huenneke, L. F. 1989. Distribution and regional patterns of California grasslands. Pp. 1–12 in L. F. Huenneke and H. A. Mooney (eds.). *Grassland Structure and Function: California Annual Grasslands.* Kluwer, Dordrecht.

Huenneke, L. F., and H. A. Mooney (eds.). 1989. *Grassland Structure and Function: California Annual Grasslands.* Kluwer, Dordrecht.

Huntley, B., and T. Webb III (eds.). 1988. *Vegetation History.* Kluwer, Dordrecht.

Innis, G. S. (ed.). 1978. *Grassland Simulation Model.* Springer-Verlag, New York.

Jackson, L. E. 1985. Ecological origins of California's Mediterranean grasses. Journal Biogeography 12:349–361.

Johnson, H. B., and H. S. Mayeux. 1992. Viewpoint: a view on species additions and deletions and the balance of nature. Journal Range Management 45:322–333.

Kemp, P. R., and G. J. Williams III. 1980. A physiological basis for niche separation between *Agropyron smithii* (C_3) and *Bouteloua gracilis* (C_4). Ecology 61:846–858.

Lauenroth, W. K. 1979. Grassland primary production: North American grasslands in perspective. Pp. 3–24 in N. R. French (ed.). *Perspectives in Grassland Ecology.* Springer-Verlag, New York.

Mack, R. N. 1986. Alien plant invasion into the Intermountain West: a case history. Pp. 191–213 in H. A. Mooney and J. A. Drake (eds.). *Ecology of Biological Invasions of North America and Hawaii.* Springer-Verlag, New York.

Mack, R. N. 1989. Temperate grasslands vulnerable to plant invasions: characteristics and consequences. Pp. 115–180 in J. A. Drake, H. A. Mooney, F. di Castri, R. H. Groves, F. J. Kruger, M. Rejmanek, and M. Williamson (eds.). *Biological Invasions: A Global Perspective.* Wiley, New York.

McAndrews, J. H. 1988. Human disturbance of North American forests and grasslands: the fossil pollen record. Pp. 673–697 in B. Huntley and T. Webb III (eds.). *Vegetation History.* Kluwer, Dordrecht.

McNaughton, S. J. 1991. Dryland herbaceous perennials. Pp. 307–328 in H. A. Mooney, W. E. Winner, and E. J. Pell (eds.). *Response of Plants to Multiple Stresses.* Academic Press, San Diego, Calif.

Mengel, R. M. 1970. The North American central plains as an isolating agent in bird speciation. Pp. 279–335 in W. Dort, Jr., and J. K. Jones, Jr. (eds.). *Pleistocene and Recent Environments of the Central Great Plains.* University Press of Kansas, Lawrence.

Monson, R. K., R. O. Littlejohn, and G. J. Williams III. 1983. Photosynthetic adaptation to temperature in four species from Colorado shortgrass steppe: a physiological model for coexistence. Oecologia 58:43–51.

Pimm, S. L. 1991. *The Balance of Nature: Ecological Issues in the Conservation of Species and Communities.* University of Chicago Press, Chicago.

Pound, R., and F. E. Clements. 1900. *The Phytogeography of Nebraska, I.: General Survey.* Botanical Seminar, Botanical Survey of Nebraska. University of Nebraska, Lincoln.

Reichman, O. J. 1987. *Konza Prairie: A Tallgrass Natural History.* University Press of Kansas, Lawrence.

Risser, P. G. 1985. Grasslands. Pp. 232–256 in B. F. Chabot and H. A. Mooney (eds.). *Physiological Ecology of North American Plant Communities.* Chapman and Hall, New York.

Risser, P. G. 1988. Abiotic controls on primary productivity and nutrient cycles in North American grasslands. Pp. 115–129 in L. R. Pomeroy and

J. J. Alberts (eds.). *Concepts of Ecosystem Ecology: A Comparative View.* Springer-Verlag, New York.

Risser, P. G., E. C. Birney, H. D. Blocker, S. W. May, W. J. Parton, and J. A. Wiens. 1981. *The True Prairie Ecosystem.* Hutchinson Ross, Stroudsburg, Pa.

Ross, H. H. 1970. The ecological history of the Great Plains: evidence from grassland insects. Pp. 225–240 in W. Dort, Jr., and J. K. Jones, Jr. (eds.). *Pleistocene and Recent Environments of the Central Great Plains.* University Press of Kansas, Lawrence.

Rydberg, P. A. 1885. *Flora of the Sand Hills of Nebraska.* Government Printing Office, Washington, D.C.

Sala, O. E., and W. K. Lauenroth. 1982. Small rainfall events: an ecological role in semiarid regions. Oecologia 53:301–304.

Sala, O. E., W. K. Lauenroth, W. J. Parton, and M. J. Trlica. 1981. Water status of soil and vegetation in a shortgrass steppe. Oecologia 48:327–331.

Sims, P. L., J. S. Singh, and W. K. Lauenroth. 1978. The structure and function of ten western North American grasslands. I. Abiotic and vegetational characteristics. Journal Ecology 66:251–285.

Soule, M. (ed.). 1987. *Viable Populations for Conservation.* Cambridge University Press, Cambridge.

Tobey, R. C. 1981. *Saving the Prairies: The Life Cycle of the Founding School of American Plant Ecology, 1895–1955.* University of California Press, Berkeley.

Tomanek, G. W., and G. K. Hulett. 1970. Effects of historical droughts on grassland vegetation in the central Great Plains. Pp. 203–210 in W. Dort, Jr., and J. K. Jones, Jr. (eds.). *Pleistocene and Recent Environments of the Central Great Plains.* University Press of Kansas, Lawrence.

Weaver, J. E. 1954. *North American Prairie.* Johnson, Lincoln, Nebr.

Weaver, J. E. 1965. *Native Vegetation of Nebraska.* University of Nebraska Press, Lincoln.

Weaver, J. E. 1968. *Prairie Plants and Their Environment: A Fifty-Year Study in the Midwest.* University of Nebraska Press, Lincoln.

Weaver, J. E., and F. W. Albertson. 1939. Major changes in grassland as a result of continued drought. Botanical Gazette 100:576–591.

Weaver, J. E., and F. W. Albertson. 1943. Resurvey of grasses, forbs and underground plant parts at the end of the Great Drought. Ecological Monographs 113:63–117.

Weaver, J. E., and F. W. Albertson. 1944. Nature and degree of recovery of grassland from the great drought of 1933 to 1940. Ecological Monographs 14:393–479.

Webb, T., III. 1988. Eastern North America. Pp. 385–414 in B. Huntley and T. Webb III (eds.). *Vegetation History.* Kluwer, Dordrecht.

Wells, P. V. 1970a. Historical factors controlling vegetation patterns and floristic distributions in the Central Plains region of North America. Pp. 211–221 in W. Dort, Jr., and J. K. Jones, Jr. (eds.). *Pleistocene and Recent Environments of the Central Great Plains.* University Press of Kansas, Lawrence.

Wells, P. V. 1970b. Vegetation history of the Great Plains: a post-glacial record of coniferous woodland in southeastern Wyoming. Pp. 185–202 in W. Dort, Jr., and J. K. Jones, Jr. (eds.). *Pleistocene and Recent Environments of the Central Great Plains.* University Press of Kansas, Lawrence.

Wigdor, D. 1974. *Roscoe Pound: Philosopher of Law.* Greenwood Press, Westport, Conn.

2

Cultural Perception and Great Plains Grasslands

PAUL A. OLSON

> From the mid-nineteenth century through the end of World War II, mechanical metaphors dominated people's ideas of beauty, progress, architecture and home furnishings, even to the point of machines becoming the objects of art and the subjects of paintings, as illustrated by the painting "Suspended Power" by Charles Sheeler. . . . Ecology developed primarily in the twentieth century and, like the Friden calculator, was a child of the machine age, which began with the Industrial Revolution but flowered between the two world wars. Mechanical metaphors and machine models dominated the science of ecology until the early 1970s, when radical changes began to occur whose effects are only now beginning to be felt and understood.
>
> D. Botkin, *Discordant Harmonies*

Successive groups of witnesses have observed Plains grasslands, and each has seen them through its own lens and within its own frame. This essay concerns the lenses or frames used by the Plains Indians, who perceived the Plains as a circle; the Coronado group, who saw the Plains in "georgic" (or traditional agricultural) terms; the explorers and explorer-painters of the early nineteenth century who represented the grasslands as if they were a "picturesque" painting; and the late-nineteenth-century settlers who turned the grasslands into epic oceans. These represent, in my view, the major stages in perceiving the grasslands, although other scholars might delineate the stages differently. In every case save for the Lakota, the mode of seeing the Plains develops somewhere else, and the successive imported stages of seeing imply increasing destruction of natural grasslands as European settlement and plow agriculture more and more dominate the vision of what the Plains can be.

The subject of cultural perception is, in a sense, redundant; we always see in terms of the frame we bring (Dansereau 1975). In the view of Benjamin Whorf and his school (Rollins 1980), "sight" tends to be based on the uses that the local culture and language assign to phenomena, on the patterns of perception encouraged by group religion, art, poetry, and popular culture (Mathiot 1979). The uses determining perception may be aesthetic or nutritional. They may be medicinal or ceremonial. Whatever the uses, when spectators from outside an area look at its "new" landscape, they see its potential in terms cultivated in the locale from which they came (Dansereau 1975). Although "science" is also a cultivated mode of seeing, the purpose of this chapter will not be to look at recent scientific examinations of the Plains. It will be to look at differences among historical prescientific and early scientific perceptions.

THE NATIVE AMERICAN STAGE

The French anthropologist Claude Lévi-Strauss (1962) has argued that in many cases the human mind endeavors to make a structural analysis of the world that surrounds it. In such a structural analysis, often embodied in myth, objects that have no direct day-to-day material use for the group have a use in filling out a logical pattern. Thus utilitarian objects are placed in structures beside their relatives as seen in the taxonomies of wild science.

It is not unusual for tribal groups to classify and assign significance or ceremonial uses to thousands of plants, although they use many fewer directly. It may seem surprising that Plains tribes in their myths give attention to only a few species of grasses. Although human beings have occupied the Great Plains area for 12,000 or more years, they have left no records of grasslands perception comparable to their animal petroglyphs marking the rock outcroppings of the area, and their records in historical time are not replete with references to varieties of grasses. Their interest covers grasslands, not grasses.

It is difficult to know how the Plains tribes originally came to their knowledge of plants. I have encountered the following standard explanations from Native American informants:

1. The plant showed a medicine man its uses in a vision.
2. The plant spoke to a human being describing its uses at a time when the people were in great need.[1]

Although the origins of plant knowledge may have been shamanistic, Plains tribes' naming systems clearly reveal a taxonomic sense in the allied vocabulary used for similar plants; for example, in Lakota grass is *peji,* bluestem is *pejisasa okihetan kinkinyan,* bluegrass is *pejiblakaka,* grama is *pejiokijata.* However, for all of the Plains tribes' interest in vision revelation by plants, grasses do not play a major role in their plant accounts. Few grasses, as opposed to forbs, seem to have been the subject of vision statements or journey revelations. The best account of Plains Indians' use of plants appears in Gilmore's *Uses of Plants by the Indians of the Missouri River Region* (1977). Gilmore makes clear that the Missouri Basin tribes used few species of grasses directly. The Plains people used sweet-grass for ceremonial purposes, a slough grass in Omaha–Ponca earth lodges, and porcupine grass in Omaha–Ponca combs. They harvested wild rice from Great Plains lakes, and the Omaha and Ponca tribes gathered bluestem for a woven material placed between earth-lodge poles to support the sod covering. They seem to have used bluestem for a few medicinal purposes also. The most important "grass" for all the Plains agricultural tribes was, of course, maize; but it was—by the time it reached the Plains—thoroughly domesticated, carefully bred by Plains peoples into subspecies having differentiated food uses, and adapted to the moister river-bottom areas. One would not call such a usage of Plains grasses exhaustive.

Yet clearly the Plains tribes had a good sense of how to manage grass rangelands, a better sense than many early European-based specialists in the fostering of range grasses. Most of the early European-based visitors to the Plains note the Plains tribes' practice of burning off areas of dead grassland to remove the cover so that the spring grasses would sprout earlier and attract bison to graze nearer the village sites. Although Plains peoples prior to contact do not seem to have distinguished many direct uses for subspecies of Plains grasses, they had a strong sense of managing the land in such a way as to enhance the harvest of game, and this sense, I believe, supports and is supported by their profound circle aesthetic, their powerful sense of the interrelationship of plants, animals, and people on the Plains (Olson 1990). The plant system, creeping and winged things, four-leggeds, and two-leggeds are all part of an interdependent whole that runs in cycles and that human beings violate at their own peril.

Plains tribal perceptions of the grasslands divide into two types: one developed by the sedentary or semisedentary agricultural tribes, such as the Pawnee, Omaha, and Wichita; and the other developed by the nomadic hunter-gatherer groups, such as the Sioux, Comanche, and Cheyenne.

Both the Plains agricultural and the Plains nomadic tribes tend to divide the sacred powers that construct the grasslands landscape into the powers of earth and heaven, female and male. Gill's (1990) argument that the Mother Earth idea was a white invention ignores the numerous female sacred powers in Plains religious accounts that clearly do not have a white origin. The powers are not called Mother Earth and admittedly have complex functions, but they are female earth powers. For some of the agricultural tribes, the dominant figure related to the earth represents a domesticated grass. For instance, the sacred figure that leads the Pawnee tribe in many of the rituals, particularly in its quest for another group to which to be "married" in the Hako ceremony, is Mother Corn (Murie 1981). Although the ceremony describes her as leading the people through rivers, over mesas and hogbacks, and to mountains, it does not describe any passage through specific native grasses, surely a prominent feature of any traversing of such varied environments (Fletcher 1904). Again, the Omaha tribe, also an agricultural group, assigns the southern moiety in the village to earth clans that are also female-stressed in their sacred system. It associates the origin of maize with the bison, also an earth force, but no sacred figure accounts for wild grasses or grasslands (Olson 1979). The agricultural tribes possibly did not represent a perception of grasslands aside from corn in their sacred stories because they were less dependent on protein from the grassland-supported bison, depending as they did on the special help of domesticated corn (or maize). Although their dependence on the bison was by no means negligible, their diet included a quantity of protein from beans, squash, and maize.

For the pastoral nomadic tribes, the mythos of the grassland was usually of a generalized sort that encouraged a holistic sense of the grasslands landscape and a concern for the management of relationships among the whole. For example, in the Lakota mythos the earth is called Maka, and Maka has, as her sacred color, green

(Jahner 1983). Through her are sustained the insects, reptiles, birds, and animals—the four-legged creatures. There is no effort to divide Maka's grasses or to make her production in any sense species-specific.

Maka, in the Lakota sacred narratives gathered by J. R. Walker (1980, 1983), derives from the rock, Inyan, that originates all things.[2] He takes the power that he has and gives it to a disk, naming the disk Maka, or earth. Maka thus becomes a sacred power permanently attached to the now powerless rock. The rock appears to symbolize the inert soils and rocky substrata of the grasslands, and the related Maka nourishes stems and seeds or grains (Jahner 1983). As a great green disk, she must represent the Lakota perception of the great green round disk of Plains grassland, stretching to the circle of the horizon. Since the center of the circle is where the individual or the tribe is, the circle is a perceptual circle.[3]

The apparently related account in the Walker materials of the discovery of the four directions tells of the construction of a Plains-like area. At the center is a household dominated by Wohpe, a close relative or alter ego of Maka (Jahner 1983). The great circle of terrain stretching away from her central tipi includes a sort of mountain region in the west where Wakinyan-Heyoka—the sacred thunder power—lives, a northern frost-bound region where Yata claims his direction, the lake and sand region to the east where Yanpa puts his mark, and the forested region to the south where Okaga fixes his direction. Direction keeping is obviously crucial to an open grasslands hunter-gatherer or pastoral nomadic people. The great circle does not speak of grasslands at all save by implication. That is, the center and western part of the circle are clearly open. In contrast, the environments to the east and south are marked by distinctive features that make travel difficult for the four brothers who must mark the original directions. Most of the remaining landscape near the center of the circle appears to be open, a place where things far away can be seen. At the center of the circle where Wohpe, the companion of Okaga, the sacred fertile direction of the south, rules her household, the only grass mentioned is the sacred sweet-grass that Okaga, Wohpe's favorite, carries.

The absence in Lakota sacred story of analytic separation of species of grasses may strike one as strange in a culture that completely depended on a supply of grasses adequate to support the bison on which the people survived. However, the Lakota creation system works less to separate and analyze than to look at wholes. The creation story is the story of a circle, Inyan, broken to form a series of related circles of which Maka is the first and Wakinyan, or the thunder, is the next. The grasses as well as the insects, reptiles, and animals are the result of a collaboration of these circle powers. In the case of grasses, Maka, the earth power, collaborates with Wakinyan, the lightning-and-thunder power (Jahner 1983). Since the fundamental processes are collaborative and balancing ones, it is natural that the representation of grasslands landscape should be holistic rather than analytic. The general direction of Lakota environmental practice was to "manipulate" a few plants ritualistically and to walk lightly on most of the earth. Lakota accounts of how their plant use came about ignore any analytic or narrowly utilitarian training of perception, since dream-vision did the job. This is not to say that medicine men and older women may not, in fact, have given their apprentices training in gathering and using the species that appeared to have utility.

The Lakota holy man Black Elk, whose sense of the Plains was probably formed in the 1860s, supplements the perspective of the Walker materials. At least once he attributes personhood to the grass when he remarks, concerning the period of his group's breaking of camp in the spring of 1880, that "this was the time when the grasses were showing their tender faces" (DeMaillie 1984). The personhood of the grass, like that of everything else—four-leggeds, two-leggeds, human beings, and all other living things—is seen as part of a universal life. Plants, four-leggeds, and so forth are organized into something like tribes, as Baird Callicott has demonstrated (Olson 1990). Yet the personhood and individuality of the species and tribe is subordinated to the circle that governs all things:

> The power (i.e. sacred power) won't work in anything but circles. . . . Even the birds and their nests are round. You can't take the birds' eggs and put them in a square nest. . . . The mother bird just won't stay there. We Indians are relatives to the birds. Everything tries to be round—the world is round. (DeMaillie 1984)

In a revision on the Inyan–Maka relationship, Black Elk describes the earth as grandmother and the growing plants as mother (DeMaillie 1984).

Outside of Native American myth, one does not find such a close sense of relatedness of human beings to grasslands.

THE CORONADO–SPANISH STAGE

The Lakota analysts of the Plains are in their home; the European spectators are away from theirs when they come to the Plains. They domesticate the landscape by seeing it as potential plowland, potential Europe. For religious reasons having to do with the Christian sense of the absolute transcendence of God, European-based people have never represented grass as having personhood. One does not find in the mythos of early European explorers of the Plains the emphasis on holism of the Lakota. Yet the earliest European representation of the Plains has some commonalities with Lakota representation. The first European-based exploration of the Plains grasslands comes in the early 1540s with Coronado and his band. The Spanish had obviously seen semiarid regions similar to the Plains in Spain, although most of these areas had, by Coronado's time, been turned to wheat growing or were heavily grazed by sheep. Furthermore, the Spanish had also encountered some arid and semiarid regions in traveling through the pueblo region of New Mexico prior to their entry into the Plains.

Coronado's assistant, Castañeda, gives the fullest rendering of what the group saw or, rather, what it half saw and half made up. It created its own mythos of the Plains out of its georgic sense, its understanding of Renaissance Spanish agriculture. One of the first accounts comes in Castañeda's narrative of the trip of Coronado from the base of the Sangre de Cristo Mountains near modern Pecos pueblo to central Kansas. Some of the account, save for its mention of the crossbow shot and its slight interest in analytic dichotomies (short/tall grass, fresh/salt lakes), could have been created by a Lakota person:

> The opposite cordillera could not be seen, nor a hill or mountain as much as three estados (levels) high, although we traveled 250 leagues over them. Occasionally there were found some ponds, round like plates, a stone's throw wide, or larger. Some contained fresh

> water, others salt. In these ponds some tall grass grows. Away from them it is all very short, a span long and less. The land is the shape of a ball, for wherever a man stands he is surrounded by the sky at the distance of a crossbow shoot. There are no trees except along the rivers which there are in some barrancas (ravines). These rivers are so concealed that one does not see them until he is at their edge. They are of dead earth, with approaches made by the cattle in order to reach the water which flows quite deep. (Hammond and Rey 1940)

Castañeda's "bowl" is not greatly different from the Lakota disk or circle; the physical features observed are generalized like those in the Pawnee Hako ceremony. However, the noting of long and short grasses, of trees along the rivers, and of lakes "round like plates" and a stone's throw across sounds more like the rhetoric of detached observation that goes with explorer literature than anything as at home with the environment as Black Elk. With Coronado, we are in a world where folk taxonomies begin to separate things by difference, as is the case in Virgil's *Georgics* or European agricultural manuals of the Renaissance.

As the expedition approaches Kansas, Castañeda observes that "this country is very similar to that of Spain in the varieties of vegetation and fruits" and mentions, among the wild plants, grapes, nuts, mulberries, ryegrass, pennyroyal, wild marjoram, and flax. Flax does not do the Indians "any good because they do not know how to use it" (Hammond and Rey 1940, p. 263). Castañeda, of course, sees the flax as a source for linen, a use that grassland Plains tribes, with their abundance of bison hides, could eschew (actually the Plains tribes did use flaxseed as a flavoring in cooking [Gilmore 1977]). What Castañeda means by ryegrass is not clear. The other plants that Castañeda identifies as useful are largely flora from the woodland corridors of creeks and rivers, plants very similar to those that had been domesticated in Spanish agriculture. It is the agriculturists' perception that leads Castañeda to refer to the bison as cattle in the first quote. In short, Castañeda's perception of plant and animal life is governed by his experience of Renaissance domestic agriculture in Spain and by the search for immediate agricultural utility in the wild areas he saw.

Castañeda's first account of the Plains does not seem wide-eyed, as it assimilates much of what he sees to Spanish agriculture. However, in a later section, he admits that the Plains grasslands are astonishing and that he earlier "suppressed and canceled things" that he had seen but that he dared now "to set them down [because] many of the men are still living who saw them all and . . . will vouch for my story" (Hammond and Rey 1940, pp. 278–280).

Castañeda's first interest is in the tracklessness and durability of the grasslands. He notices that 1500 persons, 1000 horses, 500 cattle, and 5000 sheep passed over the grassland when Coronado's group traveled and "left no more trace when they go through than if no one had passed over" (Hammond and Rey 1940, p. 279). He notes the wiriness of the High Plains short grasses: "although the grass was short . . . trampled it stood up again as clear and straight as before" (p. 279). Castañeda had not experienced traveling through grasslands as tough as buffalo or blue grama grass, for his was the Spanish world of wheat and oats and domesticated Spanish grasses, of tilled fields and roads. Consequently, he is astonished that the tramping of the expedition makes no trail and that there is a necessity to mark it (for the rear guard) with "bones and dung" (p. 279).

The grassland bison also trouble him, for they seem a chimera composed of the domestic desert and grassland animals known to Spain and North Africa—sheep's wool in the back, he-goat beard, lion's mane, camel's hump, bull-like general appearance (Hammond and Rey 1940, p. 279). The herds are so huge that "there was no one who could count them" (p. 280).[4]

Any explanation of why the Spanish do not settle the Plains after the Coronado expedition must look to a colonization ideology more interested in golden cities than in agriculture, for Coronado's narratives provide no "great American desert" myth. The Spanish know that the land is like the farmland they have known. Their georgic perspective is understandable. The Spanish had seen real desert farther south and west in Mexico and had encountered semiarid grasslands in their own country. They knew that the Plains were potential agricultural land; perhaps they knew this because they saw the Plains soil, grasses, and forbs directly and did not, like many later explorer groups, follow wooded creek and river corridors. I would argue that the later open-range rancher owes something to Coronado and the later Spanish–Mexican ranchers to the south of the Plains who developed the love of open ranges interspersed with small agricultural gardens.

THE NORTHERN EURO-AMERICAN STAGE

By the late eighteenth and early nineteenth centuries, Europeans had learned to be "tourists" in the modern sense, travelers questing for the scenic and "picturesque" in Italy, in the Alps, along the Rhine, or in the Lake District (Andrews 1989). They took their sense of the "picturesque" primarily from an Italian and French painter, Salvator Rosa, and a French one, Claude Lorrain. Lesser influences came from the Dutch seventeenth-century landscape painters and, later, from the Düsseldorf and Barbizon schools. As Hussey and Manwaring show, a picturesque painting displays two modes: (1) the quiet mode, in which a placid landscape is viewed from a dark prospect—the near landscape beneath being composed of meandering water, broad fields, ruins or castles, and golden light and the far landscape of a bluish purple vagueness (Figure 2.1); (2) the sublime mode, in which rugged, rocky landscape surrounded by a wild sky centers on exotic human beings caught in some turmoil (Manwaring 1925; Hussey 1927; Andrews 1989). Claude Lorrain typifies the first kind of landscape, Salvador Rosa the second, and it is Lorrain's sort of flatland that dominates the perception of the Plains. In addition, interest in the portrayal of common life led, in the late eighteenth and early nineteenth centuries, to a revival of interest in Low Countries landscape or the landscape-with-peasants-and-cows paintings in the Barbizon school (Hussey 1927; Bermingham 1975). Even in Lorrain and Rosa one often finds peasants and their animals posed in the foreground (Figure 2.1).

The first systematic American expedition into the Plains grasslands area, that led by Lewis and Clark, expresses a perception of the grasslands that looks at the Plains taxonomically, one pointedly directed toward agricultural and industrial colonization, but it also points to the myth of the picturesque. In his diary entry for July 4, 1804, Clark, for example, wonders at the beauty of a Plains grassland scenery, in what is now modern Kansas, unattended by "civilized" European-based peoples:

> The Plains of this countrey are covered with a Leek Green Grass, well calculated for the
> sweetest and most norushing hay—interspersed with * (copses) of trees, Spreading their

Figure 2.1. Claude Lorrain, *Pastoral Landscape*. (Reproduced with permission of the Detroit Institute of Art. Accession number 41.10)

> lofty branches over Pools Springs or Brooks of fine water. Groops of Shrubs covered with the most delicious froot is to be seen in every direction, and nature appears to have exerted herself to butify the Senery by the variety of flours Delicately and highly flavered raised above the Grass, which Strikes and profumes the Sensation, and amuses the mind throws it not Conjectering the cause of So magnificent a Senerey * (several words illegible, crossed out) in a Country thus Situated far removed from the Sivilised world to be enjoyed by nothing but the Buffalo Elk Deer and Bear in which it abounds & * (page torn) Savage Indians. (Moulton 1983, 2:346–347; spelling Clark's)

For Clark something is wrong with a providence that provides the resources of leek green grasses and lofty branches over pools and leaves to "savages". Similarly, Lewis writes from Fort Mandan in 1805 that there exists "no other objection to it [i.e., the country between the Platte and Fort Mandan] except that of the want of timber, which is truly a very serious one," a grassland deficiency that he attributes to the fires set in the grasses by "the natives" (Jackson 1978, 1:223–224). He has no sense of the management uses of fire. Both he and Clark perceive the Plains as an unattended garden, awaiting the appearance of European-based people to cultivate the land, grow timber for building, and impose a European-based regimen.

Yet Lewis and Clark's landscape is not fully the domain of taxonomic science. Although Baconian inductive science had captured the Royal Society in the seventeenth and eighteenth centuries, Lewis and Clark are not full-fledged scientists but half scientists and half aesthetes–travelers. Clark writes on August 1, 1804, concerning the Council Bluffs area, "What a field for a Botents *(botanist) and a natirless (naturalist)," clearly defining himself as neither of these. Lewis had served as Jefferson's secretary and must have acquired, from his mentor and from the general climate of opinion, an interest in dividing and classifying, but neither he nor Clark do very detailed analyses. When Clark is traveling in the Missouri River timbered corridor, he distinguishes strata, proceeding away from the river lowland wet soil. The strata begin with lowland cottonwood and willows; then come "high bottom" cottonwood, walnut, ash, and hackberry; then "high land" covered with oak, ash, and walnut; and finally the grassy prairies. Near present Rulo, Nebraska, Clark sees "rich weeds" (or richweed) (Moulton 1983, 2:369–370). Near present Brownsville, he sees timothy with lamb's quarters, and cocklebur and richweed (Moulton 1983, 2:377). East of modern Williston, North Dakota, Lewis sees, in a landscape dominated by a blue grass, other forbs and bushes resembling "sage, hysop, wormwood, southernwood," and two species of artemisia that he cannot identify (Moulton 1987, 4:35). Near Musselshell, Montana, he observes the landscape, aside from the piney cedared hills, covered with "a short grass, arromatic herbs, and the prickley pear" (Moulton 1987, 4:170). The scientific and commercial interests of Lewis and Clark dictate an account of the relationships among grasses (often of unknown species), forbs, tree corridors, and soil and moisture characteristics, but the observations are neither very detailed nor very scientific. They are the observations of colonizers.

Colonization, however, also requires the aesthetic sense. Lewis and Clark write before many Americans had encountered the typical Romantic perception of a nature filled with the "presence of something far more deeply interfused," as Wordsworth puts it. The landscape has not yet become the domain of pantheistic Romantic poets such as Wordsworth and painters following his path. The early Romantics saved their belief in some sort of deity, after their encounter with the eighteenth-century skeptics, by positing that the human mind reaches the divine through nature, through sublime landscapes interpreted by sunsets, mists, storms, and snow. Although Lewis and Clark are not Romantics, they do, I think, show the influence of the more pedestrian cult of the picturesque that excited travel and tourism in the eighteenth century.

The popularity of Italian landscape art, particularly of Claude Lorrain and Salvator Rosa, led to the cult of the picturesque in art, with its emphasis on seeing landscape from prospects, in golden light, land with wild rocky outcroppings or distant watercourses curving sinuously around. So fashionable was seeing the landscape under the aspect of Rosa and Lorrain that English landscape poetry in the second half of the eighteenth century seems to be read from their canvases (Hussey 1927; cf., Andrews 1989). Travelers seeking the picturesque often carried about with them golden glass mirrors framed with a wooden frame. They took these to high prospects so that they could look down on Nature laid out as in a Lorrain landscape (Andrews 1989).

Clark's perception of landscape seems to have been particularly affected by the myth of the picturesque; Lewis's less so. On July 19, 1804, Clark comes out of the timbered corridor and looks out on "boundless Prarie" except for the timbered Platte and the creeks near it. He observes, "This prospect was So Sudden and entertaining that I forgot the object of my prosute? and turned my attention to the Variety" of the infinite grasslands. The term "prospect" is a "picturesque" technical term for the view below (Hussey 1927, pp. 1–17; Andrews 1989, pp. 17–23, 55).[5] The distant landscape—with river or water—represents a perfect Lorrain view (Figure 2.1). Near Council Bluffs, Clark sees "the most buttiful prospects imagionable"—a plain, a river with bottomlands and interspersed timber, the river meandering as in an Italianate Lorrain.

It may be, however, that this aesthetic response and the utilitarian responses to grasslands cited earlier are part of a single stance that assumes that European-based people have the Plains spread out before their feet literally and metaphorically.[6]

The picturesque perception of the Plains grasslands appears in the paintings of two later explorer-painters whose modes represent distinct phases in the picturesque movement—what may be called the landscape-as-garden and the landscape-as-God phases. Clark belongs to the landscape-as-garden phase. The landscape-as-garden mode of examining grassland landscape appears also 25 years later in the work of Karl Bodmer, one of the first of the significant explorer-painters, a man torn between precise observation of grasslands and picturesque representation. Bodmer grew up as a student of the master of the picturesque, Claude Lorrain, and also studied the seventeenth-century Dutch and Italian landscape painters (Goetzmann 1984). Thus he had been well indoctrinated in the clichés of picturesque perception—the search for prospects, meandering rivers, rock outcroppings, and, in the case of the Netherlands painters, detailed pictures of animals and peasant life. The master of the expedition with which Bodmer traveled, Prince Maximilian of Wied, though, had studied with the "German Linnaeus," Johann Friedrich Blumenbach, Alexander von Humboldt's teacher. He was thoroughly at home in the precise identification of plants and animals.

Bodmer's superb paintings of the Great Plains display little interest in grasslands as species-specific areas, however. His Plains animals—elk, antelope, bison—are as precise in their rendering as Dürer rabbits. His trees are not the conventionalized trees of a Sir Joshua Reynolds, nor are they drawn from the European windless landscape of a Lorrain or from the domestic quietude of the Dutch painters. They represent, in some cases, identifiable Plains species and always Plains tree forms. But the Bodmerian grasslands landscape from Missouri to western Montana is always of about the same color, a light blue green with some tan flourishes recalling the "brown sauce" school of European painters. One can identify no species varieties in the grasses. Forbs do not exist save for one sketch of a cactus (Goetzmann 1984, Plate 194). Bodmer's focus often goes to the Plains equivalents of the European picturesque. The painting takes on a castle-on-the-Rhine or Gothic church appearance, but includes rocky outcroppings on the Missouri or Native Americans beautifully painted but separated from their landscapes and work as if they were the peacocks in the painter's later Barbizon paintings (Figure 2.2). Bodmer does catch the great grassy distance of the Plains, something of the western Plains blue green color produced by blue grama grass and buffalo grass. He shows no interest in speciation, forbs, or bison grazing patterns. Although Bodmer's paintings have a wholly documentary intent and do accurately document the life of fauna and tribe, the life of the grasses remains a blur, as if the grasses that made up the Plains were too vast collectively and too inconsequential individually to document with the eye of an Audubon.

The leader of Bodmer's expedition, Prince Maximilian of Wied, *did* collect plants and grasses from the Great Plains. However, his grasslands interest came to as little as Bodmer's, but for other reasons. The two boats in which he sent the greater portion of the Plains plants he collected came to grief, and the samples did not reach Germany. A few plants sent separately did arrive at their destination, and to understand those the prince worked with professional botanists to identify their species relationships.[7]

Figure 2.2. Karl Bodmer, *White Castles on the Missouri*. (Reproduced with permission of the Joslyn Art Museum, Omaha, Nebr. Accession number NA32)

Figure 2.3. Albert Bierstadt, *Platte River, Nebraska.* (Reproduced with permission of Masco Corporation, Taylor, Minn.)

Bodmer's lack of interest in the specificity of grassland species is understandable, but he also set the stage for the later depreciation, in painting and writing, of specific understanding of grasslands ecology.

The penchant for moving the Plains into the realm of the picturesque is completed, as far as painting is concerned, in the work of Albert Bierstadt in the 1850s to 1870s. Bierstadt gives us the landscape-as-God. In the 1850s, Bierstadt began as a student of the Düsseldorf school of painting, a school that emphasized design over color, and he continued throughout his life to emphasize magnificent vistas at the expense of precise color or plant observation. He traveled to America and made three trips west in the 1850s to 1870s, making sketches for his monumental pictures of the Plains and mountains that set the style of modern calendar art. When he painted the Platte in 1863, he showed it full of rocky outcroppings and surrounded by hogbacks and miniature mountains that come straight from Lorrain. In the foreground, pioneers instead of shepherds move with their animals (Figure 2.3). Bierstadt's Plains pictures from farther west often set the Rocky Mountains in the distance and show Indians or covered-wagon settlers in the plains beneath the foothills, dwarfed by enormous mountains in the distance. Clearly Rosa is somewhere in the background. However, Bierstadt has been captured by the Romantic sense of divinity in landscape. His Plains reach into an infinite distance, and above them plays a light suffused by the presence of divinity.[8] His 1871 painting of covered wagons moving into a ''sublime'' sunset surely apotheosizes the journey west as both a religious and a physical journey (Figure

Figure 2.4. Albert Bierstedt, *The Overland Trail*. (Reproduced with permission of the Anschutz Collection, Denver, Colo.)

2.4). The divinity is, however, distant, unknowable, not the near-at-hand Maka of the Lakota.

THE SETTLEMENT MYTH EPIC STAGE

The fourth phase in the development of Plains perception comes in the settlement period. It turns the grasslands alternately into a garden and a seascape, especially the seascape of the epic. These two formulations alternate in settler-era representations of Plains grasslands much as the two aspects of Wittgenstein's duck/rabbit advance and recede as we focus on one or the other (Wittgenstein, 1976). The seascape grassland generally reflects a sense of estrangement, a desire to assimilate the new experience of vastness and emptiness to the seascape known in Europe or European poetry. Northern European or Eastern American lettered peoples take refuge in an old mode of seeing when they encounter the Plains—the Scandinavian and classical seagoing epics. And working-class people from Europe sometimes recall pictures and stories of the hardships of seamanship. Indeed, the name ''prairie schooner'' reflects the common perception of grasslands as seas. In contrast, the natural garden or georgic representation obviously looks at the grasslands as properly converted to domesticated agriculture.

One may illustrate the appearance of both perceptions in the journal of Mollie Dorsey Sanford, a relatively unlearned woman who was 18 years old in 1857 when her family came from Indiana to eastern Nebraska. Mollie's first glimpse of the Plains tallgrass prairie, what she calls the high prairie, comes on May 4, 1857:

> (Father) took us all out for a ramble out to the high prairie where we had a fine view of the country. The landscape was beautiful. The prairies look like a vast ocean stretching out until earth and sky seem to meet. We found many varieties of our cultivated flowers, growing in wild profusion. Father says we will love our new home. (Sanford 1980, p. 21)

The perceptual movement of the passage illustrates precisely the two faces of pioneer perception. First the landscape is a vast ocean, presumably because of the rolling hilliness of the area and the vast distances of the horizon, and perhaps because of the waving of big bluestem. Simultaneously, the flowery forbs, analogous to domesticated flowers, suggest the analogies between this landscape and the known landscapes of domestic agriculture and horticulture. Sanford later says that fate has led her ''into a sea of perplexities and cares'' (p. 81). Later in 1860 when she travels west of Fort Kearny and approaches Denver on the High Plains, she sees the landscape near the Platte as composed of trees ''like apple trees,'' soft grass, a sparkling river, and mountains in the distance, and she calls the landscape the ''promised land.'' The georgic or garden myth implies a promised land running with milk and honey because of the area's apparent suitability for domestic agriculture.

Such popular perception of grasslands reappear in subtler form in the novel or epic–novel of settlement that is the central genre of Great Plains literature (Olson 1981). For example, in Owen Wister's *The Virginian* (1902), when the narrator first ''strolls out on the plains,'' he sees the Plains' sagebrush, antelope, and prairie dogs and mixes these with memories of his experience—all of which make him feel that he is ''swimming slowly at random in an ocean that was smooth'' (1964, pp. 37–38). Later the narrator and the Virginian repeatedly see wild Wyoming scenes as gardenlike in their beauty and arrangement.

Again, Willa Cather, who came to relatively unsettled short and mid-length grasslands near Red Cloud, Nebraska, in pioneer times, creates in *My Ántonia* (1918) a child narrator, Jim Burden, an alter ego to her pioneer self. Jim comes to the Plains grasslands at night. He sees the half-visible trackless lands at night, without ''fences . . . creeks . . . trees . . . hills . . . or fields'' as ''slightly undulating'' (1954, p. 7). Then when he first goes out into the native grasslands (probably little bluestem pasture), he feels ''that the grass is the country, as the water is the sea'' (p. 15). Cather then introduces a Homeric touch in her hero's perception of the sealike grasslands, the ''red of the grass,'' ''the color of wine stain'' (Cather 1954, p. 15; Olson 1981, pp. 276–284). We have returned to seafaring in Homer's wine-red or wine-dark sea—the first ''pioneers'' in the Western world, the Achaians, hovering above Cather's pioneers.

The garden is not far away, however. In the same passage, the narrator walks with her grandmother to see the fall pumpkin and potato garden (Cather 1954, p. 16). Shortly thereafter, Cather projects the garden image on untilled grasslands when the narrator kills a large snake in a prairie-dog town and indicates how the snake seemed like ''the ancient, eldest Evil'' (p. 47). Perhaps because the untamed garden of the prairie appears as a fallen Eden, the novel moves inexorably toward the replacement of native grasslands by agriculture. Eden returns only when domestic agriculture comes, when Ántonia's orchard appears as a "walled" garden surrounded by hedges—like the walled Eden of Gustav Doré's engravings (Doré 1937, p. 15). In that garden, apple trees planted and watered by Ántonia substitute for grasslands, and everything is good (Cather 1954). In Cather's georgic epic, the untilled garden is a fallen Eden, and the act of garden tillage, planting, and watering to replace the grasslands becomes supremely redemptive. Ántonia as redeemer and epic heroine has only to stand in the orchard, to put her hand on the little crab-apple tree and look up at the fruit "to make you feel the goodness of planting and tending and harvesting at last" (Cather 1954, p. 353).

The settler mythos allows for numerous permutations of the sea–garden myth. On the sea side, Ole Rölvaag's hero, Per Hansa, comes into the South Dakota grasslands driving a schooner, steering by the stars, and dragging a rope behind the schooner. He is any Norwegian sailor trying to make certain that he sails straight toward his destination in the grasslands seas (Rölvaag 1955; Hahn 1979). In contrast to Per Hansa, who sees the inland seas as a challenge, Per Hansa's wife, Beret, sees the grasslands as demonic—troll-possessed, without place to hide, beyond the reach of church and God. Trolls often came from watery places in the country, and Beret also sees her monster-filled sea–plain as heaving with evil. But her husband, his friends, and their pastor see the grasslands as "the pastures of Goshen" (Rölvaag 1955, p. 28). They are the "[1] and of Canaan" (p. 2), a potential garden, a fairy place containing a royal palace with "a snow-white picket fence around a big, big garden" (p. 109). Rölvaag participates in the common myth wherein epic people conquer the dangerous, trackless seas of the grasslands and replace them with fenced, innocent, domesticated places. We are obviously far from Maka.

The four stages in the myth of the grasslands (the Lakota circle, the Spanish georgic form, the explorers' picturesque landscape, and the dangerous seascape turned garden) represent increasingly sharp rejection of wild grasslands in favor of domesticated agriculture. The movement goes away from comfort with undomesticated grasslands and toward their transformation. The Lakota circle does not assume significant transformation of the grasslands—only the notion that human beings must find their way around the directions. Maka and the grass-consuming bison are human beings' protecting powers. All things have personhood and power derived from the original rock. The Spanish myth infers that bison may become cattle and tallgrass prairie, flaxland. By the time of Lewis and Clark and the explorer-painters, the Plains grasslands have become a wild place seen from a distance requiring the domestication of their meandering rivers, hogbacks emulating castles, and wild mountains on the fringes. Finally, the sea–garden myth celebrates the destruction of grasslands—their conversion to something domestic and "useful."

Each stage after the first mythologizes the destruction of the wild climax grasslands ecology of the Plains. Ultimately, of course, we do not know that the domestication or the myths that celebrate its coming were useful in any permanent sense. We know of the destruction of perhaps 60 million bison and of the relatively stable climax grasslands in the area. We know of soil erosion and water pollution and pesticide–herbicide rain. Although it is clear that the replacement system can produce more protein consumable by human beings over the short haul, it is not clear that the new system can long endure. In its construction, myth has played a major role. Clearly the retention of the Plains grasslands will require the construction of new stories celebrating what we hope to save.

NOTES

1. These observations are based on interviews that I did with Otoe, Winnebago, Santee, Lakota, and Omaha elders, which are now in the possession of the University of Nebraska Center for Great Plains Studies.

2. I recognize that some scholars have questioned Walker's accuracy. However, most of the material that follows was given by George Sword, a respected holy man.

3. Black Elk makes this point; it is not clear to me that the Walker material narrators shared it.

4. For a similar account, see Jaramillo's narrative (Hammond and Rey 1940, p. 305).

5. Later some picturesque aestheticians recommend a low viewpoint instead of a high (Andrews 1989, p. 61).

6. I omit from this discussion the "Great American Desert" image liberally discussed elsewhere; it is obviously generally without much basis in grasslands reality (Lewis 1965, pp. 1–11; Allen 1975, pp. 3–11).

7. I have this information from Paul Schach, who is preparing a translation of Bodmer's writings.

8. Bierstadt's use of the sun's dazzling effects to stand for the obscure distances in which divinity hides was understood in his time. An 1864 reviewer of Bierstadt's *Rocky Mountains* said:

> We have studied the picture under all circumstances of external light, in dark and in bright weather, in bad and in good positions. It needs advantages as little as any picture we ever saw. Its sunlight seems self-supporting. . . . Human skill and patience, audacity and tireless enterprise of knowledge, have pioneered their way, all these thousands of dangerous slow miles into the very vestibule of virgin Nature. But the Holiest of Holies locks its door against them. The inmost, topmost spirit of things closes the gates of sight behind it and retires into the silent bosom of the Heavens. Skill cannot fly nor patience climb to the opalescent cradle of that glacier-stream. The King of the World is denied his subject's grandest confidence, and stands dwarfed in the valley, mutely, hopelessly gazing whither Nature is closeted alone with God! (Hendricks, n.d., p. 147)

LITERATURE CITED

Allen, J. L. 1975. Exploration and the creation of geographical images of the Great Plains: comments on the role of subjectivity. Pp. 3–11 in B. W. Blouet and M. Lawson (eds.). *Images of the Plains*. University of Nebraska Press, Lincoln.

Andrews, M. 1989. *The Search for the Picturesque: Landscape Aesthetics and Tourism in Britain, 1760–1800*. Aldershot: Scholar.

Bermingham, P. 1975. *American Art in the Barbizon Mood*. Smithsonian Institution Press, Washington, D.C.

Cather, W. 1954. *My Ántonia*. Houghton Mifflin, Boston.

Dansereau, P. M. 1975. *Inscape and Landscape: The Human Perception of Environment*. Columbia University Press, New York.

DeMaillie, R. J. (ed.). 1984. *The Sixth Grandfather: Black Elk's Teachings Given to John G. Neihardt*. University of Nebraska Press, Lincoln.

Doré, G. 1937. *The Bible in Pictures*. Wise, New York.

Fletcher, A. C. 1904. The hako: a Pawnee ceremony. Annual Report of the Bureau of American Ethnology 22:313–368.

Gill, S. D. 1990. Mother Earth: an American myth. In J. A. Clifton (ed.). *The Invented Indian.* Transaction, New Brunswick, N.J.

Gilmore, M. R. 1977. *Uses of Plants by Indians of the Missouri River Region.* University of Nebraska Press, Lincoln.

Goetzmann, W. H. 1984. *Karl Bodmer's America.* University of Nebraska Press, Lincoln.

Hahn, S. 1979. Visions and reality in *Giants in the Earth.* South Dakota Review 17:85–101.

Hammond, G. P., and A. Rey. 1940. *Narratives of the Coronado Expedition, 1540–1542.* University of New Mexico Press, Albuquerque.

Henricks, G. n.d. *Albert Bierstadt.* Abrams, New York.

Hussey, C. 1927. *The Picturesque.* Putnam, London.

Jackson, D. D. (ed.). 1978. *Letters of the Lewis and Clark Expedition, with Related Documents, 1783–1854.* 2 vols. University of Illinois Press, Urbana.

Jahner, E. (ed.). 1983. *Lakota Myth.* University of Nebraska Press, Lincoln.

Lévi-Strauss, C. 1962. *La pensée sauvage.* Plon, Paris.

Lewis, G. M. 1965. Early American exploration and the Cis–Rocky Mountain Desert, 1803–1823. Great Plains Journal 5:1–11.

Manwaring, E. W. 1925. *Italian Landscape in Eighteenth Century England.* Oxford University Press, New York.

Mathiot, M. 1979. *Ethnolinguistics: Boas, Sapir and Whorf Revisited.* Mouton, The Hague.

Moulton, G. E. (ed.). 1983–1990. *The Journals of the Lewis and Clark Expedition.* 6 vols. University of Nebraska Press, Lincoln.

Murie, J. R. 1981. *Ceremonies of the Pawnee.* Smithsonian Institution Press, Washington, D.C.

Olson, P. A. 1979. *The Book of the Omaha: Literature of the Omaha People.* Nebraska Curriculum Development Center, Lincoln.

Olson, P. A. 1981. The epic and Great Plains literature. Prairie Schooner 55:263–286.

Olson, P. A. 1990. *The Struggle for the Land: Indigenous Insight and Industrial Empire in the Semi-Arid World.* University of Nebraska Press, Lincoln.

Rollins, P. C. 1980. *Benjamin Lee Whorf: Lost Generation Theories of Mind, Language, and Religion.* Popular Culture Association, Ann Arbor, Mich.

Rölvaag, O. E. 1955. *Giants in the Earth: A Saga of the Prairie.* Trans. L. Colcord. Harper & Row, New York.

Sanford, M. D. 1980. *Mollie: The Journal of Mollie Dorsev Sanford.* University of Nebraska Press, Lincoln.

Walker, J. R. 1980. *Lakota Belief and Ritual.* University of Nebraska Press, Lincoln.

Walker, J. R. 1983. *Lakota Myth.* University of Nebraska Press, Lincoln.

Wister, O. 1964. *The Virginian: A Horseman of the Plains.* Airmont, New York.

Wittgenstein, L. 1976 *Philosophical Investigations.* Trans. G. E. Anscombe. Blackwell, Oxford.

3

Personal Journal

KEITH JACOBSHAGEN

In one sense, it is curious that we are so limited in our perception of space and time. We inhabit the midpoint of the physical universe as we know it today—the smallest known objects, atoms, are billions of times smaller than we are, and the largest object, the universe, is that many times larger than we are. It is intriguing to wonder if this could be an artifact of our own intellect. If we were as small as atoms, would we still see approximately equal distances up and down the size scale?. . . . Why do we have such difficulty in perceiving and understanding the portions of our spatial and temporal world just beyond the close-by and the near term?

O. J. Reichman, *Konza Prairie*

I began keeping a journal in the late winter of 1965. I had just graduated from the Kansas City Art Institute and was facing a future of vast possibilities. The question of what I was to do with my life loomed large.

The idea of keeping journals stemmed from that question and from a mixture of dread and fascination for the journey ahead. There was a strong sense that my life up to a point was filled mostly with instructional information and was primarily without insight.

At first I explored and documented my everyday life. I filled pages with hours of observations of prosaic small matters. (Figure 3.1). But this period of writing was a kind of warm-up to the questions I would ask myself later as I moved into the struggles of a young artist trying to solve the visual problems of observing a vast planar space.

To a certain extent the question of aesthetics is secondary to the activity of art making. In 1972 the abstract artist Barnett Newman was quoted as saying, ''Aesthetics is for me as ornithology must be for the birds.'' Yet the problems confronted by an artist living on the Plains are not all formal ones of composition and color. There is a constant gnawing of the mystery of perception and the interpretation of the beautiful in one's life. The ensuing modest chapter is a compilation of observations made in my journal over the past 10 years. The ideas are not completely my own—I do not live in a vacuum. But there are always complications, in my mind, when the written word tries to somehow fill in the spaces where visual language would be the most powerful means of expression. Nevertheless, the questions are honest ones asked by

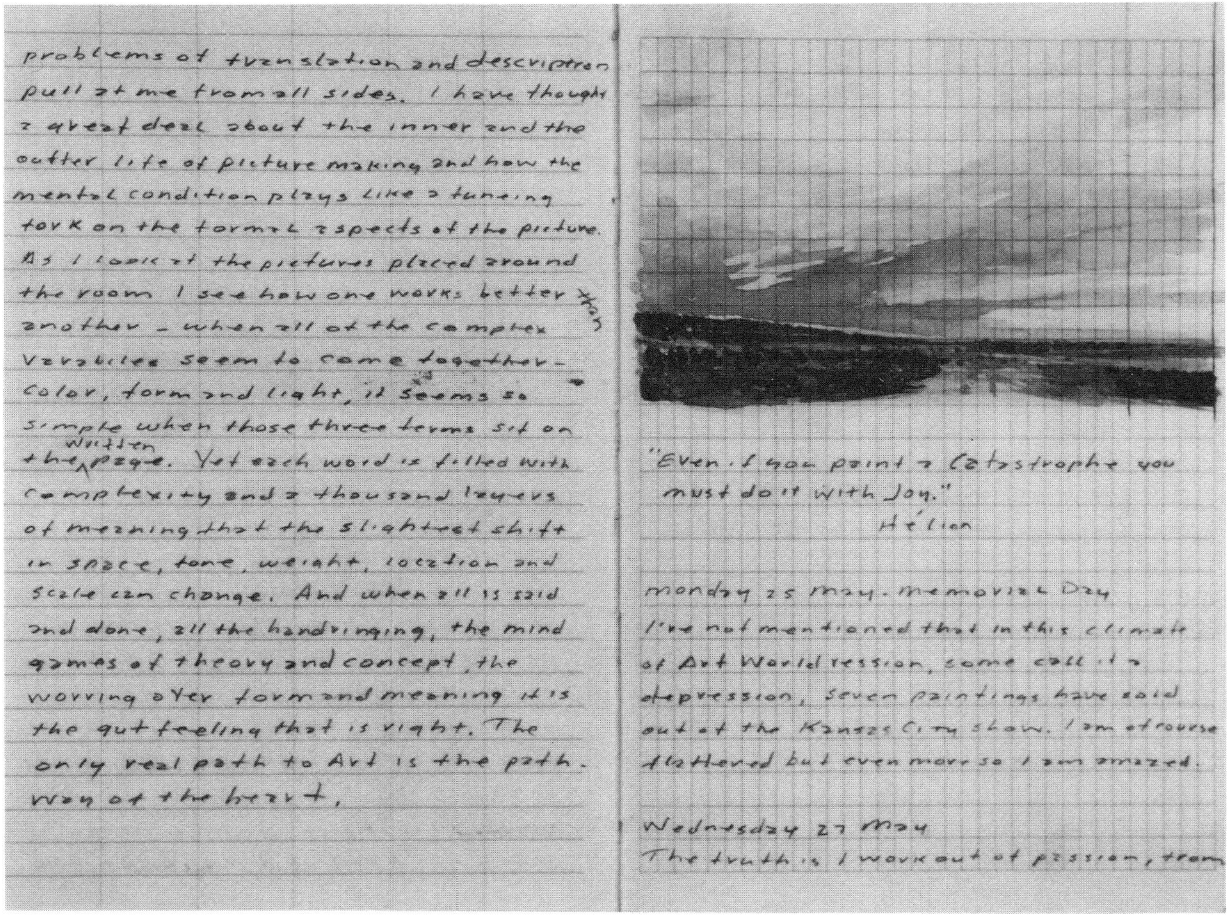
problems of translation and description
pull at me from all sides. I have thought
a great deal about the inner and the
outer life of picture making and how the
mental condition plays like a tuning
fork on the formal aspects of the picture.
As I look at the pictures placed around
the room I see how one works better than
another - when all of the complex
variables seem to come together -
color, form and light, it seems so
simple when those three terms sit on
the written page. Yet each word is filled with
complexity and a thousand layers
of meaning that the slightest shift
in space, tone, weight, location and
scale can change. And when all is said
and done, all the handringing, the mind
games of theory and concept, the
worrying over form and meaning it is
the gut feeling that is right. The
only real path to Art is the pathway of the heart,

"Even if you paint a Catastrophe you
must do it with Joy."
Hélion

monday 25 may. memorial Day
I've not mentioned that in this climate
of Art World ression, some call it a
depression, seven paintings have sold
out of the Kansas City show. I am of course
flattered but even more so I am amazed.

Wednesday 27 may
The truth is I work out of passion, from

Figure 3.1. Keith Jacobshagen, Personal Journal, 1991–1992. (Reproduced with permission of the artist) an artist who has a strong desire to understand his vision, a vision that seems to be always turned toward the low sweeping elegance of the horizon.

• Driving west at early evening after two hours of drawing on the Platte River. The soft curve of Highway 6 out of Ashland moves in a subtle downward path to the flat land toward Lincoln. It reminds me of the elegant simplicity of the minimalist paintings of Frank Stella or Newman. It is about straight lines that live by the nature of their parallel shapes and modest forms, forms that merge in the coming darkness and articulate the hard edge of the horizon line. The question is always asked by the curious travelers who have crossed the Plains at Interstate speeds, ''How can you live here without the mountains, the ocean, the woods?'' But what they are really speaking to is their desire to ''get it'' right away. The sublime of this place that we call the prairie is one of patience and looking. There is no quick fix, as there is with the monumental vertical of the mountains or the dark mysteries of timberlands. The coming to grips with the prairie, for the artist at least, has to do with a long and expansive relationship, one that seems, at first, without parameters. One that speaks of the giving up of ego and arrogance, of seeing as opposed to just looking. If one is to understand the beauty of this place, the old answers just won't do.

• At first it seems that the aesthetic of the Plains is somehow about its flatness, its long stretch unbroken by hills or the walls of tree lines. What it is about, of course, is its space, its vastness, its insistence on bringing home to you your meager place in trying to somehow understand your place. It is a space so vast in its presence, so unlimited in its conceptual parameters, that to come to an understanding (the graphic and formal two-dimensional understanding of the artist), one must, in a word, make a leap of faith. For how can the artist explain a distance so deep into the heart of the picture plane that if you could pull yourself through the rectangle, like Alice through the looking glass, it would only touch the beginning of an idea that could be understood. One is held at the threshold of this space, poised at its beginning and embraced by its visual mystery.

• There is no fat on this land. In the throws of deep winter without snowfall, the fields stretch bone dry, a brittle ocher color in the late afternoon sunlight. I take to the high ground to make a drawing. Looking west toward the silhouetted shapes of the Greenwood elevators, the space before me is as sensuous and visually loaded as one could ask for. It is late in the day, and the sky has gone from a milky midday cerulean of thin light to a deep icy cobalt blue that peeks through warm goose-down gray broken clouds that have begun to glow from a low sun. It occurs to me, as I begin blocking in my drawing, the issue here is always about the big elemental forms and their relationship to the occupation of the small forms that sit on the seemingly smooth plain. What is so compelling about this place, what is in part the visual seduction of the Plains, is the ability to see, without great interruption, how each element, silo, grain elevator, tree line, roadway, pond, farmhouse, railroad line sits and occupies its place in the scheme of things. The slightest rise in the land suddenly gives one a view, modest in subject as it may be, that can take the breath away in its subtle elegance. It is not the heroic space one finds at the rim of the Grand Canyon;

it is instead a distance that is at once human in its visual and emotional scale—a scale in both the land and its relationship to sky that embraces, shelters, and signals the visual journey forward without intimidation. For me, it always has implicit in it the outer expression of an inward journey.

• Most problems of visual language to some extent concern themselves with placement and arrangement. It is this not-so-simple problem of design that can give a kind of silent universal meaning to specific facts. What is this visual mystery of the flatlanders' space that can so compel an artist to description, invention, and the sometimes unrestrained desire for making seen those things seen by the artist? There can be no doubt in anyone who has spent time roaming the Plains that the deep space of the land as it rushes toward the horizon and its subsequent visual meeting with the great arching bowl of the sky is a point of convergence, a focus toward a core of a mental and spiritual journey. The horizon is an anchor for the placement of those earthbound objects that give veracity to their positions. That a small patch of prairie willow or goldenrod may have a visual importance equal to the perpendicular mass of a grain elevator or and electrical power station speaks to the ocular democracy of objects and their perceived meanings, meaning unobstructed by a closed space or confused by a competition of visual scale. The visual language of the Plains is made complex only by its accumulation of signs and symbols and the mystery of interpretation that they may hold as they are layered one after another across the artist's picture plane.

• If there is an aesthetic value to the prairie it is in its unbounded clarity, its sense of freedom and its independence from ruthless clichéd visual systems that so stereotype the beautiful. These things are embraced by the extraordinary effect of weather on all forms of life and living. In other words, to talk about an aesthetic value is to speak of an organic constantly changing form, a form in flux whose visual meaning changes as often as the weather and the seasons. Our ideas about the Plains function in close symbiosis with our needs, fears, and the concepts we carry deep within us. Our visual perception of the prairie is altered and distorted by factors sometimes unknown to us, but determine how our eyes interpret the prairie as seen by us.

II

GRASSLAND ECOLOGY

4

The Physical Environment of Great Plains Grasslands

THOMAS B. BRAGG

At sunset I saw a thick smoke rising at the foot of the mountain toward the Indian's camp, and soon after perceived the plains on fire. The weather was cloudy at dusk, and the wind blew strong from the N., causing the flame to make rapid progress; at ten o'clock it had extended as far as Salt river, presenting a dismal and lurid appearance. We could plainly distinguish the flames, which at intervals rose to an extraordinary height, as they passed through low spots of long grass or reeds. They then would cease their ravages for a few moments, soon afterward rise again with redoubled fury, and then die away to their usual height. The sight was awful, indeed, but as the wind was from us, and the fire was on the S. side of Park river, we had nothing to dread. If this fire spreads all over the country, we shall be hard up for provisions, as there will be no buffalo; nothing can stop its fury but snow or rain (near Park River, North Dakota, 1 December 1800).

A. Henry and D. Thompson, *New Light on the Early History of the Greater Northwest*

From head-high prairie cordgrass in eastern Nebraska valleys to carpet-short buffalo grass in western Kansas plains, the North American prairies represent a diversity of life that we are only now beginning to understand and enjoy. This great diversity reflects thousands of years of selection for plant and animal characteristics that allow them to survive in the fire-frequented, climatically variable conditions that support grasses and that are variously affected by an array of geologic and climatological conditions. This chapter focuses on three principal factors—soil, climate, and fire—each of which is an important part of any explanation about differences in the grasslands of North America.

Each grassland differs from others and each contains its own diversity, a consequence of nonuniform conditions throughout its range. Some differences provide sharp contrasts, such as those between the tallgrass and shortgrass prairies, while other differences are more subtle, such as those between north-facing and south-facing slopes within the Nebraska sandhills prairie. To explain these differences, we must first look to the nonliving, or abiotic, environment, for it is in an understanding of features such as soil and climate that we can find explanations for differences between and within prairies.

The physical boundary between different grassland types is rarely, if ever, distinct. Instead, grasslands grade from one type into another, arranging themselves along soil, moisture, and temperature gradients. The tallgrass prairie of the more moist eastern part of the Great Plains, for example, also extends into the drier mixed-grass and shortgrass prairie regions occurring along river valleys where moisture conditions are more suitable for it. The mixed-grass prairie itself is considered by some to be only a broad region of intergrading of the tallgrass prairie to the east with the shortgrass prairie to the west.

Despite this continuum and intermingling of prairie types, it is useful to define specific communities, thus allowing us to describe and try to better understand them. In this chapter, I describe the abiotic environment of specific grassland types with the understanding that any distinct boundaries implied are artificial and not identifiable in the natural landscape. For discussion purposes, I have divided the Great Plains grasslands into six grassland types: tallgrass prairie, tallgrass prairie–forest transition zone, northern mixed-grass prairie, Nebraska sandhills prairie, southern mixed-grass prairie, and shortgrass prairie (Figure 1.1).

SOIL: AN ECOLOGICAL SUMMATION

Soil is an aggregation of weathered minerals and decaying organic material that covers the earth in a thin layer and in which plants can grow. This aggregation reflects the sum of a multitude of ecological activities, including those of plants and animals, climate, topographic condition, and original geologic material, that have interacted over long periods of time. The result of these interactions is shown in the various soil types that characterize the North American grasslands (Figure 4.1). A detailed description of the geologic background of the grasslands and their biotic and chemical dynamics is given in Collins and Glenn (Chapter 7) and Seastedt (Chapter 8). Only a summary of particularly salient points regarding soils is presented here.

The North American prairies have been formed on a variety of geologic material. Many soils in the Great Plains reflect the effects of the Cretaceous sea that once covered much of the region from the present Gulf of Mexico to the Arctic Ocean. Subsequent geologic processes of erosion, deposition, and mountain-building resulted in the varied topography of the region and in some unique features such as the Nebraska Sand Hills, in which the sandy substrate supports a more productive grassland than expected given the amount of rainfall received. Superimposed over many of these earlier geologic deposits are additional deposits such as glacial till and windblown soil, some of which is called loess. Loess deposits, for example, ultimately formed into the deep, rich prairie soils that extend throughout portions of the Great Plains.

The classification of soils into groups provides an efficient means by which to describe a soil. Of the 10 soil orders, 5 are relevant to grasslands (Akin 1991). A brief description here provides some indication of grassland soil variability. *Mollisols* are characterized by a dark color reflecting a large proportion of decomposed plant and animal matter. This soil order dominates most Great Plains grasslands as well as bunchgrass steppe prairies of the western grasslands. *Entisols,* generally young soils without clearly developed soil horizons, characterize the soils of the sandhills prairie of west-central Nebraska, portions of the northern mixed-grass prairie of eastern Mon-

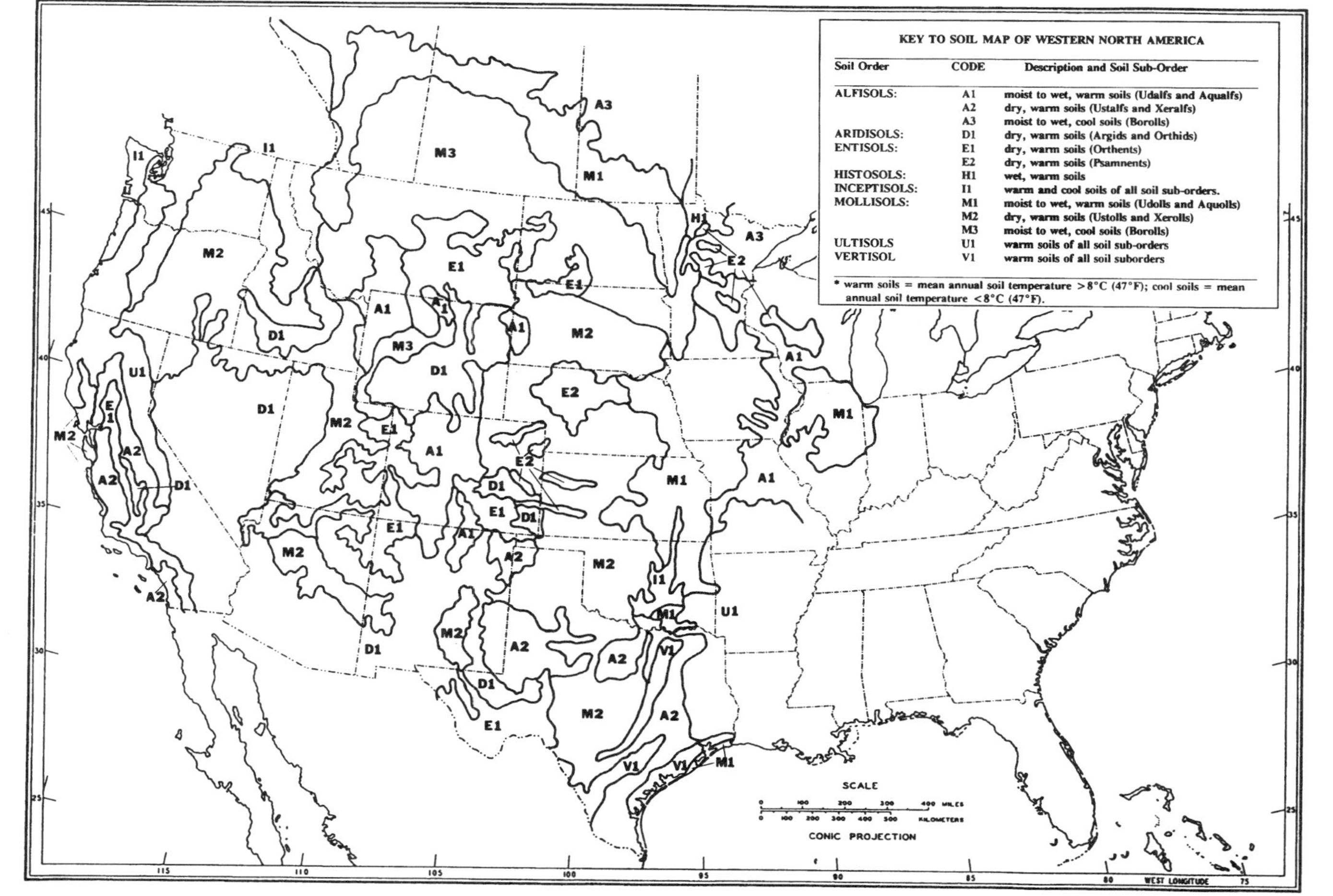

Figure 4.1. Major soil regions of North America. (Based on *National Atlas Map: Soils—1985*, with permission of the U.S. Department of the Interior, U.S. Geological Survey, and Leggett [1968])

tana, and small pockets throughout portions of the western Great Plains. *Aridisols,* characterized by a lack of water and low organic matter, dominate the desert grasslands and are an important secondary soil of the bunchgrass prairie of the northwestern United States. *Alfisols,* characterized by only shallow occurrence of organic matter, dominate many of the shortgrass areas of central and western Texas and eastern New Mexico as well as California grasslands. *Vertisols,* characteristically dark clay soils that feature wide, deep cracks when dry, dominate the tallgrass areas of eastern Texas.

CLIMATE: THE GRASSLAND'S ENVIRONMENTAL CHAMBER

Climate is defined as the sum of weather (i.e., meteorological) conditions that characterize the atmosphere over a long period of time at a given point on the earth's surface. It incorporates the elements of temperature (including radiant energy), moisture (including precipitation, humidity, and cloudiness), wind (including storms), pressure, evaporation, electrical phenomena, and other elements, such as chemical atmospheric phenomena. Here I focus primarily on climatic elements, temperature, and precipitation (rain, snow, sleet, hail) that most directly affect vegetation. A detailed discussion of the climate of North America and the factors affecting it are given by Bryson and Hare (1974).

Precipitation

Differences in climate across the North American continent reflect both global and regional weather patterns. Global patterns are those affected by differential heating of the earth's surface that results in rising air masses over the equator and in rising and descending air masses at various more northern and southern latitudes. These, in combination with the rotation of the earth on its axis, cause the broad, global patterns of weather movement across the earth's surface. In North America, the prevailing weather is borne on generally westerly winds exhibiting seasonal modification by arctic airstreams from the north and tropical airstreams from the south (Figure 4.2). The clashing of these air masses, particularly in the summer, results in the variable weather and storm patterns that characterize the Plains region.

On a regional scale, surface features further modify the broad air-mass movements in North America. The first feature important to the Great Plains grasslands is the series of north–south mountain ranges (collectively called the Western Cordillera) that dominate much of the western portion of the continent. The westerly air masses that move over the Pacific Ocean are saturated with moisture when they reach the North American continent. As they move inland, they are obstructed first by coastal mountain ranges and then by successive tiers of mountains ending with the Rocky Mountains on the eastern extreme of the cordillera. As the air masses reach each of these obstructions, some are diverted north or south while others are forced up over portions of the mountains as they continue their eastward movement. At each mountain range, rising air cools and, since cool air holds less water, the excess falls as precipitation on the windward side of the mountain range. Thus precipitation on the west-facing, windward slopes supports the majority of the forests of the mountain ranges of western North America. Once over the mountain range, the cool, dry air

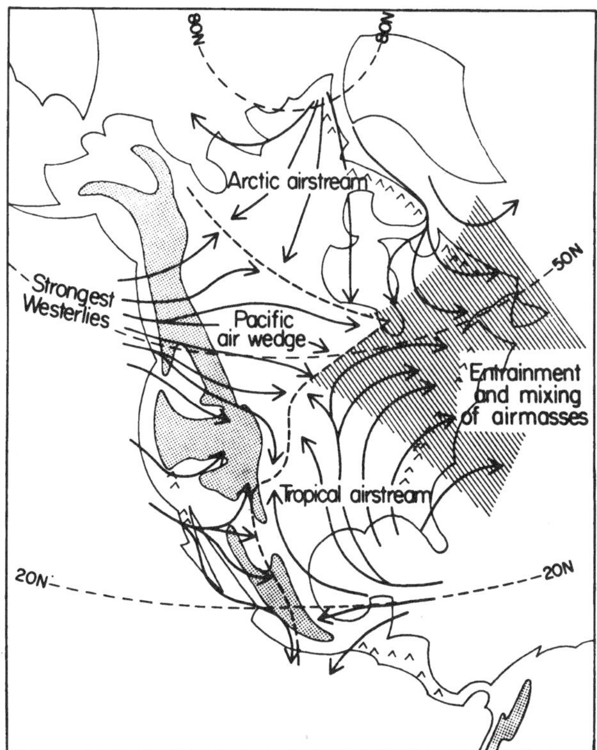

Figure 4.2. Surface flow across North America, illustrating the movement of Pacific air across the Western Cordillera and the midcontinent clashing of the westerlies and the tropical airstream that results in storms throughout the Great Plains. Airflow is based on July surface winds. (From Bryson and Hare [1974], with permission of Elsevier)

descends and begins to warm and, being rather dry, picks up surface moisture. These conditions result in dry conditions on the leeward (downwind) side of mountains in an area called the rainshadow. The rainshadow effect of the Cascade Range results in conditions that support bunchgrass steppe grassland of eastern Washington State. Similarly, California grasslands are a product of the rainshadow effect of the Coast Range. The rainshadow effect of the entire Western Cordillera also explains the dry conditions that prevail along the eastern flank of the Rocky Mountains and that define the western boundary of the Great Plains, the region once referred to as the Great American Desert. It is within the western portion of the rainshadow that the driest conditions occur and in which the shortgrass and desert grasslands predominate.

The second relevant geologic feature is the broad, flat plain that extends from the mountainous West across the central part of the continent. As the air mass descends the Western Cordillera and moves farther east across the broad, flat plains of the central interior, it increases in temperature and can hold more moisture. The broad plain also allows for unobstructed movement of the arctic and tropical air masses into the central region, creating unstable summer climatic weather conditions. Increased moisture in eastward-moving air masses, in combination with the humid tropical air masses moving up from the south, particularly in the summer, provides more rainfall to support the mixed-grass prairie of the central Great Plains and still more for the tallgrass prairie of the eastern Plains. Thus is formed the moisture gradient, increasing from west to east (Figure 4.3), and the variable climate that characterizes the various prairie types in the Great Plains.

Average annual precipitation across the Great Plains grasslands ranges from about 300 millimeters (12 inches) in the dry western shortgrass and northern mixed-grass prairies to about 600 millimeters (24 inches) in the northern portion to more than 1000 millimeters (40 inches) in the southern portion of the tallgrass prairie (Figure 4.3). Two-thirds of this rainfall occurs during the summer growing season, generally from April through September. This considerable variability in precipitation between grassland types suggests that climate works with other factors in explaining the occurrence of a particular grassland type. Other factors, of which fire is often significant, are likely to be at least equally as important.

Temperature

Temperature is another important factor in explaining the distribution of grassland types. The general temperature gradient that dominates North America shows an increase from northwest to southeast. In the Great Plains grassland region, the January gradient ranges from 15°C in southern Texas to −15°C in south-central Canada. In July, the gradient ranges from 30°C to 20°C. This gradient has important effects on biota, especially affecting geographic distributions of some groups of plants. An important, broad distributional effect may be seen in the north–south gradient of two groups of grasses, one referred to as cool-season and the other as warm-season. Although there are exceptions, cool-season plants (most of which are referred to as C_3 plants) are most efficient at photosynthesizing when temperatures are cool, whereas the warm-season species (C_4 plants) are most efficient in comparatively warmer conditions (Black 1971). These differences in photosynthetic pathway, in large part, explain the gradient of plants within the Great Plains from domination by warm-season species in the warmer, southern regions to domination by cool-season species to the north (Figure 4.4). A similar gradient occurs with elevation, where cool-season species predominate in cooler, higher altitudes and warm-season species are more prevalent in lower, warmer elevations.

Potential Evapotranspiration

Potential evaporation is another characteristic of climate that is related to the distribution of different grassland community types. When plants grow, they take up water through their roots, use it in various cellular activities, and then release it as water vapor into the atmosphere through pores in their leaves. The release of water vapor from plants to the atmosphere is called transpiration. This process explains why plants contribute greatly to changes in soil moisture. Transpiration coupled with evaporation of water directly from the soil surface is called evapotranspiration. Under ideal soil moisture conditions, and depending on temperature, a maximum amount of soil moisture can be lost to the atmosphere by way of evapotranspiration. This maximum is termed potential evapotranspiration. The relevance of this concept for understanding

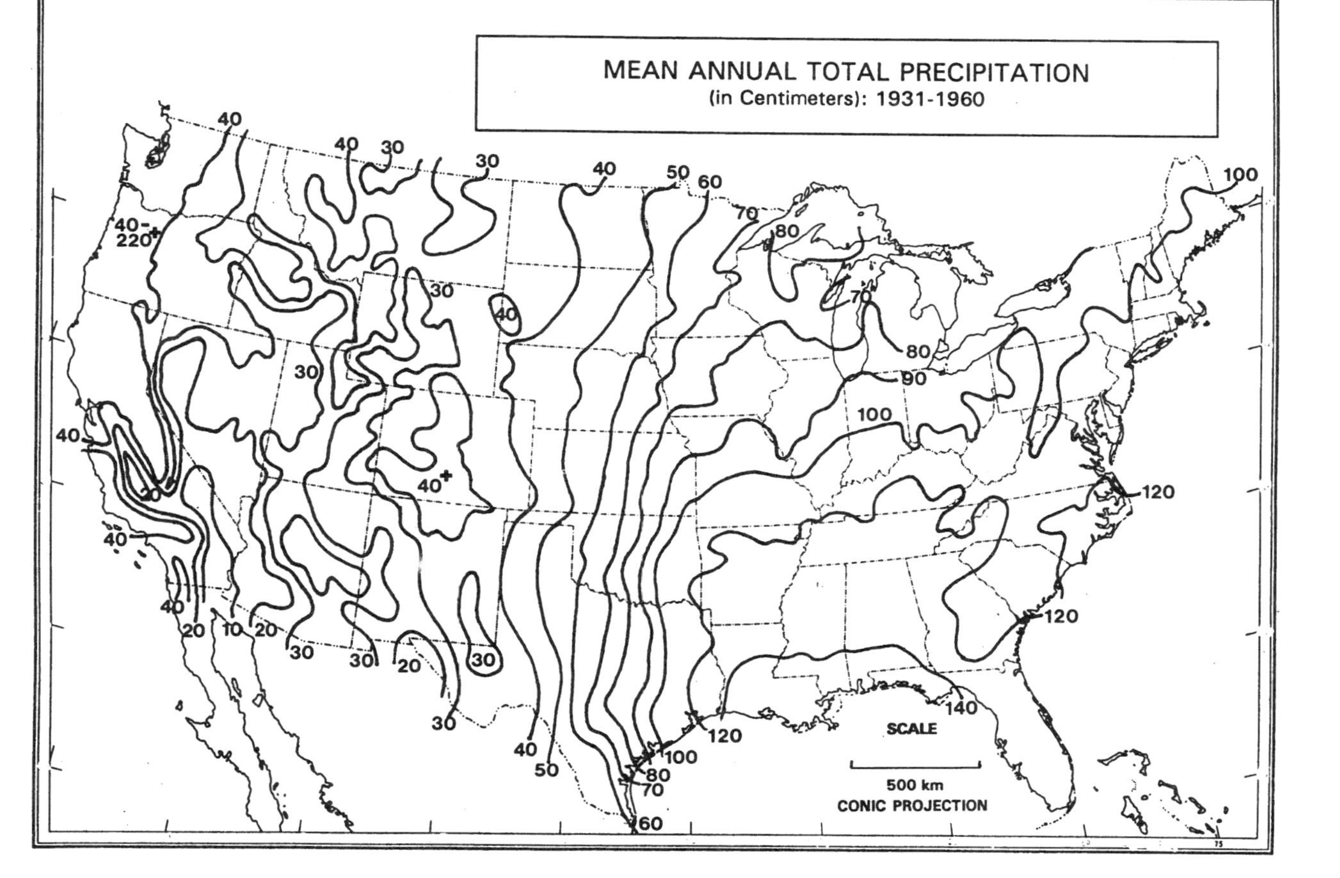

Figure 4.3. Mean annual total precipitation, illustrating the gradient of moisture from lower in the western Great Plains to higher in the east. (From Bryson and Hare [1974], with permission of Elsevier)

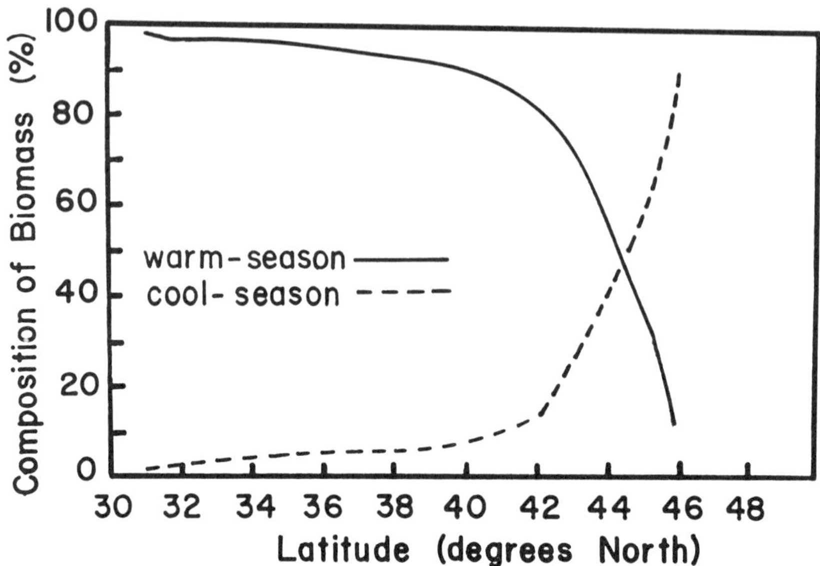

Figure 4.4. Changes in relative composition of warm-season and cool-season plants, primarily grasses, across the North American grasslands from the southern to the northern latitudes to illustrate the effect of climate on plant species composition. (From Sims [1988], with permission of Cambridge University Press)

plant systems is that it can be compared with actual rainfall to determine if the amount of water needed to support plant life (potential evapotranspiration) is greater or less than that available to plants (e.g., from precipitation). The use of the ratio between these two measures to consider differences in climate was proposed by Thornthwaite (1948). He developed a simple mathematical calculation of a moisture index (I_m) that reflects the relationship between precipitation (P) and potential evaporation (PE):

$$I_m = 100 * (P - PE)/PE$$

Areas with values of I_m greater than 1 are considered moist areas that have the potential to support tree growth. Such areas include the forests of eastern North America. Climatic regions where I_m is less than 1 designate the dry regions, including the shortgrass and mixed-grass prairies. The tallgrass prairie and the interdigitation of grassland and forest (the grassland–forest transition zone) occur where I_m fluctuates around 1. This zone is approximated by a line that extends from western Minnesota south through eastern Nebraska and Kansas and into central Oklahoma and Texas. Roughly, this line delineates the western boundary of the tallgrass prairie, and it approximates the 700-millimeter (24-inch) precipitation zone (Figure 4.3). Thornthwaite's moisture index, however, only partly explains the presence of North American grasslands, which result as much from the effects of factors such as periodic drought and fire (summarized by Collins and Wallace 1990) as they do from moisture conditions alone.

Drought

In the Great Plains in particular, one notable characteristic of grassland climate is its great year-to-year variability. To survive in such a variable environment, organisms must be adapted to surviving occasional periods of greater than normal stress. This stress is often manifested as one of water shortage or drought, where drought represents the sum of decreased precipitation, increased evaporation, and increased water runoff that combine to reduce soil moisture and thus to stress the plant community. In general, grasslands are found in areas of North America in which periodic droughts occur (Weaver 1968; Wilhite and Hoffman 1979). Of the North American grasslands, the occurrence of the most severe drought conditions is often highest in the shortgrass and southern mixed-grass prairies of the central Great Plains (Kansas, Nebraska, Oklahoma, and central Texas). For example, for the period 1931 through 1960, these grasslands in general experienced drought for 20 to 25% of the months. Drought was measured as the difference between actual evapotranspiration and the adjusted normal evapotranspiration (Bryson and Hare 1974).

Drought can have significant effects on species composition of grasslands. In a series of studies by John Weaver and others (Weaver 1968) in the Great Plains, the composition of tallgrass, mixed-grass, and shortgrass prairies was noted to change substantially during six years of severe drought from 1934 to 1940. Vegetative cover shifted from continuous to spotty with many areas of bare soil where plants had once occurred. Plants rarely flowered and were regularly in wilted condition. These conditions substantially affected the composition of native prairies. The response of individual species, however, varied depending on factors such as where the plants were growing, whether they were grazed, and whether dust deposited from the great "black blizzards" was deep enough to cover growing points. Most species were adversely affected by the drought, but some less so than others. In general, species most adversely affected included little bluestem and Indiangrass. Sideoats grama, blue grama, buffalo grass, and western wheatgrass were among those better able to withstand or to take advantage of drought conditions (Weaver 1954). Dust burial was a particular problem for short-statured species such as buffalo grass and blue grama.

FIRE: THE KEEPER OF THE PRAIRIE

Fire is a factor in any terrestrial ecosystem in which there is dry fuel, an ignition source, and oxygen. Forests, grasslands, chaparral, and marshes are among the many plant communities in which fires have occurred for thousands or hundreds of thousands of years. Fire occurs in a wide variety of plant communities, but it is particularly important in grasslands because without fire, most grasslands would ultimately succeed to forests or shrublands (Sauer 1950). The flammability of grasslands is attributed to numerous characteristics of both the individual plants that dominate and the plant community as a whole.

Evidence for fire's prevalence and extent in the historic Great Plains region includes both frequent observations of prairie fires by early travelers (see page 49) and the absence of trees at the time of European settlement in areas capable of supporting them (i.e., much of the eastern portion of the Great Plains). One example of the extent of such fires is indicated by a report on a fire in 1885 that started in western Kansas, jumped the Cimarron River, and burned across the north plains of Texas, a distance of 282 kilometers (175 miles) (Haley 1929). It is difficult to envision a fire of this magnitude today or to know if such fires were possible when great herds of bison grazed down the grass throughout portions of the Great Plains. However, large fires predating European settlement clearly occurred and probably were not uncommon.

The role that fire played in the natural grassland community differed both for each grassland and for any given set of conditions of a burn. Important variables of fire that explain such different responses include the season or time of year in which burning occurs and the number of years between fires, or fire frequency. The importance of these variables is apparent when one considers certain seasonal, biological phenomena such as when ground-nesting birds are raising young, when insect larvae emerge from protection belowground, and when plants are flowering and producing seeds. Fires at any particular time of year may have detrimental effects on these or other biological phenomena, whereas fires at other times may be less consequential. Further, fires at any one time would be beneficial to some species and detrimental to others.

Fire season and frequency rely, to some extent, on the natural fire ignition. Lightning is the most obvious cause of natural grassland ignition. Native Americans, however, are considered to be equally as natural in this chapter. The rationale for this consideration is based on two historical realities:

1. Much of the northern Great Plains was glaciated as recently as 9000 years ago.
2. Humans had arrived in North America perhaps as many as 30,000 years ago.

Thus Native Americans have occupied large portions of today's North American grassland region since the grassland's development. The frequency and timing of Native American fires, therefore, is as important as a consideration of lightning.

Fire ignition occurred from spring through fall. For example, from 1535 through 1890, Native Americans used fire frequently both in dormant seasons (e.g., April and October), when there is little green plant matter, and during the growing season (e.g., July through August), when green plant matter may have prevailed (Moore 1972). Further, lightning ignitions were probably common during these times, as evidenced by a present-day average of 40 lightning storms that sweep the Great Plains from May through September each year (Figure 4.5) (Bryson and Hare 1974). Of these, 75% occur during the summer months (July and August), and many are likely to have resulted in grassland fire ignitions (Figure 4.6). Thus both lightning and Native American fire information suggest fires throughout the year, with a high probability of frequent summer fires in most grasslands. However, with respect to summer fires, they would have occurred when plants were green and when dry plant matter would be only a small part of the total fuel available for burning. Thus summer fires would have burned less completely and with lower intensity. Consequently, while available evidence suggests frequent summer fires, I have observed them to be smaller in extent than those occurring during the dormant seasons, with the fire often extinguished by rain that follows.

The frequency of fires and the season in which they occur affect the way a fire burns, which is reflected in temperatures that occur during burning. The temperatures that occur with burning is another variable of fire that requires some understanding

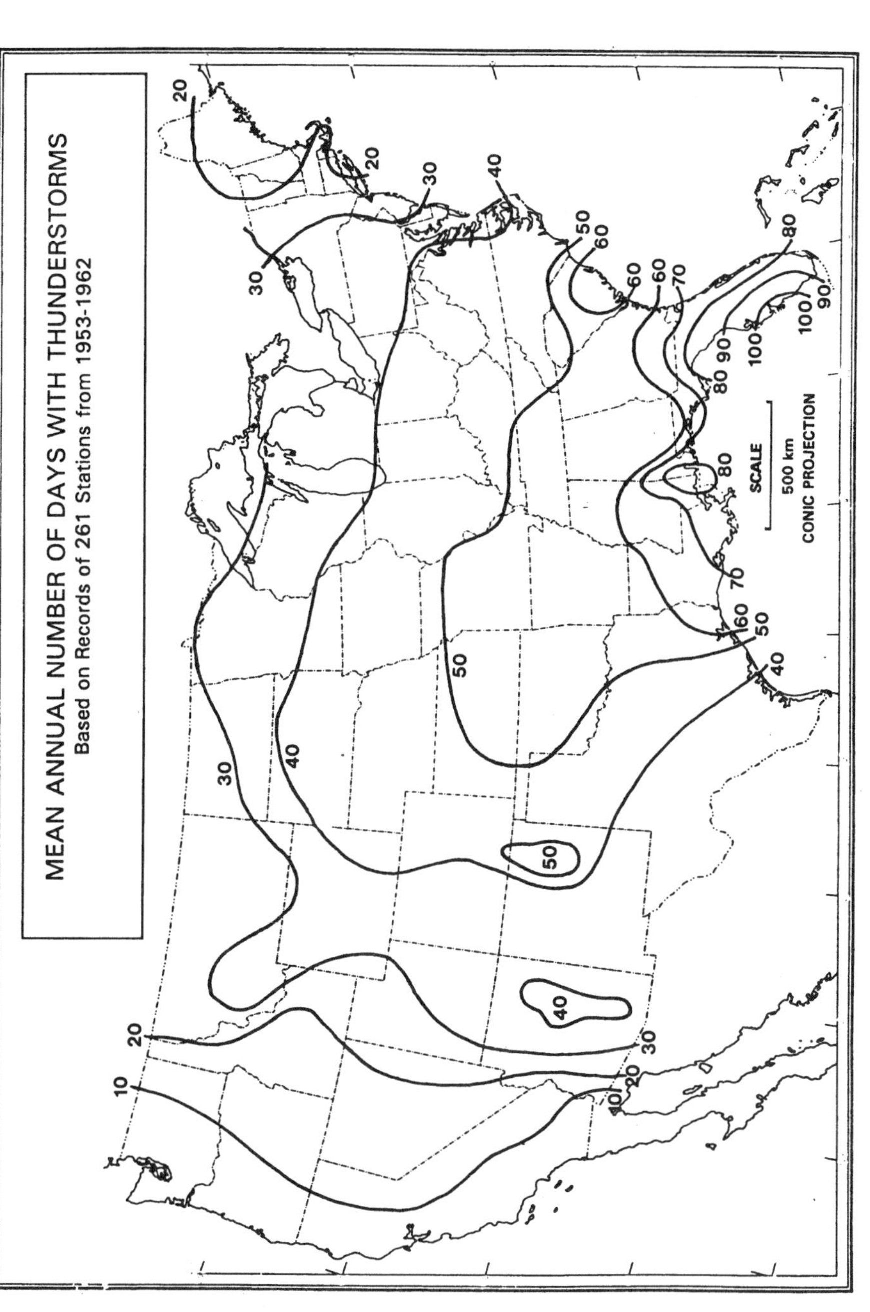

Figure 4.5. Mean annual number of days with thunderstorms, illustrating the gradient in potential ignition from lower in the northern Great Plains to higher in the south. (From Bryson and Hare [1974], with permission of Elsevier)

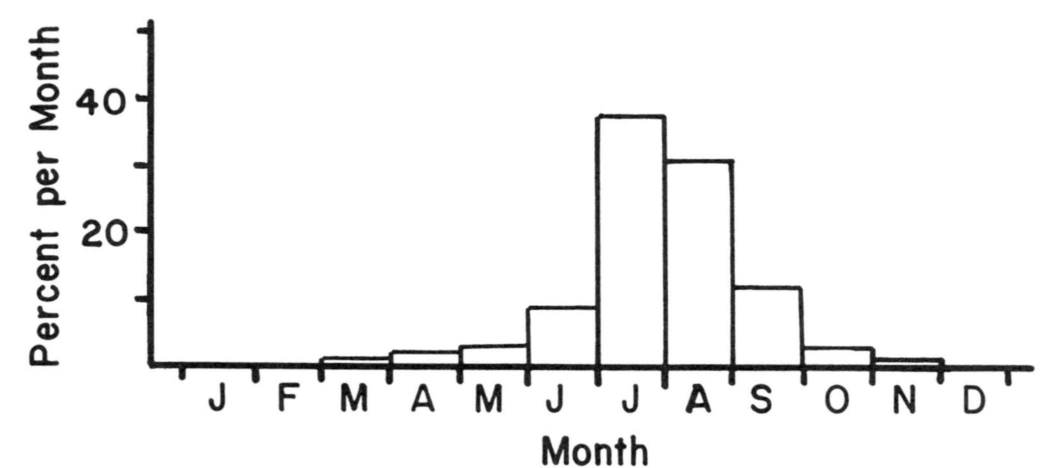

Figure 4.6. Seasonal occurrence of lightning-caused fires in Nebraska, illustrating the high frequency of summer ignitions. Values represent averages for the period 1971 through 1975 when the number of lightning-caused fires averaged 138 per year. (From Bragg [1982], with permission of the Ecological Society of America)

Table 4.1. Fire behavior variables ($\pm$ standard error) from headfires and backfires in a grazed, Oklahoma tallgrass prairie

	Fire type	
Variable	Headfire	Backfire
Fire-line intensity[a] (kilowatts/meter)	1167 ± 445	97 ± 16
Rate of fire spread[b] (meters/minute)	12 ± 6	1 ± 0.2
Residence time[c] (seconds)	250 ± 114	255 ± 75
Temperature (°C) at		
60 cm[d]	274 ± 83	78 ± 14
30 cm	307 ± 83	270 ± 68
soil surface	216 ± 59	215 ± 81

[a]Measure of heat energy generated per meter of fire line.

[b]Speed with which the fire line moved downwind.

[c]Time that flames occurred at one place.

[d]Height above the soil surface.

Source: Data from Bidwell and Engle (1991), with permission of the U.S. Department of Agriculture, U.S. Forest Service.

because of its various effects on plants and animals. Average soil surface temperatures of grass fires in the Great Plains have been found to vary from 102 to 388°C (215–730°F), although extremes of 83 to 682°C (182–1260°F) have been reported (Wright and Bailey 1982). These differences may reflect differences in the amount of plant matter, since there is a direct relationship between the amount of fuel and the temperature of the fire. High temperatures, however, rarely last more than a few minutes, because the fire front in a grassland rapidly consumes the fine fuel. Even within a small area, fire temperatures can vary significantly (Gibson et al. 1990; Hickey 1992). Importantly, soil temperatures below 10 millimeters (0.4 inches) in depth rarely reach lethal levels during a fire; thus soil organisms are not immediately affected.

An additional factor affecting fire temperature is whether fires are headfires, which move rapidly with the wind, or backfires, which move slowly into the wind but burn more completely. Bidwell and Engle (1991) reported headfires to be 10 times more intense (Table 4.1) and to travel 10 times more rapidly than backfires. The differences between backfires and headfires may explain, at least in part, different effects on animals. Rapidly moving headfires, for example, might skip over small animals in the litter or crouched on the ground, but they may also overtake others trying to flee from the advancing flame. Backfires, on the contrary, move more slowly but burn more completely and thus would be particularly destructive to immobile organisms, such as eggs, young in nests, and larvae, or invertebrates that are aboveground. Plants also respond differently to headfire/backfire conditions. Bidwell et al. (1990) reported that grass production was 21% greater in the headfired area than in the area burned with a backfire on grazed tallgrass prairie.

Effects of Fire on Great Plains Grasslands

The effects of fire on grasslands and on individual species within grasslands are extremely variable, so much so that studies often appear inconsistent. This variability is a consequence of differences between numerous factors, including species composition; specific burning conditions; climatic conditions before, during, and after a burn; inherent genetic variability within species; soil conditions; grazing history; and postburn grazing intensity. In addition, the interpretation of fire's effects is complicated by the use of different sampling procedures and of short-term results for drawing long-term conclusions. Generalizations about fire's effects, therefore, need to be considered carefully before using the information for decision making. Nevertheless, it is useful at this point to note that fire does affect grasslands in various ways even though the degree of effect and direction may not always be certain. The following discussion is intended only to provide indicators of potential consequences, although some are more certain than others.

Fire in the Tallgrass Prairie and Grassland–Forest Transition Zone. The tallgrass prairie varies considerably across its range but, in general, is dominated by warm-season species, such as big bluestem, switchgrass, Indiangrass, prairie cordgrass, asters, leadplant, and sunflower. Some cool-season species, such as porcupine grass, are also common. Substantial knowledge has accrued on the role of fire in the tallgrass prairie, in part because substantial parts of the ecosystem remain (e.g., the Kansas Flint Hills) and in part because there has been an increasing interest in preserving the small remnants where the once-extensive ecosystem is mostly gone (e.g., in Iowa, Missouri, and Illinois). The remnant tallgrass prairies occur mostly within the grassland–forest transition zone. Good summaries for the region are found in Daubenmire (1968), Vogl (1974), Wright and Bailey (1982), Sims (1988), and Collins and Wallace (1990).

Lightning-caused fires in the region are most prevalent during the summer. For example, 66% of such fires in Nebraska occurred from July through August despite accompanying rainfall (Figure 4.6). Further, Moore (1972) noted that most fires in the tallgrass prairie area occurred from March through December, being most common during July and August. Fires in this region could have occurred at any time of the year, with a high frequency of summer ignitions, but due to the abundance of dry plant matter, larger fires most likely occurred during the dormant seasons of spring and fall.

This prairie type is located in the eastern portion of the Great Plains where precipitation is sufficient to support trees. Historically, therefore, one of the major effects of fire is the prevention of woody plant invasion. Within 10 years of fire suppression, there can be a noticeable increase in woody plants (Figure 4.7), which ultimately will eliminate the prairie plants (Gehring and Bragg 1992). Once established, many of these woody species, particularly those that can resprout, are extremely difficult to remove. Problem woody invaders include shrubs such as rough-leaved dogwood and smooth sumac and trees such as eastern red cedar, ash, elm, and some oaks (Bragg 1974). Only eastern red cedar can be comparatively easily eliminated using fire; the other species resprout from roots.

In addition to preventing woody plant invasion, fire affects the nonwoody component of the prairie, although the effects vary by season. Mid- to late-spring burning, for example, increases the productivity of the tallgrass prairie in general (Table 4.2), although it is more visible in some species, such as big bluestem, than in others, such as Indiangrass (e.g., see Risser et al. 1981). Fires in this season also increase forb

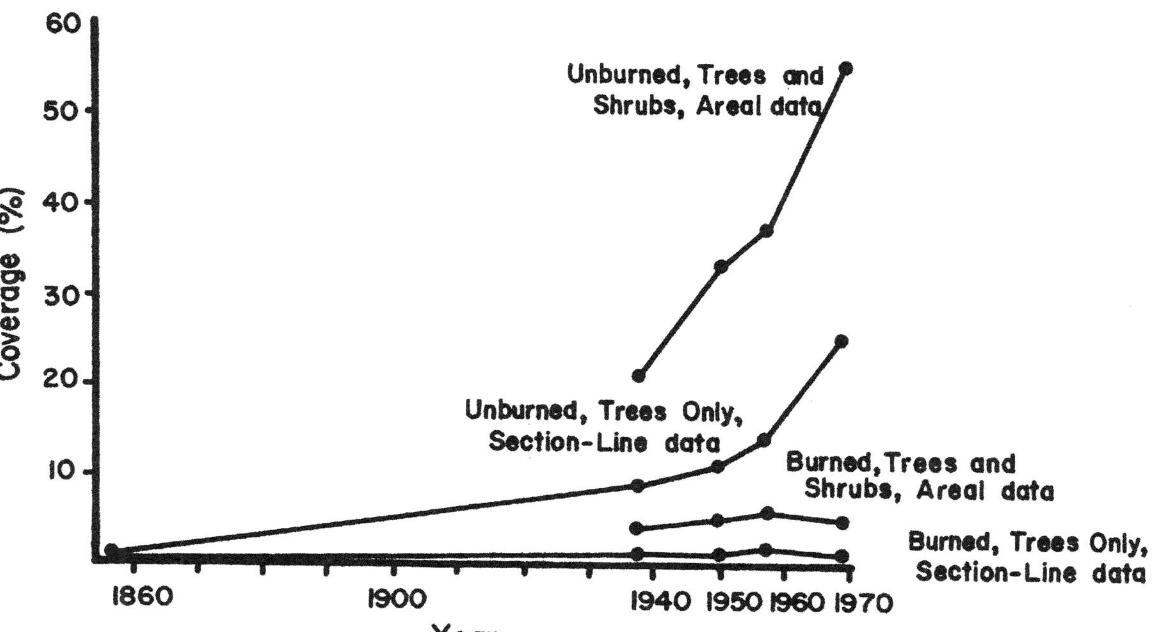

Figure 4.7. Average percent cover of woody plants on burned and unburned prairie sites from 1856 through 1969, illustrating the increase of woody plants in the absence of burning. Data are based on section-line notes made in 1856 and from aerial photographs taken from 1937 through 1969. (From Bragg and Hulbert [1976], with permission of the Society of Range Management)

Table 4.2. Postburn plant production for representative North American prairies

Prairie type and state or province	Grazing status	Postburn plant production (kg/ha) and season of burn				Reference
		Spring	Summer	Fall/winter	Not burned	
Northern mixed-grass prairie						
South Dakota	Grazed	1618 (Apr)	—	—	1775	Gartner et al. (1986)
South Dakota	Grazed	2178 (May)	—	2461 (Mar)	2495	Gartner et al. (1978)
Bunchgrass steppe (Aspen Parkland)						
Alberta	Lightly grazed	1365 (May)	—	1145 (Oct)	1180	Bailey and Anderson (1978)
Tallgrass prairie						
Illinois	Ungrazed	6480 (Apr)	—	—	3890	Old (1969)
Iowa	Ungrazed	5436 (Mar)	—	—	4201	Ehrenreich (1959)
Kansas	Ungrazed	3400 (Apr)	—	—	1701	Hulbert (1969)
Nebraska (reestablished)	Ungrazed	3147 (Apr)	2580 (Jul)	1755 (Oct)	1288	Bragg (unpublished)
Sandhills prairie						
Nebraska	Ungrazed	—	—	2076 (Oct)	2687	Morrison et al. (1986)
Nebraska	Ungrazed	—	2400 (Jun)	—	2120	Bragg (1978)
Nebraska	Grazed	800 (May)	—	—	1110	Bragg (1978)
Southern mixed-grass prairie						
Nebraska	Ungrazed	5000 (Apr)	—	—	5250	Nagel (1983)
Kansas	Lightly grazed	1205 (Mar)	—	1570 (Nov)	2379	Hopkins et al. (1948)
Oklahoma	Ungrazed	—	—	2160 (Nov)	1320	Adams and Anderson (1978)
Shortgrass prairie						
Kansas	Lightly grazed	711 (Mar)	—	2148 (Nov)	3028	Hopkins et al. (1948)
Kansas	Lightly grazed	1674 (Mar)	—	—	4424	Launchbaugh (1964)
Annual grassland						
California	Moderately grazed	—	3536 (Jul)	—	5507	Stechman (1983)

Note: Values shown are for biomass produced during the growing season immediately following burning. The month of burning is indicated after each value.

species such as tall gayfeather and white aster. Both the production of seedlings (mostly in biennials and short-lived perennials) and the number of flowering and nonflowering grass stems of some species are increased by late-spring burns. While favoring warm-season species, late-spring burns adversely affect cool-season, native grasses and forbs such as porcupine grass and purple coneflower. Henderson (1992), for example, reported that frequent (8 out of 10 years) late-spring burns result in a substantial decline in species diversity that primarily represents effects of cool-season species. Late-spring burns, however, are also detrimental to non-native, cool-season plants such as smooth brome, musk thistle, and Kentucky bluegrass, which are presently invaders into remnant prairies, particularly in the northern portion of the tallgrass prairie region.

Fires in the summer, fall, and winter result in different effects than those in the spring. Dormant-season fall and winter fires remove litter, which, with no regrowth to shade the soil surface, allows the soil to warm more rapidly in the spring. This gives cool-season species, including some introduced plants, such as sweet clover, an advantage over warm-season species, which do not initiate growth until late spring. Winter fires also tend to reduce the productivity of the dominant, warm-season species, including big and little bluestem and Indiangrass. These general results suggest that, over time, winter burning may cause a shift in the species composition to cool-season plants. Late-summer burns are also damaging to warm-season species, because they kill the aboveground parts at the time when the plants are growing most rapidly. As previously discussed, fires occurred most frequently at this time of the year before European settlement. Research on the long-term effect of summer burns is only beginning, and more time is needed before we will have a better idea of the importance of summer burns, particularly in understanding how fires during this season affect long-term biotic diversity.

Because it increases productivity, controls non-native invaders, and does not adversely affect total species composition over the short term, spring burning is the prevalent form of management in the tallgrass prairie. As with summer burning, however, additional research is needed to evaluate the long-term impact of regular spring burning (but, see Henderson 1992). Among the more important questions to address is whether spring burning alone can maintain the diversity of the prairie over many years. In addition, there is increasing evidence that fire must be used along with large-animal grazing in order to maintain the long-term diversity of tallgrass and other prairies (Collins and Glenn, Chapter 7; Wallace and Dyer, Chapter 9).

Fire in the Northern Mixed-grass Prairie. The northern mixed-grass prairie extends from South Dakota, in the north-central United States, into Saskatchewan in south-central Canada. This grassland is dominated by cool-season species, such as western wheatgrass and needle-and-thread. As with other regions in the Great Plains, Native Americans are believed to have been a dominant force in fire initiation. Literature references from 1673 to 1920 indicate that most Native American–caused fires in this prairie type occurred from March through May, with a peak in April, and from July to November, with a peak in October (Higgins 1986a). Fires were used mostly as a tool for activities such as hunting and food gathering. Of lightning-caused fires, over 70% occur during July and August (Higgins 1984). The frequency of lightning-ignited grassland fires varies considerably from an average of 6 fires per year in the more humid eastern part of North Dakota to 25 fires per year in the drier west.

Fire studies in this prairie type indicate that, in normal precipitation years, both dormant-season and growing-season burns generally decrease total plant production (Table 4.2), although production of dominant species may increase. Effects are extremely variable, but, in general, dormant-season spring burns are most beneficial to warm-season species, such as blue grama, and less so to cool-season species, such as needle-and-thread. Fires during the growing season, however, are detrimental to both. Fall and winter burns favor cool-season over warm-season species. In drought years, summer burns dramatically reduce productivity. Burns before warm-season species start growing suppress Kentucky bluegrass (Wright and Bailey 1980; Steuter 1987; Higgins et al. 1988).

Fire in the Nebraska Sandhills Prairie. The Nebraska sandhills prairie is centrally located in the Great Plains. Dominant species include sand bluestem, prairie sandreed, and sedges. The Nebraska sandhills prairie is often listed as a tallgrass prairie because taller grasses dominate in areas; however, it is more accurately described as a separate grassland type because it is an extensive, intact plant community in which the soil and the flora differ distinctly from those of surrounding grasslands (Bleed and Flowerday 1990).

The sandy nature of the soils, the absence of extensive interior roads for fire control, the large areas in which cattle are grazed, and the slow rate with which woody plants are invading have prevented fire from being used in management of the sandhills prairie. Historically, fires may have occurred in the region every four to five years, as indicated for the period 1850 to 1900 using fire-scarred trees along the northern border of the prairie (Bragg 1986). This period of time predates extensive settlement of the area, but the fire frequency needs to be considered carefully because it was during the middle years of this period that the once-ubiquitous bison herds were decimated. Loss of bison may have meant an increase in fuel for burning, since there were no other large grazers in sufficient numbers to consume the accumulating plant matter. The season of pre-European–settlement fires is unclear, although early references to fire in this region indicate both dormant-season and growing-season fires were likely.

Fire research in the sandhills prairie is limited but indicates that this grassland, like others, is adapted to fire (Bragg 1978). There is an initial decline in plant productivity following burning (Table 4.2), but the decline may not persist for more than a year. Also, as in other grasslands, variables such as the season of burning have different effects on different species. Burning in any season causes a decline in species such as sand lovegrass and sand muhly, whereas sand dropseed increases with burning, particularly with summer burns. Total species composition declines the year following burning, but it recovers to above preburn amounts thereafter. There is also clear evidence that along the margins of the Nebraska Sand Hills, woody plants are invading into the prairie (Steinauer and Bragg 1987), presumably because of fire suppression since settlement. A commonly expressed concern for burning in this ecosystem is the potential effect on soil movement; however, there is no evidence of fire initiating such a response to any significant degree (personal observations).

Fire in the Southern Mixed-Grass Prairie. Although this grassland may be considered only an intermediate plant community occurring between the tallgrass prairie and the shortgrass prairie, here we consider it an identifiable ecosystem because of its considerable extent and its distinctive characteristics. Extending from southern Nebraska south through central Oklahoma, the southern mixed-grass prairie is dominated by little bluestem. In contrast to the northern mixed-grass prairie, where cool-season grasses are co-dominants, here the warm-season grasses sideoats grama and blue grama prevail.

Direct evidence of the natural fire regime of the southern mixed-grass prairie is lacking, but fires were reported by early travelers in the region. The seasonal aspect of burning can only be guessed, but available information shows lightning storms to occur from March through October, peaking from May to August. Native Americans of the region appear to have used fire most frequently in July and August as well (Moore 1972); thus, fires may have been common in mid- to late summer.

Little research has been conducted on burning in this type of grassland, and most such studies are based on wildfires rather than on controlled burning (Launchbaugh 1973; Nagel 1983). Most of the dominant grass species appear to be adversely affected by burning, particularly in dry years, but several forbs increase, including western ragweed. While adversely affected, most grasses, except sideoats grama and Kentucky bluegrass, appear to be somewhat tolerant of fire, although it usually takes two to three growing seasons for the dominants to recover to preburn cover. Spring burning in the southern mixed-grass prairie appears to be more detrimental than fall burning, although both effectively reduce Kentucky bluegrass. Summer fires appear to be the most detrimental, causing all plants to decrease. Limited information suggests that burning in areas where litter is substantial may be beneficial for large grazers by increasing plant production and palatability of coarse grasses. Because of slow litter accumulation, however, fires in the southern mixed-grass prairie probably did not occur more frequently than every 5 to 10 years.

Fire in the Shortgrass Prairie. Natural fires were reported in the shortgrass prairie by early travelers through the region; thus fire should be considered a natural component of the environment. Plants of the shortgrass prairie of the west-central Great Plains are generally adversely affected by fire during dry years (Table 4.2), with the dominant plants, such as buffalo grass and blue grama, requiring up to three growing seasons to recover to preburn conditions (Wright and Bailey 1980). In normal-to-wet years plant production decreases, but recovery is more rapid for dominant grasses as well as other species. Forbs are generally less adversely affected by dormant-season (early spring and fall) fires than by midspring fires when they may have already initiated growth.

Within the shortgrass prairie region, where clay soils dominate, there are islands of sandy soils on which occur taller grasses, such as sand bluestem and switchgrass. Burning on these sandy soils is less deleterious to the vegetation and may increase productivity even though some species, such as little bluestem, are adversely effected. Fire effects in these sandy enclaves may be similar to those in the Nebraska sandhills prairie.

Whereas burning may affect vegetation in the shortgrass prairie in a manner that humans may consider deleterious, that does not mean that fire has been unimportant in the development of the plant community; for example, natural fires may have supported a vegetative composition that differs from the one seen today. Although burning may not be an advisable range-management practice, it is useful to consider the effects of fire from an ecological perspective. As with the southern mixed-grass prairie, slow litter accumulation probably prevented fires from occurring more frequently than every 5 to 10 years.

Effects of Fire on Animals and Other Biota

The effect that fire has on animals and other organisms (e.g., fungi and bacteria) are as variable as the effects on plants, depending on factors such as the conditions of the burn, the time of year burning occurs, and the behavioral and physical limitations of the organism. In general, effects can be divided into two categories: direct effects (effects occurring during burning) and indirect effects (effects resulting from changes in the habitat).

Direct, fire-caused mortality is primarily related to the mobility of organisms, although any animal can be killed under certain conditions. Organisms most susceptible to being killed during a fire include young animals, eggs in nests, and insect larvae that are aboveground and that cannot move or whose movement is too slow to escape a moving fire front. Among the organisms least likely to be killed in a fire are large animals and birds. Direct effects, however, are of less consequence to the animal community as a whole than are the longer-term, indirect effects that alter factors such as food availability, protection from predators, and habitat conditions (e.g., increased soil temperatures that affect soil organisms). The consequences of fire on an organism's habitat may persist for at least a year.

Large Mammals. In general, large animals are able to move rapidly enough to escape fires, although there are historical references to many bison being killed in a single fire event, apparently in areas with substantial fuel in which they became trapped. These instances are most likely to have been the exception among the 30 million bison that roamed throughout the Great Plains. One explanation for the small effect of fire on large grazers may have been their ability to move when they saw or smelled an approaching fire. In addition, the grazing and trampling effects of large numbers of bison would have left little fuel to carry a fire, thereby halting an advancing fire. The latter explanation suggests an important interaction between bison and fire that is the current focus of some grassland research.

The principal, indirect effect of grassland fires on large animals is primarily an increase in the quantity of available forage. There appears to be little effect of burning on the quality (e.g., nutrients) (Wright and Bailey 1982), although Ohr and Bragg (1985) reported significant differences in some nutrients; calcium and magnesium were lower and potassium, phosphorus, zinc, and manganese were higher in grasses from burned areas. In addition to improving forage for animals, fire removes standing, dead plant matter, making new growth more accessible to large grazers. For example, little bluestem grows in clumps that in the absence of burning retain many dead plant stalks; these unburned clumps are often avoided by cattle.

Whether by quantity or quality, the indirect effect of burning on herbivores is

Table 4.3. Average percent abundance of common small mammals (ranked from fire-positive to fire-negative) in burned and unburned prairie from autumn 1981 to autumn 1987

	Percent abundance	
Species	Burned	Not burned
Deer mouse	65.7	40.3
Thirteen-lined ground squirrel	5.3	4.3
White-footed mouse	6.4	11.5
Western harvest mouse	6.1	17.1
Prairie vole	4.2	5.9
Southern bog lemming	1.5	2.4
Elliot's short-tailed shrew	7.7	16.6

Note: Data from the Konza Prairie Research Natural Area, a tallgrass prairie located south of Manhattan, Kansas.

Source: Kaufman et al. (1990), with permission of the University of Oklahoma Press.

indicated by increased weight gains by cattle on spring-burned pasture. Plant regrowth following burning is so great an attraction that burned areas are often overgrazed while unburned areas are grazed only lightly. In addition to bison, elk, pronghorn, and rabbits concentrate on burned areas in North American grasslands (Lewis 1973; Evans and Probasco 1977). The attraction of large herbivores to recently burned areas was used by Native Americans to facilitate hunting (Moore 1972).

Small Mammals. Small mammals are generally unable to run from fire and thus must otherwise be adapted to survive in a fire-frequented environment. The abundance of small mammals in prairies, however, is evidence of their behavioral, physical, or reproductive capacity to survive in a fire environment. Most studies indicate that direct, fire-caused mortality is low (e.g., see Kaufman et al. 1990). Detailed searches following burning rarely find many dead small mammals, although it is not uncommon to find empty aboveground nests of small mammals. As with larger animals, the effects of fire are primarily indirect. Fires in ungrazed areas are generally detrimental to small mammals that eat foliage, use surface nests constructed of plant matter, or hunt in the litter layer for insects. In the Great Plains grasslands, this group includes the prairie vole and the short-tailed shrew. Fire also removes cover, thus exposing small mammals to increased predation. Fires have a positive effect on small mammals that eat seeds or insects in areas that contain little litter, such as the deer mouse and the meadow jumping mouse. The effect, whether positive or negative, is largely a consequence of the animals surviving the fire but moving or staying depending on their habitat requirements (Kaufman et al. 1990) (Table 4.3).

Birds. The direct effect of fire on bird populations varies. Hawks and purple martins, for example, are attracted to fire and use the fire front to locate prey (Lehman and Allendorf 1989). For other bird species, fire can be detrimental if, for example, it occurs during nesting or hatching of ground-nesting birds. Kruse and Piehl (1986), however, found that 69% of nests survived a mid-June fire in a northern mixed-grass prairie in North Dakota. They also suggested that many birds have time to complete hatching and nesting either before or after the time during which fires are most common.

Indirect effects of fire on birds act primarily through alteration of the foraging and nesting habitat, changes in vegetation structure, and its effect on food sources

Table 4.4. Individual species observations by treatment and sample period

| | Treatment and month sampled | | | |
| Species | Burned | | Not burned | |
	June	July	June	July
Sharp-tailed grouse	2	1	0	0
Killdeer	2	0	0	0
Upland sandpiper	50*†	24*†	0*	6*
Western kingbird	2	0	0	1
Horned lark	7	0	0	0
Common yellowthroat	0	1	4	2
Bobolink	0	0	1	0
Western meadowlark	4†	22‡†	8	10‡
Yellow-headed blackbird	3	3	10	0
Red-winged blackbird	17	6*	19	24*
Brown-headed cowbird	18	12	15	11
Savanna sparrow	0	0	1	4
Grasshopper sparrow	0*	8*	41*	48*
Vesper sparrow	0	0	0	2
Clay-colored sparrow	0	0	2	9
Chestnut-collared longspur	10	0	0	0
Unidentified	0	2	2	7
Totals	115	79*	103	123*

Note: Each number represents the total of five sample days for four transects per treatment. Data were collected in 1983 at the Samuel H. Ordway, Jr., Memorial Prairie, a northern mixed-grass prairie located west of Leola, South Dakota.

*Significantly different (P < 0.05) between treatments.

†Significantly different (P < 0.05) between sample periods within a treatment.

‡Significantly different (P < 0.10) between treatments.

Source: Modified from Huber and Steuter (1984), with permission of the North Dakota Natural Science Society.

(Kirsch and Kruse 1973; Risser et al. 1981; Kruse and Higgins 1990). Tester and Marshall (1961), for example, indicate that bobolinks, savannah sparrows, and Le Conte's sparrows all avoided recently burned grassland. Cannon (1979), however, noted that lesser prairie chickens in western Oklahoma moved the location of a lek (mating site) to a burned area, thus indicating that fire may be important in maintaining conditions for the reproductive behavior of this species. Similarly, Huber and Steuter (1984) noted that sharp-tailed grouse, a related species, was present on burned areas but noticeably absent from unburned areas. They also recorded the response for 15 other species in northern mixed-grass prairie, including the upland sandpiper and the western meadowlark, which favor burned areas, and the grasshopper sparrow, which favors unburned areas (Table 4.4). As with the sharp-tailed grouse, bobwhite quail were also noted to move to burned areas within a month of burning (Seitz and Landers 1972). In addition to these effects of burning in general, differences in response were found to be associated with the season in which burning occurred. Higgins (1986b), for example, indicated that duck nesting success was significantly greater with fall than with spring burns.

Reptiles and Amphibians (Herpetofauna). Reference on the effects of fire on reptiles and amphibians in grasslands is limited (Mushinsky 1985). As with other small animals, these organisms escape from fire by retreating belowground when possible. However, if they are caught aboveground, their physiological and morphological status makes them particularly vulnerable to being killed in a fire, perhaps because their body temperature rapidly reaches a lethal level. While some reptiles (lizards and snakes) die as a direct effect of burning, most others survive (Erwin and Stasiak 1979). Northern prairie skink, for example, survived a spring fire in an eastern Nebraska tallgrass prairie (personal observation). Similarly, 72% of common skinks in a New Zealand tussock grassland survived a September fire (Patterson 1984). As with mammals, young herpetofauna are particularly susceptible to fire. For example, a large number of young ringneck snakes were killed in a slow-moving, spring fire in an eastern Nebraska tallgrass prairie (personal observation).

Although little research has focused on fire and the herpetofauna of the North American grasslands, some effects are suggested by studies in a west-central Florida, slash pine community that has a grass-dominated understory. Burning in this community resulted in increased diversity and abundance of amphibians and reptiles over control plots; some fire frequencies were better than others for maintaining high diversity. These responses may, in part, have been caused by changes in food source as well as in the maintenance of the nonwoody habitat in which some reptiles do best.

Invertebrates. Invertebrates include a wide variety of organisms. Some reside above the soil and in or above the plant litter (e.g., butterflies and spiders), others live belowground (e.g., earthworms), and still others may be above- or belowground, depending on activities (e.g., ants) and the nature of their life cycle (e.g., beetles and beetle larvae). Aboveground organisms are more susceptible to direct fire mortality than are those remaining belowground, because lethal subsurface soil temperatures are rarely reached in grasslands. The effect of fire on soil organisms, therefore, is primarily a response to postburn conditions. The lack of litter, for example, results in increased soil temperature and reduced soil moisture. Increased plant growth, particularly in the tallgrass prairie, also results in more water lost through increased photosynthesis. The increased growth, however, also results in increased root production, thereby providing more organic matter for consumption by subsurface organisms. Ahlgren (1974), Risser et al. (1981), Warren et al. (1987), Anderson et al. (1989), and Seastedt and Ramundo (1990) summarize many of the effects of burning on the various categories of invertebrates in Great Plains grasslands. The following discussion is largely based on these references.

In general, the effects of burning on invertebrates are quite variable. As a group, insects do not appear to experience a severe decline following fire, suggesting that they have adaptations that allow at least some individuals to survive periodic burning. Often this survival is the result of immigration from nearby, unburned areas. Inappropriately timed fires, however, can result in the extirpation of butterflies and other prairie insects in isolated areas (Panzer 1988), such as typifies many nature preserves. Particular sensitivity to timing, frequency, and homogeneity of prescribed burns, therefore, is essential when managing habitats for invertebrates. Fire effects, however, vary for different invertebrate groups and for specific grassland types.

Grasshoppers and leafhoppers have been shown to increase with burning in tallgrass prairies in Kansas, Illinois, and Minnesota, perhaps because of lush growth following burning (Knutson and Campbell 1976). However, they were found to be

Table 4.5. Percentages of early (first through third) instar grasshopper nymphs belonging to individual species classified as either grass or forb feeding

	Percent of grasshoppers captured by watershed			
	1	2	4	U
Grass feeders				
Orphulella speciosa	66	56	13	4
Phoetaliotes nebrascensis	25	24	35	64
Other species	2	1	2	1
Forb feeders				
Melanoplus keelerii luridus	4	13	15	6
Melanoplus scudderi	3	6	28	20
Other species	tr[a]	1	8	5
Total N[b]	422	1147	513	396

Note: Data were collected from June to September 1982 at the Konza Prairie Research Natural Area, a tallgrass prairie located south of Manhattan, Kansas. Watersheds 1, 2, and 4 were burned in April 1982 but, from 1972 to 1981, had been burned every year (watershed 1), every other year (watershed 2), and every fourth year (watershed 4). Watershed U was not burned in 1982 and had not been burned since spring 1973.

[a]Less than 0.5%.

[b]Total number of individuals from which percentages were calculated.

Source: Modified from Evans (1984), with permission of Munksgaard, Copenhagen.

unaffected by burning in a sand prairie in Illinois. Varying fire conditions differently affected the response of these organisms to burning. Early-spring burns in tallgrass prairie, for example, resulted in more grasshoppers than did late-spring burns. Fire every four years appeared to result in greater grasshopper diversity than did more or less frequent fires (Table 4.5) (Evans 1984).

Ants have also been found to increase with burning (Table 4.6), averaging populations more than one-third larger in burned than in unburned areas. This group of invertebrates spends much of its time belowground; thus at least some of each colony survive a fire. The combination of their colonization habit, their scavenging behavior, and the general adaptation of ants, as a group, to the hot, dry conditions of a burned area make them one of the animals that most rapidly and substantially increases in population size in burned areas.

Beetles also appear to increase with burning. Although there may be an initial drop in population numbers caused by fire mortality, that is followed by rapid recolonization. For example, scarab beetle larvae (mostly *Phyllophaga* spp.) in the soil were two times greater in burned than in unburned tallgrass prairie (Seastedt 1984). Other studies, however, noted lower beetle populations with burning (Table 4.6) (Bertwell and Blocker 1975), apparently a consequence of differences in sampling technique. Other explanations for this difference may be that there is considerable variability in fire behavior and fire temperature, even within a small area (Hickey 1992), or that it only reflects the large fluctuations that characterize year-to-year invertebrate population sizes.

Spiders are drastically reduced by fire in most areas, although not all species are equally affected (Table 4.6). In southwestern Wisconsin, however, spiders were found to increase following a prairie fire, presumably because they were able to find refuge

Table 4.6. Invertebrates collected from native prairies from March through May 1928, following a March burn of part of the study area

Organisms	Combined number of species	Number of individuals by treatment	
		Burned	Not burned
Beetles	16	316	427
Spiders	14	21	265
Segmented worms	4	240	63
Moth (cutworm) larvae	5	80	175
Species including ants, wasps, bees, etc.			
Ants	4	2480	1200
All others	6	21	14
True bugs	6	250	76
Centipedes	2	20	82
Millipedes	2	40	50
Species including flies or mosquitoes			
Cranefly larvae and adults	—	36	25
All others	10	23	30
Springtails	4	30	130
Nymphs of species including grasshoppers, crickets, etc.	—	10	4
Nymphs of species including aphids, cicadas, hoppers, etc.	—	47	8
Mites	—	4	1
Total		3618	2550
Total minus ants[a]		1138	1350

Note: Surface invertebrates were collected by visually locating organisms in 1 meter^2 (11 foot^2) prairie plots. Subsurface invertebrates were collected from 0.1 meter^2 (1 foot^2) by 8 centimeters (3 inches) deep prairie soil both by visual searches and by screening of soil. The burned study site was a native, tallgrass prairie located 2 kilometers (1 mile) west of Seymour, Illinois. The prairie that was not burned was located 7 kilometers (4 miles) east, about 2 kilometers (0.5 mile) east of Bondville. Both sites were in a railroad right-of-way.

[a]Ants are separated from the total because of their large numbers, which mask the general decline in invertebrate populations.

Source: Extracted from Rice (1932), Table II with permission of the Ecological Society of America.

during the burn and find abundant food in the postburn prairie (Riechert and Reeder 1972).

Centipedes (a predatory group) and millipedes (a group that eats dead plant matter) are among surface dwellers that generally are reduced in numbers by burning in grasslands (Table 4.6). Their decline may be a consequence of direct mortality during burning because most of their activities are in the surface litter, which often reaches lethal temperatures or is consumed in a fire. Centipedes are often abundant in unburned areas, which may reflect the greater number of insect prey where litter is abundant.

Prairie soil fauna are largely affected indirectly by burning (Anderson et al. 1989; Seastedt and Ramundo 1990). Mites and springtails are a major part of the total soil fauna that are involved in the essential process of decomposition. The numbers of these invertebrates are reduced by burning (Table 4.6). In forests, their decline appears to be a direct consequence of increased soil temperature during burning. In grasslands, however, where soil temperature is not substantially increased, the decline of these groups is most likely a consequence of a postburn increase in soil temperature caused by increased sunlight and decreased soil moisture. Earthworm populations have been reported to be higher in previously burned areas (James 1988). This is apparently in response to increased belowground resources resulting, after some time, from increased plant and root growth in response to burning. Additional factors affecting earthworm populations include soil temperature and soil moisture conditions. Rice (1932), for example, reported fewer earthworms on burned areas in which soil moisture was substantially lower than in unburned areas. Where soil moisture remains adequate, the effect of fire is probably negligible.

Other Organisms. Fire also affects other organisms, but only limited research has been conducted in grasslands (Annala and Kapustka 1982; Seastedt and Ramundo 1990). Effects on fungi were reported by Shearer and Tiffany (1989), who found less fungal foliar diseases (leaf spot and mildew) on plants in burned areas than in unburned areas. Without burning, the number of diseased plants increased for at least two years. Vesicular-arbuscular mycorrhizal (VAM) fungi, commonly found associated with many prairie plant roots, increase the uptake of moisture and nutrients, thus enhancing the plant's growth. Gibson and Hetrick (1988) noted that burning frequency affected both individual VAM species and the VAM community as a whole.

Annala and Kapustka (1982) noted that burning did not significantly affect microbial activity in an Ohio prairie; however, Sambol (1981) found a significant decline in the number of bacteria/actinomycetes, fungal propagules, and bacterial endospores the day following burning. These numbers returned to preburn conditions one week after the burn, indicating increased microbial activity. During the remainder of the year, numbers in unburned plots were stable, while those in burned plots fluctuated greatly. The increase in microbial activity most likely also affects nitrogen fixation, which increases with burning and thus compensates for losses of nitrogen caused by burning (Ojima et al. 1990).

In a reestablished tallgrass prairie in Nebraska, Kragskow (1982) found that burning increased both algal abundance (+34%) and diversity (+27%), although different groups responded differently to burning; among those increasing with fire were green algae (Chlorophyta) (+29%) and blue-green algae (Cyanophyta) (+32%). Opposite results were recorded by Johansen and Rayburn (1989) in an arid grassland in eastern Washington. There, cryptogams, which form a protective crust over the soil, were killed by burning. Cryptogams include lichens, mosses, algae, and other plantlike organisms that do not form true flowers or seeds. Re-formation of cryptogamic crusts was estimated to take up to 10 years for the algae and 20 years for lichens and mosses. The differences in these two studies on algae reemphasizes the point that fire affects different grasslands in different ways.

Effects of Fire on the Grassland Environment

The effect of fire on the grassland environment often occurs in a site-specific manner. Wells et al. (1979) and Wright and Bailey (1982) summarize much of the results of research on the effect of fire on the physical environment.

Soil Organic Matter and Soil Chemistry. Dead plant and animal matter is consumed by many different organisms, adding humus (partially decomposed organic matter) to the soil until the nutrients are ultimately returned in their elemental state to be taken up again by plants. This process of decomposition is essential to completing the cycling of nutrients in any ecosystem. The main effect of fire on organic matter is that it compresses the longer process of biological decomposition into a very short time span in a process termed heat oxidation (burning). The product of grassland burning is the same as the product of decomposition except for (1) losses caused by runoff of ash and nutrients when heavy rains immediately follow a fire or (2) losses of nitrogen and sulfur to the atmosphere during burning, such as are known to occur in forests. Nitrogen present in rain offsets some of the losses of nitrogen that occur during burning. Long-term losses of nutrients during burning have not been demonstrated in grasslands. Further, of the generally short-term studies conducted to date, significant changes in soil chemistry resulting from annual burns in tallgrass prairie have not been reported. Frequent burning, however, does result in severe nitrogen limitation for plant production, but primarily through its effect on microbial activities involved in nitrogen fixation (Seastedt and Ramundo 1990).

Litter. Undecomposed plant matter (litter) is consumed in most burns, releasing moisture to the air and nutrients to the soil. Additional effects of litter loss with burning differ from one grassland to another. In the more moist tallgrass prairies, removal of litter stimulates plant growth because excessive litter accumulation retards growth of the dominants (Hulbert 1969). This stimulative effect probably reflects increased soil temperatures following spring burning, which advances plant growth by as much as 10 days. The adverse response to litter accumulation appears to be a unique feature of the tallgrass prairie (Table 4.2) (Knapp and Seastedt 1986). In dry grasslands, fire reduces plant production, and several years are required before plants recover to pre-burn levels. Whether this decline is ''good'' (i.e., a result of a natural event) or ''bad'' (e.g., for cattle grazing) depends entirely on the perspective of the observer.

The removal of litter has several additional effects, one of which is to expose the mineral soil surface to sunlight. This results in differences in surface temperature averaging 10°C (18°F) higher in burned than in unburned areas. Soil moisture also decreases primarily because of increased evaporation in the absence of litter, which, when present, insulates the soil against evaporation and protects the soil surface from heating, which increases evaporative losses. In addition, particularly in the moister grasslands (e.g., tallgrass prairie), the increased plant growth draws more moisture from the soil. Yet another factor is that snow accumulation may affect soil moisture. For example, in the northern mixed-grass prairie of southwestern Saskatchewan, an August, lightning-caused fire was found to reduce soil-moisture recharge, apparently because, in the absence of litter, snow was blown off the burned area. The effect of this inability to hold snow, which would have melted and recharged soil moisture, was noticeable for the following two years (DeJong and MacDonald 1975).

Offsetting the loss of water resulting from litter removal, the amount of rainfall reaching the soil surface can be greater in a burned than in an unburned area, making more soil water available with burning (Gilliam et al. 1987). Litter acts to intercept and hold moisture, both preventing the moisture from reaching the soil surface and increasing the likelihood of evaporation. This effect is most likely to be substantial only in areas of low rainfall, such as the shortgrass prairie. Where rainfall is higher, such as in the tallgrass and northern mixed-grass prairies, the water-holding capacity of litter is probably rapidly saturated, and additional water reaches the soil surface where it can percolate below the soil surface and thus be stored.

In general, differences in soil moisture that result from burning persist throughout the year of the burn and even longer where litter accumulation is slow. Differences in soil temperature and soil moisture also affect microbial activity, including nitrogen fixation, since the activity of microorganisms is reduced in dry soils.

An additional effect of litter removal is the exposure of bare soil to the impact of rainfall. Raindrop impact disrupts the surface soil structure and increases the possibility of soil erosion. Despite this effect, however, soil losses following burning of grasslands have been found to be negligible, although evidence is from grasslands on other continents (Wright and Bailey 1982). Even for the sandy soils of the Nebraska sandhills prairie, erosion following spring, summer, and fall burns was too slight to measure. Fall burns, which would have exposed the bare soils to several months of winter winds, did appear to cause some very limited soil erosion, but only in the year immediately following burning (personal observation). The principal explanation for grassland's resistance to erosion appears to be that the plants are not killed; thus the extensive root system holds soil against erosion until plant canopy is reestablished, often shortly after burning. Slope steepness, however, would increase the potential for soil erosion in some grasslands (Wright et al. 1976).

Additional effects of soil exposure by fire include a slight, if any, reduction in infiltration and percolation rates and a slight increase in pH in the upper few centimeters that persists for only one to two years.

Smoke Emissions. Smoke produced by grassland fires is an important consideration, particularly near populated areas. Few studies have been conducted on smoke from grassland fires (Martin et al. 1979), although numerous studies have been done on forests (Sandberg et al. 1979; Smith 1990; Ward and Hardy 1991). One principal grassland fire study was conducted by Boubel et al. (1969) on cool-season grasses from Oregon. Results indicate that 7 grams of particulates and 46 grams of carbon monoxide were produced for each kilogram of grass fuel burned (16 and 101 pounds per ton, respectively). This is considerably less than the product of wood-burning stoves, in which up to 50 grams of particulates and up to 300 grams of carbon monoxide are produced for each kilogram of wood burned (110 and 660 pounds per ton, respectively). In addition, there is considerably less fuel to burn in a grassland than there is in a forest. For the highest-productivity grasslands, such as the tallgrass prairie, the amount of plant matter for burning (fuel load) is about 7500 kilograms per hectare (3 tons per acre). For a mixed-conifer stand with some downed trees, the fuel load is about six times greater (42,000 kilograms per hectare, or 17 tons per acre). These estimates of plant matter are based on fuel models (Model Numbers 3 and 10) used by the U.S. Department of Agriculture Forest Service to assist in planning prescribed burns (Anderson 1982).

Management and Conservation Considerations

Of the principal abiotic forces that shape the grasslands (soil, climate, and fire), fire is the one that can be most effectively regulated. Fire has been a constant force in most of the North American grasslands just as in many other terrestrial ecosystems, superimposing its effects over soil and climatic regimes although it is often associated with, or at least encouraged by, drought. Today, the occurrence of natural fires is greatly diminished because of fire control, roads and cultivated fields that act as fire breaks, and the absence of large areas of grassland throughout which a single lightning ignition can expand. Consequently, prescribed burning along with grazing management (or lack of grazing) are the principal tools available for maintaining native prairies.

The use of fire as a tool in North American grasslands started with the Native Americans. Today its use is widespread in managing the remaining prairies. The natural fire regime (e.g., fire frequency, season, and variability) provides the basis for managing a grassland to maintain the biotic diversity of a natural system. However, grasslands are also managed for other purposes, such as cattle grazing, that may have a different objective for which the natural fire regime is not best suited. Because of this potential difference, it is most important to define the specific management objective before application of prescribed fire.

When assessing the objective of prescribed burning, either for the maintenance of natural grasslands or for human-oriented management objectives, it is important to realize that fire effects differ both between grasslands and within a single grassland. Whether these effects are "good" or "bad" depends solely on point of view. Spring is presently the most common season for prescribed burning, although current research on fires in other seasons may someday alter this timing somewhat, at least for some management objectives. Summer fires, for example, occurred naturally; thus occasional application of summer prescribed burns may be essential for managing a grassland to duplicate pre-European–settlement fire conditions. Fires in summer, however, may reduce annual plant production; thus summer may be a poor time to burn for those interested in maximizing plant production for haying or for raising cattle. Fire is the appropriate tool for both objectives, although the season of burning most likely will differ.

In addition to differences in the season of conducting prescribed burns, there are differences in the number of years between burns, which vary both for management objectives and for specific grassland types. In general, where precipitation is limited, such as in the western and central grasslands, a long-term decline in grass production occurs when burning is more frequent than every 5 to 10 years. This fire frequency may be best for natural fire management of grasslands, such as the short- or mixed-grass prairies, although fire exclusion may be best for other purposes. In the more eastern and more moist grasslands, such as the tallgrass prairie, burning could occur every 1 to 3 years without adversely affecting productivity. However, a frequency of from 1 to every 2 years is likely to cause a decline in forbs and a dominance by a few grasses. This may be advantageous to cattle grazing, but it is not the best management for maintaining diverse natural areas.

Prescribed burning is the process of applying fire in order to accomplish some objective. It involves careful planning that takes into account various considerations, including fire season, fire frequency, smoke management, and erosion control. Often overlooked, smoke production is a consideration that has become an increasing concern in light of environmental pollution and suggested global warming. Headfires usually produce more smoke than do backfires (Sandberg and Ward 1981) because of incomplete combustion, particularly of surface litter. However, where smoke emissions do not adversely affect populated areas, headfires are preferable to backfires in grasslands. Backfires more completely consume surface litter and are more likely to adversely affect resident surface and subsurface fauna. In the historic fire environment, headfires almost certainly affected a considerably greater portion of a burn than did backfires.

A few words of caution about prescribed burning: First, managing an area, whether for conservation, cattle, or other objective, may require that fire be applied when only limited information, particularly on long-term effects, is available. It may be necessary, for example, to burn to prevent further woody plant invasion or to reduce excessive litter accumulation that hinders plant growth. Under these conditions, it is advisable to apply fire with caution and to monitor effects until the consequences of a specific fire plan are more fully understood. Second, it is unwise to conduct a burn without having had personal experience in the field with experienced individuals, because grass fires can be unpredictable and dangerous. Historic bison kills are only one stark reminder of the consequence of being in the wrong place at the wrong time with fire as a companion.

For grasslands, there are several references available that provide explanations of the basic concepts of planning and conducting prescribed burns. Wright and Bailey (1982), Higgins et al. (1989), and Masters et al. (1990) provide some basics.

LITERATURE CITED

Adams, D. E., and R. C. Anderson. 1978. The response of a central Oklahoma grassland to burning. Southwestern Naturalist 23:623–632.

Ahlgren, I. F. 1974. The effect of fire on soil organisms. Pp. 47–72 in T. T. Kozlowski and C. E. Ahlgren (eds.). *Fire and Ecosystems*. Academic Press, San Francisco.

Akin, W. E. 1991. *Global Patterns: Climate, Vegetation, and Soils*. University of Oklahoma Press, Norman.

Anderson, H. E. 1982. *Aids to Determining Fuel Models for Estimating Fire Behavior*. General Technical Report INT-122. U.S. Department of Agriculture, Forest Service, Ogden, Utah.

Anderson, R. C., T. Leahy, and S. S. Dhillion. 1989. Numbers and biomass of selected insect groups on burned and unburned sand prairie. American Midland Naturalist 122:151–162.

Annala, A. E., and L. A. Kapustka. 1982. The microbial and vegetational response to fire in the Lynx Prairie Preserve, Adams County, Ohio. Prairie Naturalist 14:101–112.

Bailey, A. W., and M. L. Anderson. 1978. Prescribed burning of a *Festuca-Stipa* grassland. Journal Range Management 31:446–449.

Bertwell, R. L., and H. D. Blocker. 1975. Curculionidae from differently managed tallgrass prairie near Manhattan, Kansas. Journal Kansas Entomological Society 48:319–326.

Bidwell, T. G., and D. M. Engle. 1991. Behavior of headfires and backfires on tallgrass prairie. Pp. 344–350 in S. C. Nodvin and T. A. Waldrop (eds.). *Fire and the Environment: Ecological and Cultural Perspectives*. General Technical Report SE-69. U.S. Department of Agriculture, Forest Service, Southeastern Forest Experiment Station.

Bidwell, T. G., D. M. Engle, and P. L. Claypool. 1990. Effects of spring headfires and backfires on tallgrass prairie. Journal Range Management 43:209–212.

Black, C. C. 1971. Ecological implications of dividing plants into groups with distinct photosynthetic production capacities. Advances Ecological Research 7:87–114.

Bleed, A., and C. Flowerday (eds.). 1990. *An Atlas of the Sand Hills*, 2nd ed. Conservation and Survey Division, Institute of Agriculture and Natural Resources, University of Nebraska, Lincoln.

Boubel, R. W., E. F. Darley, and E. A. Schuck. 1969. Emissions from burning grass stubble and straw. Journal Air Pollution Control Association 19:497–500.

Bragg, T. B. 1974. Woody plant succession on various soils of unburned bluestem prairie in Kansas. Ph.D. Dissertation, Division of Biology, Kansas State University, Manhattan.

Bragg, T. B. 1978. Effects of burning, cattle grazing, and topography on vegetation of the choppy sands range site in the Nebraska sandhills prairie. Pp. 248–253 in D. N. Hyder (ed.). *Proceedings of the First International Rangeland Congress*. Society for Range Management, Denver.

Bragg, T. B. 1982. Seasonal variations in fuel and fuel consumption by fires in a bluestem prairie. Ecology 63:7–11.

Bragg, T. B. 1986. Fire history of a North American sandhills prairie (Abstract). Program of the IV International Congress of Ecology. Syracuse University, Syracuse, N.Y.

Bragg, T. B., and L. C. Hulbert. 1976. Woody plant invasion of unburned Kansas bluestem prairie. Journal Range Management 29:19–24.

Bryson, R. A., and F. K. Hare (eds.). 1974. *World Survey of Climatology.* Vol. 11: *Climates of North America.* Elsevier, New York.

Cannon, R. W. 1979. Lesser prairie chicken responses to range fires at the booming ground. Wildlife Society Bulletin 7:44–46.

Collins, S. L., and L. L. Wallace (eds.). 1990. *Fire in North American Tallgrass Prairies.* University of Oklahoma Press, Norman.

Daubenmire, R. 1968. Ecology of fire in grasslands. Advances Ecological Research 5:209–266.

DeJong, E., and K. B. MacDonald. 1975. The soil moisture regime under native grassland. Geoderma 14:207–221.

Ehrenreich, J. H. 1959. Effect of burning and clipping on growth of native prairie in Iowa. Journal Range Management 12:133–137.

Erwin, W. J., and R. H. Stasiak. 1979. Vertebrate mortality during the burning of a reestablished prairie in Nebraska. American Midland Naturalist 101:247–249.

Evans, E. W. 1984. Fire as a natural disturbance to grasshopper assemblages of tallgrass prairie. Oikos 43:9–16.

Evans, K. E., and G. E. Probasco. 1977. *Wildlife of the Prairies and Plains.* General Technical Report NC-29. U.S. Department of Agriculture, Forest Service, North Central Forest Experiment Station, St. Paul, Minn.

Gartner, F. R., J. R. Lindsey, and E. M. White. 1986. Vegetation responses to spring burning in western South Dakota. Pp. 143–146 in G. K. Clambey and R. H. Pemble (eds.). *The Prairie— Past, Present and Future.* Proceedings of the Ninth North American Prairie Conference. Tri-College University Center for Environmental Studies, North Dakota State University, Fargo.

Gartner, F. R., R. I. Butterfield, W. W. Thompson, and L. R. Roath. 1978. Prescribed burning of range ecosystems in South Dakota. Pp. 687–690 in D. N. Hyder (ed.). *Proceedings of the First International Rangeland Congress.* Society for Range Management, Denver.

Gehring, J. L., and T. B. Bragg. 1992. Changes in prairie vegetation under eastern red cedar *Juniperus virginiana* (L.) in an eastern Nebraska bluestem prairie. American Midland Naturalist 128:209–217.

Gibson, D. J., D. C. Hartnett, and G.L.S. Merrill. 1990. Fire temperature heterogeneity in contrasting fire prone habitats: Kansas tallgrass prairie and Florida sandhill. Bulletin Torrey Botanical Club 117:349–356.

Gibson, D. J., and B.A.D. Hetrick. 1988. Topographic and fire effects on the composition and abundance of VA-mycorrhizal fungi in tallgrass prairie. Mycologia 80:433–441.

Gilliam, F. S., T. R. Seastedt, and A. K. Knapp. 1987. Canopy rainfall interception and throughfall in burned and unburned tallgrass prairie. Southwestern Naturalist 32:267–271.

Haley, J. E. 1929. Grass fires of the southern Great Plains. West Texas Historical Association Year Book 5:23–42.

Henderson, R. A. 1992. Ten-year response of a Wisconsin prairie remnant to seasonal timing of fire. Pp. 121–125 in *Recapturing a Vanishing Heritage.* Proceedings of the Twelfth North American Prairie Conference. University of Northern Iowa, Cedar Falls.

Henry, A., and D. Thompson. 1965. *New Light on the Early History of the Greater Northwest: The Manuscript Journals of Alexander Henry and David Thompson, 1799–1814.* Ed. E. Coues. Ross and Haines, Minneapolis.

Hickey, S. M. 1992. The influence of fuel bed heterogeneity on plant response to fire in tallgrass prairie. M.A. Thesis, Department of Biology, University of Nebraska, Omaha.

Higgins, K. F. 1984. Lightning fires in North Dakota grasslands and in pine-savanna lands of South Dakota and Montana. Journal Range Management 37:100–103.

Higgins, K. F. 1986a. *Interpretation and Compendium of Historical Fire Accounts in the Northern Great Plains.* Resource Publication 161. U.S. Department of the Interior, Fish and Wildlife Service, Brookings.

Higgins, K. F. 1986b. A comparison of burn season effects on nesting birds in North Dakota mixed-grass prairie. Prairie Naturalist 18:219–228.

Higgins, K. F. 1989. *Effects of Fire in the Northern Great Plains.* U.S. Department of Agriculture Publication EC-761. U.S. Fish and Wildlife Service and Cooperative Extension Service, South Dakota State University, Brookings.

Higgins, K. F., A. D. Kruse, and J. L. Piehl. 1989. *Prescribed Burning Guidelines in the Northern Great Plains.* U.S. Department of Agriculture Publication EC-760. U.S. Fish and Wildlife Service and Cooperative Extension Service, South Dakota State University, Brookings.

Higgins, K. F., D. P. Fellows, J. M. Callow, A. D. Kruse, and J. L. Piehl. 1988. *Annotated Bibliography of Fire Literature Relative to Northern Grasslands in South-Central Canada and North-Central United States.* U.S. Department of Agriculture Publication EC-762. U.S. Fish and Wildlife Service and Cooperative Extension Service, South Dakota State University, Brookings.

Hopkins, H., F. W. Albertson, and A. Riegel. 1948. Some effects of burning upon a prairie in west-central Kansas. Transactions Kansas Academy of Science 51:131–141.

Huber, G. E., and A. A. Steuter. 1984. Vegetation profile and grassland bird response to spring burning. Prairie Naturalist 16:55–61.

Hulbert, L. C. 1969. Fire and litter effects in undisturbed bluestem prairie in Kansas. Ecology 50:874–877.

James, S. W. 1988. The postfire environment and earthworm populations in tallgrass prairie. Ecology 69:476–483.

Johansen, J. R., and W. R. Rayburn. 1989. Effects of rangefire on soil cryptogamic crusts. Pp. 107–109 in D. M. Baumgartner, D. W. Breuer, B. A. Zamora, L. F. Neuenschwander, and R. H. Wakimota (eds.). *Prescribed Fire in the Intermountain Region: Forest Site Preparation and Range Improvement.* Washington State University Press, Pullman.

Kaufman, D. W., E. J. Finck, and G. A. Kaufman. 1990. Small mammals and grassland fires. Pp. 46–80 in S. L. Collins and L. L. Wallace (eds.). *Fire in North American Tallgrass Prairies.* University of Oklahoma Press, Norman.

Kirsch, L. M., and A. D. Kruse. 1973. Prairie fires and wildlife. Proceedings of the Tall Timbers Fire Ecology Conference 12:289–303.

Knapp, A. K., and T. R. Seastedt. 1986. Detritus accumulation limits productivity of tallgrass prairie. BioScience 36:662–668.

Knutson, H., and J. B. Campbell. 1976. Relationships of grasshoppers (Acrididae) to burning, grazing, and range sites of native tallgrass prairie in Kansas. Proceedings of the Tall Timbers Conference on Ecology of Animal Control by Habitat Management 6:107–120.

Kragskow, S. L. 1982. Effects of burning on soil algae in a restored tallgrass prairie. M.A. Thesis, Department of Biology, University of Nebraska, Omaha.

Kruse, A. D., and K. F. Higgins. 1990. Effects of prescribed fire upon wildlife habitat in northern mixed-grass prairie. Pp. 182–193 in M. E. Alexander and G. F. Bisgrove (technical coordinators). *The Art and Science of Fire Management: Proceedings of the First Interior West Fire Council Annual Meeting and Workshop.* Forestry Canada Northwest Region Information Report NOR-X-309.

Kruse, A. D., and J. L. Piehl. 1986. The impact of prescribed burning on ground-nesting birds. Pp. 153–156 in G. K. Clambey and R. H. Pemble (eds.). *The Prairie—Past, Present and Future.* Proceedings of the Ninth North American Prairie Conference. Tri-College University Center for Environmental Studies, North Dakota State University, Fargo.

Launchbaugh, J. L. 1964. Effects of early spring burning on yields of native vegetation. Journal Range Management 17:5–6.

Launchbaugh, J. L. 1973. Effect of fire on shortgrass and mixed prairie species. Proceedings of the Tall Timbers Fire Ecology Conference 12:129–151.

Leggett, R. F. (ed.). 1968. *Soils in Canada. Geological, Pediological and Engineering Studies.* Royal Society of Canada Special Publications No. 3.

Lehman, R. N., and J. W. Allendorf. 1989. The effects of fire, fire exclusion and fire management on raptor habitats in the western United States. Pp. 236–244 in B. G. Pendleton (ed.). *Proceedings of the Western Raptor Management Symposium and Workshop.* Scientific and Technical Series No. 12. Institute for Wildlife Research, National Wildlife Federation.

Lewis, H. T. 1973. *Patterns of Indian Burning in California. Ecology and Ethnohistory.* Ballena Press Anthropological Papers No. 1. Ballena Press, Menlo Park, Calif.

Martin, R. E., H. E. Anderson, W. D. Boyer, J. H. Dieterich, S. N. Hirsch, V. J. Johnson, and W. H. McNab. 1979. *Effects of Fire on Fuels: A State-of-Knowledge Review.* General Technical Report WO-13. U.S. Department of Agriculture, Forest Service.

Masters, R. A., R. Stritzke, and S. S. Waller. 1990. *Conducting a Prescribed Burn and Prescribed Burning Checklist.* Nebraska Cooperative Extension Publication No. EC-90-121. U.S. Department of Agriculture, Cooperative Extension Service and Institute of Agriculture and Natural Resources, University of Nebraska, Lincoln.

Moore, C. T. 1972. Man and fire in the central North American grassland, 1535–1890: a documentary historical geography. Ph.D. Dissertation, Department of Geography, University of California, Los Angeles.

Morrison, L. C., J. D. DuBois, and L. A. Kapustka. 1986. The vegetational response of a Nebraska sandhills grassland to a naturally occurring fall burn. Prairie Naturalist 18:179–184.

Mushinsky, H. R. 1985. Fire and the Florida sandhill herpetofaunal community: with special attention to response of *Cnemidophorus sexlineatus.* Herpetologica 41:333–342.

Nagel, H. G. 1983. Effect of spring burning date on mixed-prairie soil moisture, productivity and plant species composition. Pp. 259–263 in C. L. Kucera (ed.). *Proceedings of the Seventh North American Prairie Conference.* Southwest Missouri State University, Springfield.

Ohr, K. M., and T. B. Bragg. 1985. Effects of fire on nutrient and energy concentration of five prairie grass species. Prairie Naturalist 17:113–126.

Ojima, D. S., W. J. Parton, D. S. Schimel, and C. E. Owensby. 1990. Simulated impacts of annual burning on prairie ecosystems. Pp. 118–132 in S. L. Collins and L. L. Wallace (eds.). *Fire in North American Tallgrass Prairies.* University of Oklahoma Press, Norman.

Old, S. M. 1969. Microclimate, fire, and plant production in an Illinois prairie. Ecological Monographs 39:355–384.

Panzer, R. 1988. Managing prairie remnants for insect conservation. Natural Areas Journal 8:83–90.

Patterson, G. B. 1984. The effect of burning off tussock grassland on the population density of common skinks. New Zealand Journal Zoology 11:189–194.

Rice, L. 1932. The effect of fire on prairie animal communities. Ecology 13:392–401.

Riechert, S. E., and W. G. Reeder. 1972. Effect of fire on spider distribution in southwestern Wisconsin prairies. Pp. 73–90 in J. H. Zimmerman (ed.). *Proceedings of the Second Midwest Prairie Conference.* Madison, Wis.

Risser, P. G., E. C. Birney, H. D. Blocker, S. W. May, W. J. Parton, and J. A. Wiens. 1981. *The True Prairie Ecosystem.* Hutchinson Ross, Stroudsburg, Pa.

Sambol, A. R. 1981. Effects of fire on the soil microbial ecosystem in a native tallgrass prairie. M.A. Thesis, Department of Biology, University of Nebraska, Omaha.

Sandberg, D. V., and D. E. Ward. 1981. Air quality considerations in using prescribed fire for weed control. Pp. 237–247 in *Weed Control in Forest Management.* Proceedings of the 1981 John S. Wright Forestry Conference. Purdue University, West Lafayette, Ind.

Sandberg, D. V., J. M. Pierovich, D. G. Fox, and E. W. Ross. 1979. *Effects of Fire on Air: A State-of-Knowledge Review.* General Technical Report WO-9. U.S. Department of Agriculture, Forest Service.

Sauer, C. O. 1950. Grassland climax, fire and man. Journal Range Management 3:16–20.

Seastedt, T. R. 1984. Belowground macroarthropods of annually burned and unburned tallgrass prairie. American Midland Naturalist 11:405–407.

Seastedt, T. R., and R. A. Ramundo. 1990. The influence of fire on belowground processes of tallgrass prairie. Pp. 99–117 in S. L. Collins and L. L. Wallace (eds.). *Fire in North American Tallgrass Prairies.* University of Oklahoma Press, Norman.

Seitz, W. K., and R. Q. Landers. 1972. Controlled burning in relationship to bobwhite quail populations on a southern Iowa public hunting area. Iowa State Journal Research 47:149–165.

Shearer, J. F., and L. H. Tiffany. 1989. Fungus disease in relation to managing prairie plants with fire. Pp. 183–185 in T. B. Bragg and J. Stubbendieck (eds.). *Proceedings of the Eleventh North American Prairie Conference.* University of Nebraska, Lincoln.

Sims, P. L. 1988. Grasslands. Pp. 265–286 in M. G. Barbour and W. D. Billings (eds.). *North American Terrestrial Vegetation.* Cambridge University Press, New York.

Smith, W. H. 1990. *Air Pollution and Forests: Interactions between Air Contaminants and Forest Ecosystems,* 2nd ed. Springer-Verlag, New York.

Stechman, J. V. 1983. Fire hazard reduction practices for annual-type grassland. Rangelands 5:56–58.

Steinauer, E. M., and T. B. Bragg. 1987. Ponderosa pine (*Pinus ponderosa*) invasion of Nebraska sandhills prairie. American Midland Naturalist 118:358–365.

Steuter, A. A. 1987. C_3/C_4 production shift on seasonal burns—Northern Mixed Prairie. Journal Range Management 40:27–31.

Tester, J. R., and W. H. Marshall. 1961. *A Study of Certain Plant and Animal Interrelationships on a Prairie in Northwestern Minnesota.* University of Minnesota, Museum of Natural History Occasional Papers No. 8. University of Minnesota, Minneapolis.

Thornthwaite, C. W. 1948. An approach toward a rational classification of climate. Geographical Review 38:55–94.

Vogl, R. J. 1974. Effects of fire on grasslands. Pp. 139–194 in T. T. Kozlowski and C. E. Ahlgren (eds.). *Fire and Ecosystems.* Academic Press, San Francisco.

Ward, D. E., and C. C. Hardy. 1991. Smoke emissions from wildland fires. Environment International 17:117–134.

Warren, S. D., C. J. Scifres, and P. D. Teel. 1987. Response of grassland arthropods to burning: a review. Agriculture, Ecosystems and Environment 19:105–130.

Weaver, J. E. 1954. A seventeen-year study of plant succession in prairie. American Journal Botany 41:31–38.

Weaver, J. E. 1968. *Prairie Plants and Their Environment: A Fifty-Year Study in the Midwest.* University of Nebraska Press, Lincoln.

Wells, C. G., R. E. Campbell, L. F. DeBano, C. E. Lewis, R. L. Fredriksen, E. C. Franklin, R. C. Foelich, and P. H. Dunn. 1979. *Effects of Fire on Soil: A State-of-Knowledge Review.* General Technical Report WO-7. U.S. Department of Agriculture, Forest Service.

Wilhite, D. A., and R. O. Hoffman. 1979. *Drought in the Great Plains: A Bibliography.* Institute of Agriculture and Natural Resources, University of Nebraska, Lincoln.

Wright, H. A., and A. W. Bailey. 1980. *Fire Ecology and Prescribed Burning in the Great Plains—A Research Review.* General Technical Report INT-77. U.S. Department of Agriculture, Forest Service, Intermountain Forest and Range Experiment Station.

Wright, H. A., and A. W. Bailey. 1982. *Fire Ecology: United States and Southern Canada.* Wiley, New York.

Wright, H. A., F. M. Churchill, and W. C. Stevens. 1976. Effect of prescribed burning on sediment, water yield, and water quality from dozed juniper lands in central Texas. Journal Range Management 29:294–298.

5

Population Processes

DAVID C. HARTNETT
KATHLEEN H. KEELER

> That conditions of life are severe in the [tallgrass] prairie was shown conclusively by the removal of all but one species from a single square meter. . . . The remaining sunflowers or goldenrods or grass, profiting from an increased supply of water, light, and nutrients, increased in vigor, size, and production of flowers and seeds.
>
> J. E. Weaver, *North American Prairie*

POPULATIONS ARE MORE THAN JUST COLLECTIONS OF INDIVIDUALS

Within the prairies, myriad populations of plants, insects, small mammals, and other organisms form distinct levels of organization. The physical forces of weather and fire and the complex networks of species interactions control each species's structure and dynamics and determine its abundance or rarity. Variation in population dynamics and interactions over time and space determine the distribution and abundance of species, and ultimately the composition and dynamics of entire prairie communities.

To the casual observer, plant populations on prairies may seem quite static, like museums containing a diverse collection of specimens that can be seen again and again on repeated visits. Many plant populations, however, are much more analogous to a railway station than a museum, characterized by a constant flux of arrivals and departures. Furthermore, much of the dynamics within prairie plant populations occurs hidden belowground in the pools of seeds and rhizomes, obscuring the true population sizes and their dynamic nature. Even the dominant species that appear to be temporally and spatially stable are not static, but are characterized by significant population flux.

The basic processes determining the sizes and dynamics of all populations are birth, death, and migration. The size of a leadplant population in a tallgrass prairie is determined by the balance between additions of new shoots via both seedling establishment and vegetative reproduction and the death of individuals as a result of competition, herbivory, or physical factors such as drought and fire.

These basic population processes are influenced by a complex set of interacting factors. Fire affects plant populations directly by altering demographic traits and indirectly by influencing plant–herbivore interactions. Water availability, emphasized

Plate I. Keith Jacobshagen, *Dog Days Afternoon, 1989*.
(Reproduced with permission of the author, private collection)

Plate II. Representative tallgrass prairie: Konza Prairie,
near Manhattan, Kansas. (K. Keeler)

Plate III. Fire in tallgrass prairie. (T. Bragg)

Plate IV. Representative grazed shortgrass steppe. (K. Keeler)

Plate V. Historically important grasslands herbivores, grass-feeding grasshopper *(Arphia pseudonietana)*. (A. Joern)

Plate VI. Bison grazing at Yellowstone National Park in March. (A. Joern)

Plate VII. Representative prairie soil and the vegetation it supports, Pawnee Classification: (a) aboveground: vegetation associated with soil type; (b) belowground: soil profile. (Reprinted with permission of Nebraska Conservation and Survey)

as a major controlling influence in much of early prairie literature (Weaver and Mueller 1942; Weaver and Albertson 1944), has a large influence on plant population dynamics and densities. Populations of many of the prairies' most common species are persistent and highly resilient to environmental stresses such as drought, although their visibility can vary markedly from year to year because of the impact of these stresses on mortality and recruitment. Selective grazing by native ungulates such as bison, soil disturbance by small burrowing rodents, and mutualistic interactions such as those between plants and their pollinators or plants and mycorrhizal fungi all play key roles shaping the structure and dynamics of plant populations, but are not independent of the effects of fire and competition. The interplay of all these factors determines spatial patterns, variation, dynamics, and structure of plant populations across the prairies.

Animal populations respond to a similar set of forces. Population sizes of prairie birds, mammals, insects, and other animals are all affected not only by rates of birth and death, but also by immigration and emigration of individuals resulting from the natural dispersal of offspring as well as habitat changes caused by fire or other factors. An understanding of these processes and the environmental factors that influence them is key to understanding the dynamics of prairies and determining sound approaches to their management and restoration.

The population sizes of many species in the central grasslands, from large mammals such as bison to tiny plants, have been profoundly influenced by human activities since the nineteenth-century arrival of European settlers. The black-footed ferret has declined in number and is now the rarest mammal in North America, primarily because of human efforts to control populations of prairie dogs, the ferret's primary food. Prairie dogs have been reduced to less than 2% of the population size recorded several decades ago (Summers and Linden 1978). *Penstemon haydenii*, the only plant species endemic to Nebraska, is now endangered primarily as a result of human influences on grazing patterns. Even big bluestem, once continuous across the tallgrass prairie region, is now missing from large areas because of plowing, road building, and urbanization.

In this chapter, we provide an overview of patterns in populations of prairie plants, insects, birds, and mammals as well as the processes influencing their numbers. Examples are provided from different types of prairies to illustrate the variation in controlling influences and population patterns across the central grasslands.

PRAIRIE PLANTS DISPLAY A VARIETY OF LIFE HISTORIES AND DYNAMICS

Although many different plant growth forms and life history patterns are represented in the central prairies, the dominant species are primarily long-lived perennial herbs. Most prairie plants flower and set seed many times throughout their lives (polycarpic species). After a juvenile period of varying length, during which a plant accumulates a critical threshold level of necessary resources and a critical minimum size, it begins to flower and produce seeds (Roberts et al. 1977; Auffenorde and Wistendahl 1983). Seed production generally increases gradually in successive years with increasing size. A few species flower and produce seed only once in a lifetime (monocarpic species), often after years of vegetative growth. Yet another group, the annuals, complete their life cycle from seed to seed within a single year, but typically have many years of seeds lying dormant in the ground.

Many of the dominant perennials are long-lived (Anderson 1946; Wright and Van Dyne 1976; Gatsuk et al. 1980). The majority of dominant forbs of tallgrass prairie have life spans of 10 to 30 years (Blake 1935; Weaver 1954), whereas some species—such as the orchid *Spiranthes cernua,* the four-o'clock, the bush morning glory, and the Carolina puccoon—may live much longer (Platt 1976; Weller 1985; Keeler 1991; Antlfinger, unpublished data). The population dynamics of short-lived species may be highly responsive to year-to-year environmental variation. The very long-lived species appear to be well adapted and less sensitive to this temporal variation, however, and it is intriguing to think that many individuals we observe today may have experienced extremes such as the droughts of the 1930s and early 1950s as passing intervals of minor impact on their lifetime reproductive success. Longevities of prairie plants are especially difficult to study because, unlike trees, few species produce woody structures that can be aged and many of the plants live a substantial portion of a human life span or longer. A study of switchgrass in Kansas tallgrass prairie showed that the rate of lateral expansion of clones through vegetative reproduction was remarkably constant from year to year. This suggested that the ages of individual clones could be estimated retrospectively by measuring their diameters. Most of the individuals in the local population studied were estimated to be 14 to 20 years old, although a few of the clones that were several meters in diameter may have been several decades old (Hartnett 1991).

The causes of mortality of established prairie plants are numerous and interacting. Individuals face hazards throughout their lifetimes from physical factors, such as desiccation or fire, or they may succumb to the effects of burial by rodents, herbivory from large mammals, or incessant attacks of grasshoppers, beetles, insects, or fungal pathogens. One of the major stresses for established prairie plants is the effects of competitors. Neighboring plants may decrease survivorship or reproductive success by preempting light, water, nutrients, or the services of pollinating insects. The young seedling is typically the most vulnerable stage, and mortality of prairie grasses and forbs is concentrated among the smaller and younger plants in the population. Because they interact, the actual causes of mortality are difficult to discern. For example, defoliation by herbivores may not kill a plant directly, but may do so by weakening its competitive ability (Hartnett 1990; Louda et al. 1990).

MOST PLANT POPULATION REGENERATION IS VIA VEGETATIVE REPRODUCTION

The majority of prairie perennials have two modes of reproduction. They produce new individuals via the sexual reproductive process of flowering, pollination, and seed production, and they also reproduce vegetatively, producing clones of new stems via rhizomes, runners, or similar specialized structures. These reiterated vegetatively produced shoots are referred to as ramets in clonal forbs and tillers in grasses. In the prairies, most successful reproduction and population regeneration is via vegetative reproduction, and regeneration from seed typically plays a minor role (Weaver and Mueller 1942; Tripathi and Harper 1973; Glenn-Lewin et al. 1990).

Ramet or tiller recruits differ from seedlings in several important respects. They are, in most cases, genetically identical to their parent plants, in contrast to the genetically variable offspring produced by the sexual reproductive process and outcrossing. There is some debate, however, concerning the genetic consequences of seed versus vegetative reproduction in these perennial herbs, because many populations regenerating primarily by vegetative reproduction appear to be genetically just as diverse as those regenerating via seed (Ellstrand and Roose 1987). A mixture of clones can contain abundant genetic variation. In addition, mutations within the developing buds along a rhizome system may accumulate to produce genetically variable shoots, as appears to be the case in goldenrods (Schaal 1988).

Shoots arising via vegetative reproduction can also show much greater competitive ability and higher survivorship than those arising from seed. In many clonal perennials, the rhizome or stolon functions like an umbilical cord, enabling the transport of resources from the parent plant to the offspring shoot to ensure its successful establishment until the time when its own root-and-shoot system is sufficient to successfully compete against well-established neighbors (Hartnett 1983). When one considers the strong competition and other hazards facing new plant recruits on the prairie, and the small amount of stored food reserves contained within a seed relative to the vast supply available to a ramet or tiller, it is not surprising that seedling recruitment typically makes a small relative contribution to plant population regeneration.

The patterns of vegetative reproduction in prairie communities also influence spatial patterns in the prairie vegetation mosaic. Those clonal plant populations characterized by the production of numerous short interconnected rhizomes, or runners, form well-defined clones that are easy to identify, such as those of many bunchgrasses. Other species, such as big bluestem, clover, and leadplant, produce few very long wandering rhizomes, and their interconnections may senesce after a few years. These form widely spreading populations of intermingling clones that cannot be clearly delimited. These variations in clonal growth patterns are important because they affect patterns of competition among populations, and competition is a critical process limiting both seed and vegetative reproduction in many clonal grasses and forbs (Hartnett and Bazzaz 1985; Briske and Butler 1989). The tightly compact clones are "territorial." Their dense clump of stems is not easily invaded by competitors, and only the stems growing at the clone edge experience strong competition from neighbors. In the widely spreading rhizomatous species, all stems experience competition from other species as they wander through the habitat. Plant species with long rhizomes are quite mobile over time. They can potentially sample a large area within their habitat and can be very responsive to local environmental patchiness. For many of the rhizomatous prairie grasses, the local light microenvironment they encounter as they sample their habitat may strongly influence the population dynamics of rhizomes and tillers. As sunlight is filtered through a dense prairie canopy, only a small percentage reaches the soil surface. The canopy also changes the spectral composition of light, absorbing primarily red light and reflecting or transmitting longer-wavelength far-red and infrared light. Thus the soil surface under a dense prairie canopy experiences a low ratio of red to far-red light. Rhizomes of clonal grasses and forbs may respond to red/far-red light and sense resource availability, competitor density, or changes in canopy density caused by grazing or fire, and adjust vegetative reproduction and shoot population size to local conditions (Deregibus et al. 1985). An understanding of the controls on dynamics of vegetatively reproduced shoot populations is crucial. Total vegetation productivity and community structure is determined primarily by the numbers of ramets and tillers rather than their sizes.

It is difficult to identify "individuals" in prairie populations consisting of wandering and intermingling clones, especially since connections are often underground. Consequently, some aspects of the dynamics of populations of clonal plants in prairie habitats remain somewhat a mystery and are matters of great interest. Does a single, highly competitive clone eventually expand and eliminate the weaker, such that populations gradually progress toward domination by a single clone covering a large area? The high genetic diversity of prairie plants suggests that this is not often the case. If so, what factors prevent eventual domination by one or a few competitively superior individuals?

In clonal prairie plants, population processes occur at two distinct levels. Patterns of birth and death of whole clones may be very different from the patterns of birth and death of shoots making up each clone. If, for example, drought stress caused a 50% morality rate of vegetative shoots, it makes a great difference whether this represents the death of 50% of the clones or 50% of the shoots within each clone. The former results in significant loss of genetic variation in the population, whereas the latter causes no change in genetic diversity.

POPULATION REGENERATION VIA SEED IS UNCOMMON

Although regeneration by seed is infrequent in prairies, it is not usually limited by seed production, which is quite high for most prairie species. Some workers have reported total seed production in tallgrass prairie as high as 20,000 seeds or more per square meter in good years, virtually a solid carpet of seeds. This is an order of magnitude higher than the seed production of nongrassland habitats (Rabinowitz and Rapp 1980; Potvin 1988), although flowering and seed production decline markedly in drought years and the viability of seeds may be quite limited. Seed production is considerably lower in less productive sandhill (2000 to 3000 seeds per square meter), mixed-grass, and shortgrass prairies, and may be low enough to limit recruitment rates in those habitats (Potvin 1988).

Although seed production may be high, regeneration via seed is a tenuous proposition at best. Successful seedling establishment is a rare event (Anderson 1946; Christiansen and Landers 1966; Rapp and Rabinowitz 1985; Glenn-Lewin et. al. 1990). The necessary conditions for successful grass or forb germination and seedling survival may occur only infrequently in time or space; hence, there is a very low probability that a given seed will reach an appropriate microsite for germination and encounter the essential favorable conditions immediately following germination. Factors such as rodent, insect, and fungal predation, induced dormancy (seed dormancy can be induced by the low-light environment created by a dense grass canopy), and plant competition are responsible for the fact that only a tiny fraction of a given year's seed crop may successfully produce new plants (Risser et al. 1981; Potvin 1984; Gurevitch 1986; Collins and Wallace 1990).

Frequent recruitment requires appropriate plant biology. For example, sand dropseed seedlings in mixed- and shortgrass prairie emerge throughout the spring and summer and well into the autumn. This may greatly increase their probability of successful establishment. If some early emerging seedlings succumb to unfavorable weather or inadequate soil moisture, other seedlings soon emerge whose timing may coincide with favorable conditions. Seedlings of prairie grasses ordinarily do not survive unless their adventitious roots become sufficiently established in moist soil to supply the necessary water and nutrients to the photosynthesizing seedling (Ries and Svejcar 1991). At this point, the seedling can be completely autotrophic (growing on its own photosynthate and not reliant on stored food reserves in the seed) and is considered established and so a "recruit." Drought stress is probably the major cause of mortality of young seedlings, although other factors—such as trampling, grazing by cattle, predation by grasshoppers and rodents, disease, and winter killing—also take their toll. The strongly limiting influence of water availability on successful seedling establishment was once emphasized in a humorous way by Kansas range scientist Kling Anderson, who advised, "The critical factor for native grass establishment is to sow seed just prior to a long wet spell."

Next to the requirement for sufficient moisture, successful colonization of the seedlings' roots by mycorrhizal fungi is likely the most important condition necessary for successful seedling establishment. Mycorrhizae are symbiotic associations between a specific group of fungi and plant roots. Mycorrhizal fungi are ubiquitous in prairies, and almost all the dominant plant species form these associations. These fungi germinate from spores in the soil and produce threadlike vegetative structures called hyphae. Mycorrhizal colonization usually occurs when the developing roots of new seedlings or tillers contact hyphae emanating from germinated spores or from plants with established mycorrhizae. The growing hyphae establish contact with a much greater volume of soil than the plant root can alone, and the hyphae absorb mineral nutrients, transporting them into the host plant root and significantly enhancing plant growth. In return for these benefits, the host plant provides a share of its photosynthetic food production as an energy source to the fungus. Thus this is clearly a mutualistic relationship, much like that of a plant and its pollinators, and both parties benefit from the interaction. Mycorrhizae are particularly important in the acquisition of immobile nutrients such as phosphorus because their hyphae extend considerably beyond the local zones of depletion of these nutrients around roots (Hetrick 1989; Koide and Schreiner 1992). Mycorrhizae provide additional benefits for the host plant for which the mechanisms are not well understood, including reduced water stress and increased disease resistance. Mycorrhizal activity must influence competition among seedlings, between seedlings and established adults, and among established adults, and thus can also exert significant influence on mortality and reproduction of established plants. Because of differential plant species dependencies and growth responses to mycorrhizae, any changes in the activity of these fungi caused by fire or other factors can alter interspecific competitive relationships and relative abundances of different plant species (Hetrick et al. 1989).

Mycorrhizal associations may influence plant competition and population dynamics in additional ways. The hyphae of mycorrhizal fungi may form interconnections among the root systems of neighboring plants and provide a direct pathway for transport of nutrients from plant to plant, much like intershoot translocation via rhizomes in clonal plants mentioned earlier (Chiariello et al. 1982; Newman 1988). Interplant mycorrhizal connections add another dimension to the role of mycorrhizae in increasing seedling survivorship and altering the balance of competition among neighboring plants (Newman 1988). In fact, many of the responses we observe in plant populations to factors such as fire, drought, and grazing may be indirect effects, mediated belowground through changes in mycorrhizal fungi, which direct competitive interactions. These numerous interdependent effects underscore the notion that the prairie is molded by a complex network of population interactions, above- and belowground.

Flower and seed "predation" by beetles, grasshoppers, weevils, and other animals is also an important process influencing reproductive success in prairie plant populations. While the eating of plants is generically called herbivory, an attack on a seed kills an independent (potential) individual, which is logically like the predation of fox on rabbit. Predation at this stage is often intense, may limit population size, and may occasionally destroy 100% of a given year's seed crop (Platt et al. 1974; Willson and Rathke 1974; Keeler 1980). Studies of various prairie forbs have demonstrated that insect predators can successfully limit seed production and/or plant recruitment (e.g., Platt et al. 1974; Kinsman and Platt 1984; Davis et al. 1987; Louda et al. 1990). Herbivory-induced reduction in seed size can significantly reduce the viability of seeds or competitive success of seedlings in many plant species (Weaver and Mueller 1942; Hendrix 1984). The periodic but unpredictable production of large numbers of flowers and seed crops may enable sporadic escape from infestation and may be the key to the regeneration success and long-term persistence of some populations. The Platte thistle of the Nebraska sandhills prairie is an uncommon perennial, despite its great similarities to weedy perennials. Louda et al. (1990) have shown that predispersal flower and seed consumption by insects, postdispersal seed loss to vertebrates, and high mortality resulting from competition with established plants cause severely limited recruitment of the Platte thistle. The insect seed predators cause major losses that are magnified in each succeeding life stage, resulting, like other prairie plants, in a very low probability that a given seed will eventually produce an adult plant.

Many of these factors that limit successful regeneration from seed may also interact in various ways. For example, seedling competition may influence populations directly by reducing survival, or indirectly by amplifying the negative effects of seed predation (Louda et al. 1990).

SEED POPULATIONS BELOWGROUND HAVE THEIR OWN DYNAMICS

For many plant populations, the number of individuals present as dormant seeds buried in the soil far exceeds the number present aboveground as established plants. Buried seeds can be seen as forming a community with its own dynamics, which are determined by the balance between arrivals from aboveground seed production and losses caused by germination, predation, fungal decay, and migration resulting from transport by ants or small mammals. To the ecologist studying seed banks, the seeds in the soil and factors regulating their abundance are of primary interest. The aboveground plants are of lesser importance—a seed's way of producing more seeds.

The dominant perennial grasses and forbs of undisturbed prairie generally do not maintain large or persistent soil seed banks (Blake 1935; Lippert and Hopkins 1950; Rabinowitz 1981). In contrast, some of the less-common short-lived and/or weedy species—such as western yarrow, little ragweed, giant ragweed, common evening primrose, mat spurge, and stickseed—are noted for their great seed longevity, some remaining viable in the soil for decades (Rice 1989). These "soil residents" maintain significant seed banks, enabling them to disperse effectively in time. They can sample a number of years in a sit-and-wait game, opportunistically germinating to produce young seedling recruits during the occasional year when conditions are favorable. For other species, there may be little or no seed bank. However long the seed bank lasts, the life span of all seeds is finite, and the supply of dormant seeds may become exhausted if favorable conditions for establishment do not follow germination at least part of the time (Lippert and Hopkins 1950).

A few studies have specifically examined prairie seed banks. Rabinowitz (1981) reported a mean density of over 6000 seeds per square meter sampled to a depth of 12 centimeters in Missouri tallgrass prairie. Earlier studies by Weaver and Mueller (1942) and Lippert and Hopkins (1950) revealed a range of 300 to 800 seeds per square meter in undisturbed short- and mixed-grass prairie sampled to a depth of 1.3 centimeters. Recalculation of the seed-bank estimates from these studies on a soil-volume–equivalent basis revealed that Rabinowitz's study of tallgrass prairie yielded seed densities almost identical to those reported by Weaver and Mueller (1942) for mixed- and shortgrass prairie sampled to the same depth. Overall, in tallgrass prairie habitats there is no close correspondence between the abundance of particular species in the seed bank and their abundances aboveground. Seed banks of shortgrass prairies in eastern Colorado are similarly dominated by early successional annuals and show poor correlation between relative population abundances aboveground and relative abundances of species in the seed bank (Coffin and Lauenroth 1989).

A persistent seed bank represents a "memory" of previous environmental conditions, containing progeny of survivors of many previous seasons of annually varying environments. A seed bank can buffer a plant population against large fluctuations in number and against the risk of local extinction. It may also retard adaptation to changing environmental conditions. Certainly, it can greatly influence the pattern of population recovery after disturbance (Coffin and Lauenroth 1989). More detailed studies of single-species seed-bank dynamics, including rates of additions and rates and causes of losses from viable seed populations, will add greatly to our understanding of the role of buried seed in the population dynamics of prairie plants and their importance in prairie maintenance and restoration.

SMALL-SCALE DISTURBANCE PLAYS A CRUCIAL ROLE IN SOME PRAIRIE PLANT POPULATIONS

In addition to the long-lived perennials that dominate most closed, undisturbed prairie communities, several shorter-lived species depend on various small- and large-scale disturbances for successful regeneration and population persistence (Platt 1975; Platt and Weis 1977). Small-scale patchy soil disturbances are created by the excavations and mound deposition by small mammals, such as pocket gophers and prairie dogs, and the diggings of ants; large-scale disturbances are generated by fire or large mammalian grazers, such as the trampling or wallowing activities of bison. Although each of these processes may cause mortality of some plants, for others they provide patches of soil free of intense competition and essential opportunities for colonization (Weaver 1968; Platt 1975; Keeler 1987; Reichman 1987). It is in these patches of disturbed soil that population regeneration via seed may be significant. Numerous species that produce a large number of small, readily dispersed seeds that germinate and grow rapidly are well adapted to colonize these disturbed patches, which are unpredictable in time and space. These species are unable to become established and compete successfully with the dominant grasses and forbs in undisturbed areas; hence, they are constrained in size by the patterns and frequency of disturbed sites. These species are typically ephemeral, becoming extinct at a local site as it is slowly invaded by seedlings or vegetative recruits of the more competitive perennials, while dispersing and colonizing newly disturbed patches elsewhere. The continual creation of small soil disturbances and the high fluxes of recruitment and death may be critical factors in the persistence of populations of these fugitive species. Natural environmental factors or human influences that reduce or eliminate populations of small prairie mammals or other soil disturbers are likely to have detrimental consequences for some of the rarer plant species that depend on patchy soil disturbances for successful regeneration. For example, a population of stickleaf (*Mentzelia nuda*) in the Nebraska sandhills declined from 397 to 63 individuals over an eight-year period after the cessation of grazing (and the soil disturbances caused by cattle) (Keeler 1987).

Some species are well adapted to refill small openings or disturbances via vegetative reproduction, but require larger, competition-free patches for successful seedling establishment. For example, Canada goldenrod requires open patches of at least 30 centimeters in diameter for successful seedling recruitment (Goldberg and Werner 1983); however, it can rapidly recolonize smaller openings via clonal spread. The majority of seedlings and shoots that colonize disturbed patches fail, for a variety of reasons, to establish successfully. Those that do survive may persist in the disturbance for only a season, long enough to flower and disperse to newly disturbed sites, even though they are perennials that could live a decade or more.

In the shortgrass prairies, two dominant plant species, buffalo grass and blue grama, have very different responses to soil disturbances. Buffalo grass is characterized by extensive tillering and a sod-forming growth habit. It is capable of extensive lateral spread through rapid growth of long (3–4 centimeters) rhizomes, but its rather large, heavy burs, produced among its leaves or at or near the soil surface, greatly limit its seed dispersal. Like many rhizomatous species, it opportunistically colonizes local disturbed patches via vegetative reproduction. It recovers very quickly and can recolonize disturbed sites within a few years (Shantz 1917; Costello 1944). In contrast, blue grama is a slow-tillering bunchgrass with very short rhizomes (less than 1 centimeter), which limit its rate of horizontal spread. Although it is slower to recolonize disturbed sites via vegetative reproduction, it produces small, light, readily dispersed seeds on a culm well above the height of canopy leaves (Riegel 1941); hence, it can more effectively disperse to colonize other disturbed sites elsewhere. Because these two species tend to dominate many shortgrass prairies of the central and southern Great Plains, their population responses to patterns of disturbance may strongly control the overall community structure and vegetation dynamics in those habitats.

Animal-generated disturbances also create and/or modify microhabitats, allowing new species to colonize or changing the competitive relationships among existing plants. By altering microenvironmental conditions, animal disturbances may give a population a competitive advantage that it lacked in undisturbed microsites. Relict bison wallows in the Kansas Flint Hills tallgrass prairie are depressions on flat, upland soils. The compacted soils of these basins create a unique moist microenvironment in the relatively droughty, shallow-soil uplands that allows species of sedges, rushes, and other species more characteristic of lowland prairie to compete successfully. Wallows also provide open microenvironments suitable for invasion by annual colonizers. Soil disturbances may increase the persistence and size of a variety of populations and tend to increase the overall plant species diversity in these prairie habitats.

FIRE AND GRAZING SHAPE PLANT POPULATION STRUCTURE AND DYNAMICS ON A LARGER SCALE

Fire is a natural component of North American grasslands and has played a significant role in the dynamics of prairies throughout their range (see Bragg, Chapter 4 as well as other recent reviews [e.g., Anderson 1982; Axelrod 1985; Pyne 1986; Collins and Wallace 1990]). Here we focus on the influence of fire on plant and animal population dynamics and interactions. An understanding of these effects is important because the long-term effects of fire on the composition and productivity of prairies are mediated through population-level processes such as seed production, tillering dynamics, and survivorship.

The effects of fire on grass populations have been studied extensively. Spring fire in tallgrass prairies generally enhances their survivorship, growth, seed reproduction, vegetative reproduction, and recruitment (Glenn-Lewin et al. 1990); these effects account, in large part, for the stimulation of overall vegetation productivity in response to burning (Kucera and Koelling 1964; Christiansen and Landers 1966; Zimmerman and Kucera 1977; Svejcar 1990). However, considerable variation in fire effects on flowering and seed production occur among grass species and among prairie habitats (Glenn-Lewin et al. 1990). Cool-season grasses, such as Junegrass, generally show decreased flowering and densities following burning. Tillering and other population responses to burning vary greatly between wet and dry years, and year-to-year weather variation typically exerts a stronger influence on population dynamics than does fire.

Responses of forb populations to prairie fire vary considerably, and the mechanisms underlying their responses are not well understood. Frequent spring burning in tallgrass prairie affects forbs such as prairie coneflower, western ironweed, and wild bergamot by reducing the number of stems and flowering and seed production (Knapp 1984; Davis et al. 1987; Hartnett 1991). Tallgrass prairie fires also influence rhizome production, vegetative reproduction, and ramet population dynamics of the clonal forbs (Hartnett 1990). Variation among forb responses to fire is likely related to timing. For the early emerging and flowering species, spring fires may damage buds or kill plants outright (Lovell et al. 1983). Such fires in tallgrass prairie may indirectly affect later species by creating postfire environmental conditions that alter competitive abilities such that the forbs suffer greater competitive suppression by the perennial grasses (Hartnett 1991). The important influence of competition in grassland plant populations is well recognized (Harper 1977; Fowler 1981; Tilman 1988), and it is reasonable to expect that many of the effects of fire, grazing, small-scale animal disturbances, and other processes in many prairie habitats may be mediated through changing plant competitive interactions.

Fire may also indirectly influence prairie plant populations by altering the population sizes of pollinating or herbivorous insects, the abundance of seed predators, or the activities of large grazers (e.g., Knutson and Campbell 1976; Coppock and Detling 1986; Davis et al. 1987). If fire causes a delay in flowering, a plant population may be thrown out of synchrony with its pollinating insects and suffer reduced seed set. Insect populations change in abundance in response to fire as well. In prairies where very frequent fires result in greater abundance of grasses relative to forbs, grasshopper populations shift concomitantly toward greater abundances of grass-feeding species relative to forb-feeding species (Evans 1984). Gall-forming insects are another important group of insect consumers, and several dozen species of such insects form galls on prairie grasses and forbs. These and other insects are clearly reduced in abundance with frequent fire, and they increase gradually over the years when sites are protected from fire. The underlying cause of this relationship is unknown. The reduction in insect populations may be caused by the direct effects of fire on the insects, reductions in the nutritional quality of the host plant for the insect, or fire-induced reductions in the host-plant population density. Some insect populations (and levels of insect herbivory) may increase after fire (e.g., Davis et al. 1987).

Large grazers influence prairie grass populations through their effects on seed production, seedling establishment, tillering, and survivorship. These effects may vary, depending on the intensity or timing of grazing or other interacting factors such as drought and fire (e.g., Olson and Richards 1988). As discussed earlier, most of the dominant prairie grasses are clonal and thus characterized by two different levels of population structure and dynamics. Grazing can influence their demography at the whole plant level and/or the level of population of individual shoots or tillers. For example, grazing of the bunchgrass little bluestem in a Texas prairie resulted in the fragmentation of clones (clumps), decreasing their mean size but increasing the total density of clumps in the population. Within clones, grazing resulted in a stimulation of tilling, an extension in the season of tiller recruitment, and an overall increase in the number of tillers per unit area (Butler and Briske 1988). These responses varied considerably, depending on the intensity of grazing, and when herbivory was severe, tiller densities, survivorship, and reproduction all decreased (Butler and Briske 1988). Intense grazing may cause reductions in these bunchgrass populations because their basal areas (clone size) and tiller numbers are reduced to small values, and this increases their probability of size-dependent mortality caused by drought or competition. In crested wheatgrass and bluebunch wheatgrass, drought greatly limits the regrowth potential of clones following grazing via tiller production (Busso et al. 1989). Studies of other grasses have also shown decreases in tiller production and/or survivorship in grasses heavily defoliated during one or more growing seasons (Stout et al. 1980, 1981; Hall et al. 1987; Vinton and Hartnett 1992).

Other population responses of prairie grasses to grazing may be highly variable. Longevities of perennial grasses have been observed to be lengthened (Williams 1970; West et al. 1979), shortened (Canfield 1957), or unaffected by herbivory (Wright and Van Dyne 1976). Tiller recruitment and clone size are decreased by grazing in some species, increased in some species, and unaffected in others. In crested wheatgrass, heavy grazing before culm elongation has little effect on tiller population dynamics, but if it occurs after internode elongation, it results in increased overwintering tiller mortality, and reduces both the number and the height of replacement tillers produced following grazing (Olson and Richards 1988).

We emphasize again that these population responses are important because the maintenance of the production and position of perennial grasses and forbs in the prairie community depends much more on the number of individuals in their populations than on the size of individuals. Differential tiller population responses among species in response to grazing or other stress may cause reductions in some species and increases in others, influencing species composition and relative abundances within the community (Smith 1940; Neiland and Curtis 1956). Similarly, within a grass population, differential tilling and stem population regeneration among clones may influence the genotypic composition and diversity. Indeed, there is evidence that selection imposed by grazing, and differential responses among clones within grass populations can result in genetic differentiation. Western wheatgrass and blue grama plants from populations that have been historically subjected to repeated grazing show different growth forms and responses to competition or grazing compared with plants from populations that have been protected from grazing for a long time. Populations exposed to selection by repeated grazing become genetically differentiated; grazing-tolerant "ecotypes" emerge in the grazed area, while those protected from grazing for many years display low grazing tolerance, but increased competitive ability (Painter et al. 1989; Polley and Detling 1990).

Forb populations' responses to large grazers have not been well studied, but appear to be quite variable. Because cattle and bison show strong preference for grasses, forb populations are not usually directly affected by their grazing but may be strongly, if indirectly, affected by the changes in competitive interactions accompanying the removal of the grasses. They need not benefit: Some forbs in tallgrass prairie showed evidence of competitive release when growing in patches where bison intensively grazed their neighboring grasses, but other species showed decreased performance in grazed patches (Ward et al. 1991). Forb populations are likely to be significantly affected by the other activities of large animals, such as trampling and dung and urine deposition, and by grazing from pronghorn and deer. Some workers have suggested that a major influence of large grazers on prairie plant populations is an increase in seedling survival and establishment resulting from the hoof action of trampling animals and consequent small-scale soil disturbances (what range managers call the "herd effect"). However, studies comparing seedling demography in areas both subjected to and protected from grazing have yielded little evidence supporting the notion that these large animals increase seedling recruitment (e.g., Salihi and Norton 1987).

UNDERSTANDING OF PROCESSES INFLUENCING
MANY PRAIRIE ANIMAL POPULATIONS IS LIMITED

Tallgrass prairie has been so thoroughly modified or destroyed that we know very little about the natural processes that regulated the number of animals. The mammalian predators (wolves, bears) are gone; the big herbivores (bison, pronghorn) can no longer migrate across whole states. Other animals, such as prairie dogs, coyotes, and rattlesnakes, have been systematically exterminated; their current numbers and distributions are the result of human activities. Smaller animals, ones that people neither persecute nor promote (e.g., shrews, lark sparrows, and lizards), have been reduced to living on islands of native grassland scattered in agricultural, urban, and suburban areas. A few, such as cardinals, opossums, and painted-lady butterflies, have developed behaviors for living in human-created environments. How important abiotic causes of mortality, such as drought and cold, were to population sizes of prairie species, in comparison with biotic causes of death, such as predators, parasites, and disease, is now impossible to determine.

Even in shortgrass prairie, grazing with its attendant fences has greatly complicated understanding the patterns of animal numbers. Shortgrass prairie would seem less impacted by human-induced changes than mid- or tallgrass prairie, and yet the migratory grasshopper *Melanopus spretus,* which formed the hordes of grasshoppers that descended in such devastating numbers on farmers during the nineteenth century, is the only grasshopper and one of very few animal species known to have gone extinct from the Great Plains. It has not been seen since the 1920s and is believed to have been eliminated as a result of land-use changes in the High Plains of Colorado and Wyoming (Costello 1969; Joern and Gaines 1990).

Despite the devastation of the prairies, especially tallgrass prairie, there are few documented extinctions. Partly this is because the region was so vast that animals survived in sparsely populated areas. It may also be that extinction occurred before we knew what was there: Several flea hoppers from prairie marshes in Illinois are known from only a couple of museum specimens collected in the nineteenth century. The marshes are now gone, and the animals have not been found elsewhere (Opler 1981).

That climatic extremes such as drought affected the abundance of prairie animals seems certain. Historical records indicate that bison died by the thousands in sustained droughts (Roe 1951). Severe winters still kill hundreds of pronghorn. Large-scale prairie fires can destroy ground-nesting birds and mammal populations for hundreds of miles (Reichman 1987; Collins and Wallace 1990). The sizes of prairie animal populations are limited by the availability of food, water, and shelter, just like species in other regions. When those basic necessities are insufficient, there is massive death or migration.

While current evidence suggests that the variable prairie environment—fires, droughts, outbreaks of grasshoppers—drove or exacerbated variation in animal population sizes, probably a good year for one species was a bad year for another. Poor plant growth makes for near starvation for grazers; for their predators, it is a good year to catch grazers, other things being equal. Small mammal species have been shown to respond independently to environmental changes. For example, deer mice increase in number after a tallgrass prairie fire, while prairie voles and harvest mice decrease (Kaufman et al. 1990).

The number of birds in an area shifts within and between seasons, but also changes between years in response to burning and climate. Not surprisingly, migratory bird species and resident birds appear to respond to different environmental factors (Risser et al. 1981).

Invertebrate populations are also influenced strongly by climate and fire, sometimes directly, and sometimes indirectly because of changes in the plant community, predators, or competitors. Year-to-year differences in number are the rule. While many species' numbers can be explained by obvious changes in rainfall or temperature, every group includes species that do not respond noticeably to those factors (e.g., Risser et al. 1981; Joern and Pruess 1986).

The variety of responses is simply the prairie version of the principle that each species has a *unique* set of biotic and abiotic factors that limit its numbers. The species found in a prairie are a mixture of organisms that are the center, south, or east (etc.) part of their range. The animals at the southern edge of their range will flourish in a cold year; those with desert affinities, in a drought. One consequence of this is that no single management plan benefits all species. In Nebraska's sandhills prairie, good grass cover favors the prairie vole and box turtles. In bare sandy areas, Ord's kangaroo rat and the earless lizard prosper. Grazing, mowing, or other disturbance favors the latter; an ungrazed sandhills prairie tends to eliminate them.

GRASSLANDS PROVIDE SPECIAL CHALLENGES FOR ANIMALS

In contrast to forests, grasslands offer few places to hide. An animal can run (or fly) away from its enemies, but it cannot climb out of reach into a tree or hide behind the tree. One common response of prairie mammals to this challenge has been to live in large social groups; that is, animals stay together for protection. In the broad open plains, while there is no place for prey species to hide, there is also no concealment for the approaching predators. Studies show that members of a group feeding together can watch more efficiently for predators than can solitary animals. When a predator is sighted, some groups, like pronghorns, simply run away. The fastest animal in North America, pronghorns can outdistance any pursuit if they see it in time (Schmidt and Gilbert 1978). Birds take flight. Prairie dogs also vanish when a predator is seen, but they go belowground into their burrows, rather than leave the area.

Prairie dogs are rodents, a species of ground squirrel, but they have one of the most complex social systems of any rodent. Individuals live in permanent social groups called wards, cooperate with other prairie dogs, recognize and show aggression toward neighbors, and show different, more antagonistic behavior toward prairie dogs they have never seen before (Costello 1970; MacDonald and Hyngstrom 1991). They have more than a dozen different calls that are used to communicate across the prairie dog town (Costello 1970; Slobodchikoff et al. 1986, 1991). A bark warns of danger, and a "jump yip" is given when danger has passed. More impressive, the barks that warn of potential danger are distinctive: a different call is made for hawks, snakes, or coyotes (Costello 1970; Slobodchikoff et al. 1986). Recent work (Slobodchikoff et al. 1991) shows that prairie dogs gave different warning barks when they spotted individual humans, so that there was effectively individual recognition of enemies and communication of that information. Such specificity is useful, the authors suggest, because different humans (or other predators) pose different levels of threat. Recent studies have greatly added to our appreciation of the communication abilities of animals generally, but prairie dogs clearly have evolved a very complex system. Thus social-group living provides multiple watchers whose greater alertness and complex communication systems allow effective escape from predators, despite the openness of the country.

For prairie dogs, there are additional benefits to social behavior. The burrows of prairie dogs are an engineering marvel, with listening rooms for hiding close to the surface before venturing out, false tunnels for confusing the snake or ferret that runs in after a prairie dog, and passages below the living areas for safely absorbing flood waters (Costello 1970; MacDonald and Hyngstrom 1991). All this requires substantial excavation and is much easier if several prairie dogs cooperate. Social groups provide each member with a much better burrow.

Bison, too, live in large social groups, but they do not run from predators. Another defense available to animals living in a group is for the largest and strongest to defend the rest. Faced by predators, such as wolves, bison bulls put themselves between the cows and calves and the wolves (e.g., Roe 1951; Costello 1969). While wolves may be willing to attack a calf, an alert bull is almost always too dangerous to challenge. The social group offers group defense as a form of protection.

Social groups can also benefit from the knowledge of the oldest members of the group, for example, in finding food, water, and shelter. A single animal has only its own experiences to guide it; a herd has older animals that can teach or show young ones where food or water is. In a climate of extremes, such information may be critically important.

Group living by prairie animals provides many interesting behaviors and responses, but from a conservation standpoint it can be quite difficult. Most social animals have behaviors that promote mating between groups rather than within groups. For example, young male bison are expelled from the parental herd to live as a bachelor herd until they can create a herd of their own. This reduces conflicts within the group and the chance of the group developing genetic defects as a result of inbreeding. However, managing organisms that live in groups of 10 or 100 but mate between such groups is much more difficult for the park or preserve manager than managing animals that live in pairs. Keeping a viable population of, say, 16 pairs of white-tailed deer requires many fewer animals, and much less space, than keeping 16 herds of bison.

Population biology concerns the groups of plants or animals within a species that are the reproductive units and whose success or failure causes species expansion or contraction. The environment crucially shapes the types of populations that are found in an area that survive. In grasslands, long-lived clonal plants and social animals are obvious successful patterns, but much more needs to be learned. When and how seed reproduction takes place, the relative importance of seed banks, and how mycorrhizae affect plant dynamics are among the unsolved questions about prairie plants. So little is known of prairie invertebrate population biology that people may find equally fascinating adaptations to prairies when they have time to look.

LITERATURE CITED

Anderson, R. C. 1982. An evolutionary model summarizing the roles of fire, climate and grazing animals in the origin and maintenance of grasslands: an end paper. Pp. 297–308 in J. R. Estes, R. J. Tyrl, and J. N. Brunken (eds.). *Grasses and Grasslands: Systematics and Ecology.* University of Oklahoma Press, Norman.

Anderson, W. A. 1946. Development of prairie at Iowa Lakeside Laboratory. American Midland Naturalist 36:431–435.

Auffenorde, T. M., and W. A. Wistendahl. 1983. Demography and persistence of *Silphium laciniatum* at the O. E. Anderson Compass Plant Prairie. Pp. 30–32 in R. Brewer (ed.). *Proceedings of the Eighth North American Prairie Conference.* Western Michigan University, Kalamazoo.

Axelrod, D. I. 1985. Rise of the grassland biome, central North America. Botanical Review 51: 163–202.

Blake, A. K. 1935. Viability and germination of seeds and early life history of prairie plants. Ecological Monographs 5:405–460.

Briske, D. D., and J. L. Butler. 1989. Density-dependent regulation of ramet populations within the bunchgrass *Schizachyrium scoparium:* interclonal versus intraclonal interference. Journal Ecology 77:963–974.

Busso, C. A., R. J. Mueller, and J. H. Richards. 1989. Effects of drought and defoliation on bud viability in two caespitose grasses. Annals of Botany 63:477–485.

Butler, J. L., and D. D. Briske. 1988. Population structure and tiller demography of the bunchgrass *Schizachyrium scoparium* in response to herbivory. Oikos 51:306–312.

Canfield, R. H. 1957. Reproduction and life span of some perennial grasses of southern Arizona. Journal Range Management 10:199–203.

Chiariello, N., J. C. Hickman, and H. A. Mooney. 1982. Endomycorrhizal role for interspecific transfer of phosphorus in a community of annual plants. Science 217:941–943.

Christiansen, P. A., and R. Q. Landers. 1966. Notes on prairie species in Iowa. I. Germination and establishment of several species. Proceedings Iowa Academy of Science 73:51–73.

Coffin, D. P., and W. K. Lauenroth. 1989. Spatial and temporal variation in the seed bank of a semiarid grassland. American Journal Botany 76:53–58.

Collins, S. L., and L. L. Wallace (eds.). 1990. *Fire in North American Tallgrass Prairies.* University of Oklahoma Press, Norman.

Coppock, D. L., and J. K. Detling. 1986. Alteration of bison and black-tailed prairie dog grazing interaction by prescribed burning. Journal Wildlife Management 50:452–455.

Costello, D. F. 1944. Natural revegetation of abandoned plowed land in the mixed prairie association of northeastern Colorado. Ecology 25: 312–326.

Costello, D. F. 1969. *The Prairie World.* Crowell, New York.

Costello, D. F. 1970. *The World of the Prairie Dog.* Lippincott, Philadelphia.

Davis, M. A., K. M. Lemon, and A. M. Dybvig. 1987. The effect of burning and insect herbivory on seed production of two prairie forbs. Prairie Naturalist 19:93–100.

Deregibus, V. A., R. A. Sanchez, J. J. Casal, and M. J. Trlica. 1985. Tillering responses to enrichment of red light beneath the canopy in a humid natural grassland. Journal Applied Ecology 22: 199–206.

Ellstrand, N. C., and M. L. Roose. 1987. Patterns of genotypic diversity in clonal plant species. American Journal Botany 74:123–131.

Evans, E. W. 1984. Fire as a natural disturbance to grasshopper assemblages of tallgrass prairie. Oikos 43:9–16.

Fowler, N. L. 1981. Competition and coexistence in a North Carolina grassland. II. The effects of experimental removal of species. Journal Ecology 69:843–854.

Gatsuk, L. E., O. V. Smirnova, L. I. Vorontzova, L. B. Zaugolnova, and L. A. Zhukova. 1980. Age states of plants of various growth forms: a review. Ecology 68:675–696.

Glenn-Lewin, D. C., L. A. Johnson, T. W. Jurik, A. Akey, M. Leoschke, and T. Rosburg. 1990. Fire in central North American grasslands: vegetative reproduction, seed germination, and seedling establishment. Pp. 28–45 in S. L. Collins and L. L. Wallace (eds.). *Fire in North American Tallgrass Prairies.* University of Oklahoma Press, Norman.

Goldberg, D. E., and P. A. Werner. 1983. The effects of size of opening in vegetation and litter cover on seedling establishment of goldenrods (*Solidago* spp.). Oecologia 60:149–155.

Gurevitch, J. 1986. Competition and the local distribution of the grass *Stipa neomexicana.* Ecology 67:46–57.

Hall, J. W., D. G. Stout, and B. Brooke. 1987. Recovery of pinegrass after clipping (simulated grazing). Canadian Journal Plant Science 67:259–262.

Harper, J. L. 1977. *Population Biology of Plants.* Academic Press, London.

Hartnett, D. C. 1983. The ecology of clonal growth in plants. Ph.D. Dissertation, Department of Plant Biology. University of Illinois, Urbana.

Hartnett, D. C. 1990. Size-dependent allocation to sexual and vegetative reproduction in four clonal composites. Oecologia 84:254–259.

Hartnett, D. C. 1991. Effects of fire in tallgrass prairie on growth and reproduction of prairie coneflower (*Ratibida columnifera:* Asteraceae). American Journal Botany 78:429–435.

Hartnett, D. C., and F. A. Bazzaz. 1985. The integration of neighborhood effects by clonal genets of *Solidago canadensis.* Journal Ecology 73:415–427.

Hendrix, S. D. 1984. Variation in seed weight and its effects on germination in *Pastinaca sativa* L. (Umbelliferae). American Journal Botany 71: 795–802.

Hetrick, B.A.D. 1989. Acquisition of phosphorus by VA mycorrhizal fungi and the growth responses of their host plants. Pp. 205–227 in L. Boddy, R. Marchant, and D. J. Read (eds.). *Nitrogen, Phosphorus, and Sulphur Utilization by Fungi.* Cambridge University Press, Cambridge.

Hetrick, B.A.D., G. T. Wilson, and D. C. Hartnett. 1989. Relationship between mycorrhizal dependence and competitive ability of two tallgrass prairie grasses. Canadian Journal Botany 62: 2608–2615.

Joern, A., and S. B. Gaines. 1990. Population dynamics and regulation in grasshoppers. Pp. 415–482 in R. F. Chapman and A. Joern (eds.). *Biology of Grasshoppers.* Wiley, New York.

Joern, A., and K. P. Pruess. 1986. Temporal constancy in grasshopper assemblies (Orthoptera: Acrididae). Ecological Entomology 11:379–385.

Kaufman, D. W., E. J. Finck, and G. A. Kaufman. 1990. Small mammals and grassland fires. Pp. 46–80 in S. L. Collins and L. L. Wallace (eds.). *Fire in North American Tallgrass Prairies.* University of Oklahoma Press, Norman.

Keeler, K. H. 1980. The extrafloral nectaries of *Ipomoea leptophylla* (Convolvulaceae). American Journal Botany 67:216–222.

Keeler, K. H. 1987. Survivorship and fecundity of the polycarpic perennial *Mentzelia nuda* (Loasaceae) in Nebraska sandhills prairie. American Journal Botany 74:785–791.

Keeler, K. H. 1991. Survivorship and recruitment in a long-lived prairie perennial, *Ipomoea leptophylla* (Convolvulaceae). American Midland Naturalist 126:44–60.

Kinsman, S., and W. J. Platt. 1984. The impact of a herbivore upon *Mirabilis hirsuta,* a fugitive prairie plant. Oecologia 65:2–6.

Knapp, A. K. 1984. Postburn differences in solar radiation, leaf temperature and water stress influencing production in a lowland prairie. American Journal Botany 71:220–227.

Knutson, H., and J. B. Campbell. 1976. Relationships of grasshoppers (Acrididae) to burning, grazing, and range sites of native tallgrass prairie in Kansas. Proceedings of the Tall Timbers Conference on Ecology of Animal Control by Habitat Management 6:107–120.

Koide, R. T., and P. Schreiner. 1992. Regulation of the vesicular-arbuscular mycorrhizal symbiosis. Annual Review Plant Physiology Plant Molecular Biology 43:557–581.

Kucera, C. L., and M. Koelling. 1964. The influence of fire on composition of central Missouri prairie. American Midland Naturalist 72:142–147.

Lippert, R. D., and H. H. Hopkins. 1950. Study of viable seeds in various habitats in mixed prairie. Transactions Kansas Academy of Science 53: 355–364.

Louda, S. M., M. A. Potvin, and S. K. Collinge. 1990. Predispersal seed predation, postdispersal seed predation and competition in the recruitment of seedlings in a native thistle in sandhills prairie. American Midland Naturalist 124:105–113.

Lovell, D. L., R. A. Henderson, and E. A. Howell. 1983. The response of forb species to seasonal timing of prescribed burns in remnant Wisconsin prairies. Pp. 11–15 in R. Brewer (ed.). *Proceedings of the Eighth North American Prairie Conference.* Western Michigan University, Kalamazoo.

MacDonald, N. F., and S. E. Hyngstrom. 1991. Little dogs of the prairie. Nebraskaland 69:24–31.

Neiland, B. M., and J. T. Curtis. 1956. Differential responses to clipping of six prairie grasses in Wisconsin. Ecology 37:355–365.

Newman, E. I. 1988. Mycorrhizal links between plants: their functioning and ecological significance. Advances Ecological Research 18:243–270.

Olson, R. E., and J. H. Richards. 1988. Annual replacement of the tillers of *Agropyron desertorum* following grazing. Oecologia 76:1–6.

Opler, P. A. 1981. Management of prairie habitats for insect conservation. Natural Areas Journal 1: 3–6.

Painter, E. L., J. K. Detling, and D. A. Steingraeber. 1989. Grazing history, defoliation, and frequency-dependent competition: effects on two North American grasses. American Journal Botany 76:1368–1379.

Platt, W. J. 1975. The colonization and formation of equilibrium plant species associations on badger disturbances in a tallgrass prairie. Ecological Monographs 45:285–305.

Platt, W. J. 1976. The natural history of a fugitive prairie plant species (*Mirabilis hirsuta* (Pursh.) MacM.). Oecologia 22:399–409.

Platt, W. J., G. R. Hill, and S. Clark. 1974. Seed production in a prairie legume (*Astragalus canadensis* L.): interactions between pollination, predispersal seed predation, and plant density. Oecologia 17:55–63.

Platt, W. J., and I. M. Weis. 1977. Resource partitioning and competition within a guild of fugitive prairie plants. American Naturalist 111:479–513.

Polley, H. W., and J. K. Detling. 1990. Grazing-mediated differentiation in *Agropyron smithii:* evidence from populations with different grazing histories. Oikos 57:326–332.

Potvin, M. A. 1984. Factors affecting native grass seedling establishment along topographic gradients in the Nebraska Sand Hills: an experimental approach. Ph.D. Dissertation. University of Nebraska, Lincoln.

Potvin, M. A. 1988. Seed rain on a Nebraska sandhills prairie. Prairie Naturalist 20:81–89.

Pyne, S. J. 1986. These conflagrated prairies: a cultural fire history of the grasslands. Pp. 131–137 in G. K. Clambey and R. H. Pemble, (eds.). *The Prairie—Past, Present, and Future.* Proceedings of the Ninth North American Prairie Conference. Tri-College University Center, for Environmental Studies, North Dakota State University, Fargo.

Rabinowitz, D. 1981. Buried viable seeds in a North American tallgrass prairie: the resemblance of their abundance and composition to dispersing seeds. Oikos 36:191–195.

Rabinowitz, D., and J. K. Rapp. 1980. Seed rain in a North American tall grass prairie. Journal Applied Ecology 17:793–802.

Rapp, J. K., and D. Rabinowitz. 1985. Colonization and establishment of Missouri prairie plants on artificial soil disturbances. I. Dynamics of forb and graminoid seedlings and shoots. American Journal Botany 72:1618–1628.

Reichman, O. J. 1987. *Konza Prairie: A Tallgrass Natural History.* University Press of Kansas, Lawrence.

Riegel, A. 1941. Life history and habits of blue grama. Transactions Kansas Academy of Science 44:76–85.

Rice, K. J. 1989. Impact of seed banks on grassland community structure and population dynamics. In M. A. Leck, V. T. Parker, and R. L. Simpson (eds.). *Ecology of Soil Seed Banks.* Academic Press, Orlando, Fla.

Ries, R. E., and T. J. Svejcar. 1991. The grass seedling: when is it established? Journal Range Management 44:574–576.

Risser, P. G., E. C. Birney, H. D. Blocker, S. W. May, W. J. Parton, and J. A. Weins. 1981. *The True Prairie Ecosystem.* Hutchinson Ross, Stroudsburg, Pa.

Roberts, T. M., T. Robson, and P. M. Catling. 1977. Factors maintaining a disjunct community of *Liatris spicata* and other prairie species in Ontario, Canada. Canadian Journal Botany 55:593–605.

Roe, F. G. 1951. *The North American Buffalo.* University of Toronto Press, Toronto.

Salihi, D. O., and B. E. Norton. 1987. Survival of perennial grass seedlings under intensive grazing in semiarid rangelands. Applied Ecology 24:145–151.

Schaal, B. A. 1988. Somatic variation and genetic structure in plant populations. Pp. 47–58 in A. J. Davy, M. J. Hutchings, and A. R. Watkinson (eds.). *Plant Population Ecology.* Blackwell, Oxford.

Schmidt, C. A., and D. L. Gilbert. 1978. *Big Game Animals of North America: Ecology and Management.* Stackpole, Harrisburg, Pa.

Shantz, H. L. 1917. Plant succession on abandoned roads in eastern Colorado. Journal Ecology 5:19–42.

Slobodchikoff, C. N., C. Fischer, and J. Shapiro. 1986. Predator-specific alarm calls of Gunnison's prairie dog. American Zoology 26:557.

Slobodchikoff, C. N., J. Kiniazias, C. Fischer, and E. Creef. 1991. Semantic information distinguishing individual predators in the alarm calls of Gunnison's prairie dogs. Animal Behavior 42:713–719.

Smith, C. C. 1940. The effect of overgrazing and erosion upon the biota of the mixed-grass prairie of Oklahoma. Ecology 21:281–297.

Stout, D. G., A. McLean, B. Brooke, and S. Hall. 1980. Influence of simulated grazing (clipping) on pinegrass growth. Journal Range Management 33:286–29.

Summers, C. A., and R. L. Linder. 1978. Food habits of black-tailed prairie dogs in western South Dakota. Journal Range Management 31:134–136.

Svejcar, T. J. 1990. Root length, leaf area, and biomass of crested wheatgrass and cheatgrass seedlings. Journal Range Management 43:446–448.

Tilman, D. 1988. *Plant Strategies and the Dynamics and Structure of Plant Communities.* Princeton University Press, Princeton, N.J.

Tripathi, R. S., and J. L. Harper. 1973. The comparative biology of *Agropyron repens* L. (Beauv.) and *A. caninum* L. (Beauv.) I. The growth of mixed populations established from tillers and from seeds. Journal Ecology 61:353–368.

Vinton, M. A., and D. C. Hartnett. 1992. Effects of bison grazing on *Andropogon gerardii* and *Panicum virgatum* in burned and unburned tallgrass prairie. Oecologia 90:374–382.

Ward, L. E., R. Kazmaier, and D. C. Hartnett. 1991. Forb species responses to bison activity on Kansas tallgrass prairie. American Journal Botany 78:87–88.

Weaver, J. E. 1954. *North American Prairie.* Johnson, Lincoln, Nebr.

Weaver, J. E. 1968. *Prairie Plants and Their Environment: A Fifty-Year Study in the Midwest.* University of Nebraska Press, Lincoln.

Weaver, J. E., and F. W. Albertson. 1944. Nature and degree of recovery of grasslands from the great drought of 1933 to 1940. Ecological Monographs 14:394–497.

Weaver, J. E., and I. M. Mueller. 1942. Role of seedlings in recovery of midwestern range from drought. Ecology 23:275–294.

Weller, S. G. 1985. The life history of *Lithospermum carolinense,* a long-lived herbaceous sand dune species. Ecological Monographs 55:49–67.

West, N. E., K. H. Rea, and R. O. Harniss. 1979. Plant demographic studies in sagebrush–grass communities of southeastern Idaho. Ecology 60:376–388.

Willson, M. F., and B. J. Rathke. 1974. Adaptive design of the floral display in *Asclepias syriaca* L. American Midland Naturalist 92:47–57.

Williams, O. B. 1970. Population dynamics of two perennial grasses in Australian semiarid grassland. Journal Ecology 58:869–875.

Wright, R. G., and G. M. Van Dyne. 1976. Environmental factors influencing semidesert grassland perennial grass demography. Southwestern Naturalist 21:259–274.

Zimmerman, U. D., and C. L. Kucera. 1977. Effects of composition changes on productivity and biomass relationships in tallgrass prairie. American Midland Naturalist 97:465–469.

6

The Entangled Bank: Species Interactions in the Structure and Functioning of Grasslands

ANTHONY JOERN

It is interesting to contemplate an entangled bank, clothed with many plants of many kinds, with birds singing on the bushes, with various insects flitting about, and with worms crawling through the damp earth, and to reflect that these elaborately constructed forms, so different from each other, and dependent on each other in so complex a manner, have all been produced by laws acting around us.

C. Darwin, *On the Origin of Species*

Isn't it queer: there are only two or three human stories, and they go on repeating themselves as fiercely as if they had never happened before; like the larks in this country, that have been singing the same five notes over for thousands of years.

W. Cather, *O Pioneers!*

Time is God's way of making sure that everything doesn't happen at once.

Attributed to Woody Allen

Naturally occurring species in grasslands do not exist in an ecological vacuum; they confront scores of other species on a daily basis. In some cases, the interaction is a positive one for both participants (mutualism), as occurs between a flower and its pollinating fly, bee, or butterfly. Other, equally routine interactions have negative consequences for one or both participants, including competition among coexisting grasses for limited water, light, or nutrients; grasshoppers eating seedling plants (herbivory); or birds eating the grasshopper just mentioned (predation). These direct interactions determine population sizes of coexisting species and often modify spatial distributions as well. Indirect interactions are at least equally common and possibly more important. For example, a bird feeding on a seedling-feeding grasshopper exerts a positive effect on the host plant population without interacting with it directly, since the bird makes life for plant seedlings a little more secure. If that particular seedling grows and creates important habitat for the bird, both species benefit, the bird twice because it gets to eat grasshoppers as well as build a nest. Such is the nature of interactions in natural systems—straightforward and intuitive in concept, but often complex and convoluted in practice.

Networks of interactions among coexisting species provide the framework for understanding important grassland attributes, including the number of species that can coexist, the taxonomic identity of coexisting species, and the distribution of species along natural gradients, such as along a hillside or across soil type boundaries. Unfortunately, since most natural communities contain so many species, it is very difficult to precisely predict how the final balance sheet will sum up; species that seem unbeatable sometimes disappear in response to subtle, diffuse, and indirect interactions.

What "rules" allow us to predict responses to interactions and ultimately determine the nature of the grassland that we view and study? Consider the "entangled bank" in the form of a grassland from North Dakota illustrated in Figure 6.1. Prominent patterns quickly reveal themselves. Individual plant species are not uniformly distributed in space, but are relegated to specific types of habitat, such as on dry, exposed ridges or in moist depressions. Species combinations are not arbitrarily arrayed, but exist as predictable species mixes coexisting on a local scale (Barnes and Harrison 1982; Barnes et al. 1984). What underlying forces caused these patterns? These questions form the basis of this chapter.

Interactions among species, in concert with physiological responses of individuals to physical and chemical features of the local environment, determine resulting abilities of an individual to exist at a particular spot or in a specific habitat. Interactions also have dynamic, time-dependent effects when controlling ultimate population sizes. Hartnett and Keeler (Chapter 5) describe salient features of population processes that define the dynamic element of the interactions; however, another dynamic element at the community level concerns fluxes in the number and taxonomic identity of coexisting species (Collins and Glenn, Chapter 7). This chapter concerns the role that biotic interactions—interactions among species—play in determining the species abundances, distribution, and diversity in grassland communities.

While interactions are often readily observed natural history attributes between two or more organisms, such as a bird eating a grasshopper, quantitative responses at population levels are equally important. Strengths of interactions (Paine 1980) often contribute strongly to the impact of density- or frequency-dependent factors in moderating species coexistence. Density-dependence occurs when key attributes change in magnitude as density changes; lowered survival accompanying an increase in population density is an example. Frequency-dependent factors depend on responses that vary according to the relative abundance of the target organism. Insect larvae are often more susceptible to bird predation when relatively more common than when rare, even when total prey abundance remains constant. Sometimes subtle or otherwise seemingly minor interactions among species have startling impacts on grassland structure and function. Soil disturbance by pocket gophers is absolutely required to keep some "fugitive" plant species at a site, since they need disturbed soil to become established. As observers of grasslands, we typically overlook these individual responses, instead integrating in our mind's eye to create the patterns that we so greatly appreciate. My goal is to explore the nature of species interactions and suggest how they are responsible for patterns of grassland structure and function.

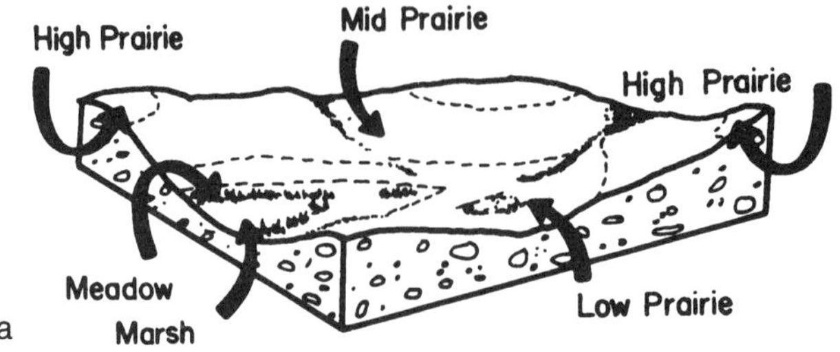

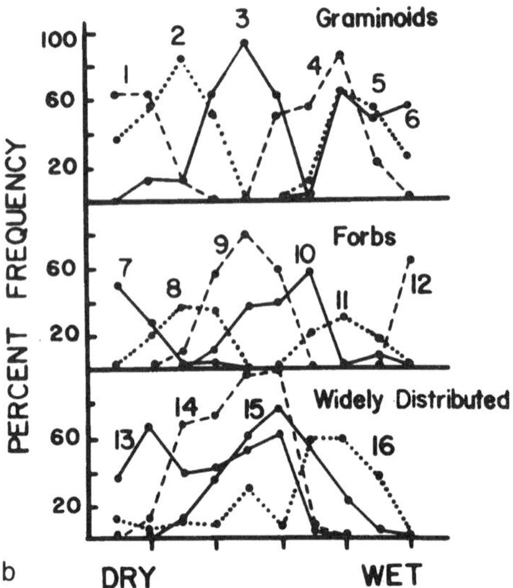

Figure 6.1. Representative grassland plant communities from Dix County, North Dakota: (a) hypothetical diagram of topography and location of plant communities; (b) distribution of 16 common plant species at this site along soil moisture gradients. Species are grouped as common graminoids, forbs, and both graminoids and forbs with widespread distributions. Note that some species are restricted to narrow bands of moisture levels (both high and low), while other species can inhabit a wide range of conditions. Species include (1) needle-and-thread grass, (2) porcupine grass, (3) big bluestem, (4) sedge, (5) sprangletop marshgrass, (6) bulrush, (7) sage, (8) candle anemone, (9) golden alexander, (10) field sow thistle, (11) field mint, (12) common bladderwort, (13) aster, (14) Richardson's muhly, (15) Baltic rush, (16) quackgrass. (From Dix and Smeins [1967], with permission of the National Research Council of Canada)

Only a few basic types of interactions actually exist, but they appear complex because of the variety of ways that a large number of species interact. I cannot reasonably cover all of them in this chapter, but I will present enough examples to provide the flavor of possibilities. Some of the points that I stress received general emphasis by most ecologists; others reflect my own special interests and expertise. The general structure of this chapter is to identify, explain, and present examples of the main interactions. Then I indicate how complex and indirect interactions can lead to unexpected community responses or cascade through the entire trophic web of a natural community. Finally, I explore how interactions will play an important role in preserving and rehabilitating natural systems, even if it is unlikely that we can fully diagnose the majority of natural species-specific interactions with care. Results from this chapter can be fleshed out more completely with examples presented in the other chapters in this book.

COMPETITION

Competitive interactions operate in an intuitively obvious fashion and can simultaneously occur among individuals of the same and different species (interspecific competition). When some important resource is limited and more than one individual or species needs the resource, further decreasing to the participants, the individuals or populations are competing. Free-market economic systems are reasonable analogues of natural competitive systems. Ecologically, competition affects demographic features such as reduced survivorship, growth, reproduction, and/or dispersal of the individuals and populations concerned (Begon et al. 1990). Ultimately, altered demographic features affect populational processes in a density-dependent or frequency-dependent fashion, leading to control of population dynamics—the change (or lack of change) in the number of individuals in the population(s). As densities or relative frequencies of individuals or species change, the strength of interactions such as competition change. Effective evaluation of competitive interactions in natural systems has been difficult, however, because other types of species interactions may superficially give the same response, but for different reasons (Connell 1990; Goldberg 1990). "Apparent competition" is an example of this type of problem, as I discuss later.

Beginning with Clements and co-workers in the 1920s (Clements et al. 1929), abundant evidence documents that plant species often compete in a variety of grassland systems. As a result, predictable mosaics of plant distributions can often be detected (Fowler 1986a; Tilman 1988; Keddy 1989; Grace and Tilman 1990).

However, competition is not always required to explain natural patterns. If resources are not limited, then competition within or among populations will not take place. Since populations will grow geometrically unless checked (much like compound interest on a bank account), constrained population growth in the absence of competition implies that populations are limited by some factor. Ready examples of alternative population controls include biotic features such as predation and disease or external, disturbing influences such as severe weather occurring at sufficiently regular intervals. Such disruptions keep population densities sufficiently below levels at which resources begin to become limited so that competition is of little consequence.

In grasslands, competition may or may not exist among species at other trophic levels, such as among insect herbivores (Strong et al. 1984) or insectivorous birds (Martin 1986; Wiens 1989; Kaspari 1991). For the most part, except for plants, insufficient evidence exists for drawing general conclusions regarding the importance of competition at other trophic levels in grassland ecosystems. However, the impacts of species in other trophic levels on plants may alter the expected outcome of competitive interactions and the resulting community structure. In this section, I summarize predominant views of how competition might be operating in constant and variable environments.

Limiting Resources and Competitive Interactions Among Plants

Clear mosaics of plant species associations and abundances typify the structure of a natural community. Often, community attributes vary in accordance with some important physical gradient, such as along a soil moisture gradient or from one soil type to another (Barnes et al. 1984; Bragg, Chapter 4; Collins and Glenn, Chapter 7). Since important resources such as water availability and mineral nutrients vary predictably along such gradients, it is not uncommon to attribute such changes in plant distribution to underlying physiological responses of the individual species, independent of the presence of other species: Either the plant can live in a specific habitat or it cannot. Since different plant species basically require the same resources (water, light, nutrients) but do better at different levels or combinations of resources, species are typically found at different points along the many naturally occurring gradients (Figure 6.1). The resulting pattern of multiple species distributions for the grassland mosaic could result from a series of independent responses of different species, a response sometimes referred to as a noninteractive view of species assembly. If true, the presence or absence of a particular species from a locality or from a portion of some environmental gradient is independent of interactions with other species. A key distinction between solely physiologically based explanations and competition lies in the importance of other individuals (same or different species) in altering levels of a limited, necessary resource—say nitrogen, water, or light.

Often, however, plants interact strongly with other plant species for limiting nutrients, as implied by the intense interplay among roots belowground, including root segregation patterns (Figure 6.2). Such segregation is consistent with the notion that these plants are competing for resources. By exploiting different regions of the soil, coexisting plants may obtain needed resources while minimizing competition. What impact do such interactions have on the resulting grassland assembly? What mechanisms are involved? Important insights have come from attempts to develop general understanding that links vegetation patterns to basic plant processes (Grime 1979; Tilman 1982, 1988; Keddy 1989; Smith and Huston 1989). To develop this framework, I begin by describing individual plant responses to resources: light and nutrients. Physiological responses to abiotic features of the environment routinely determine plant success (Barnes 1985; Knapp 1985a, 1985b; McNaughton 1991; Schimel et al. 1991). Then, following the somewhat different models proposed by Tilman (1988), Grime (1979), and Keddy (1989), I indicate how species might affect one another when they co-occur along an important resource gradient to influence the final patterns of species abundance and distribution. Finally, I describe how spatial heterogeneity, herbivory, disturbance, and temporal variation might mediate specific competitive interactions, in turn influencing species abundance patterns.

Resource Allocation and Trade-offs Within Plants

Plants obtain key resources using leaves to capture both CO_2 and light, or roots to obtain mineral nutrients and water. Each class of ''raw materials'' is required for successful photosynthesis, a process needed to provide internal resources to support resulting growth, maintenance, and reproduction in the plant. In natural systems, resources are commonly limited, with available levels determined by the balance be-

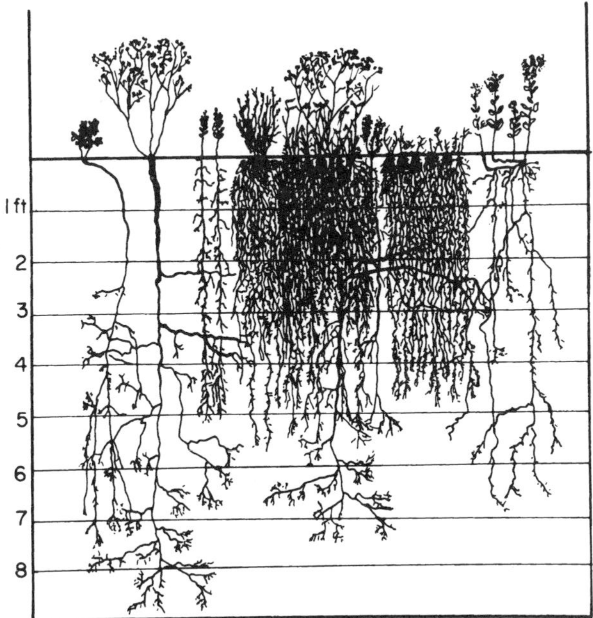

Figure 6.2. Representative growing relationships for above- and belowground components among grasses and forbs in a Kansas shortgrass plant community. Plant species include (from left to right) *Malvastrum*, scurfy pea, western ragweed, Fendler three-awn, blue grama, buffalo grass, soft goldenrod. (Redrawn from Albertson [1937], with permission of the Ecological Society of America)

tween (1) reduction in resource availability resulting from species' use and (2) resource renewal rates, primarily the result of belowground processes (Tilman 1990; Tilman and Wedih 1991a, 1991b; Wilson and Tilman 1991; Seastedt, Chapter 8). In addition, physiological responses to resource limitations direct organ growth differentially in order to increase a plant's ability to garner more of the limiting resource. For example, under limited water or nitrogen, increased root growth permits the plant to "scan" more of the soil column for these nutrients. Similarly, when a plant is light-limited, as when growing in the shade of a neighboring plant, increased leaf surface area permits increased light-harvesting capabilities, or commitment to stem tissue allows the plant to grow taller, thus increasing light availability.

Of course, there are trade-offs: Increased leaf surface area permitting increased light-capture results in increased water loss through pores (stomates) in the leaves. The good is offset by the negative; that is, there is no free lunch. Negative correlations between physiological capabilities and allocation of limited resources to life history traits such as leaves, roots, or reproductive tissue (Figure 6.3) represent fundamental physiological and energetic constraints on the capture and use of resources in the plants (Tilman 1988; Smith and Huston 1989).

Trade-offs represent compromises at the *whole plant* level involving a large number of morphological, physiological, and life history traits. In addition, trade-offs imply limited responses in allocating resources among different compartments within a plant. For example, it is unlikely that a plant can maximally allocate resources to

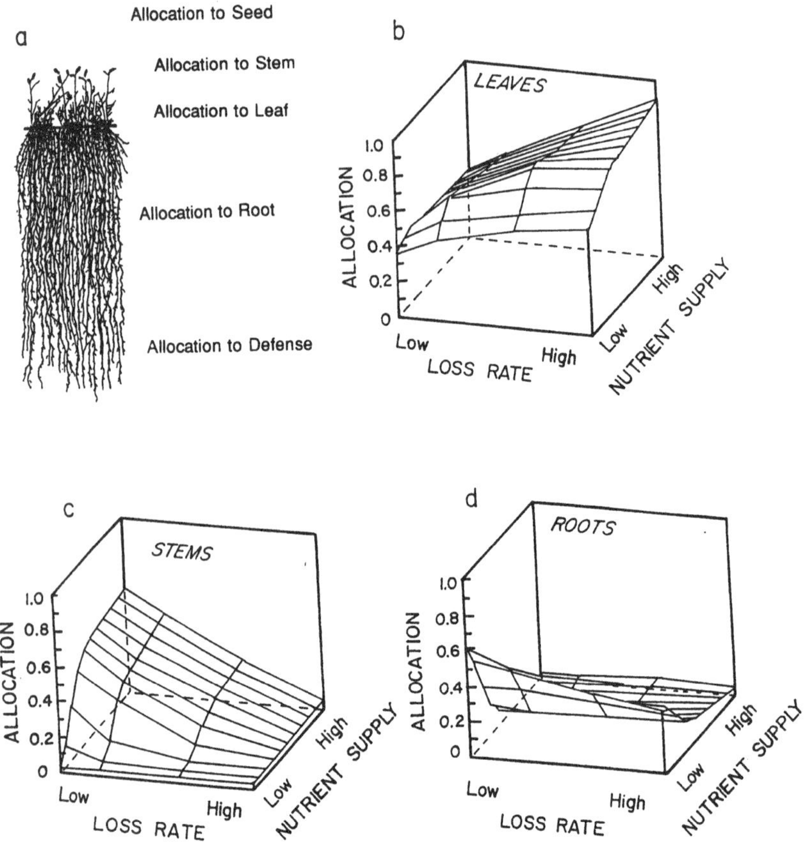

Figure 6.3. Allocation of limited plant resources among component parts to support competitively superior plant morphologies: (a) representative flow of competing allocation demands within a plant; (b–d) hypothetical, competing allocation responses of different plant organs (leaf, stem, roots) in order to provide competitively superior plant morphology based on tissue-loss rate and nutrient-supply rate. Patterns were developed with a simulation model. (Redrawn from Tilman [1988])

increased root production at the same time that it allocates maximal resources to stem and leaf production; it just cannot be done. Finally, there are compelling arguments that at equilibrium, if one exists, all key resources should be about equally limiting (Bloom et al. 1985), although it is unclear how frequently a plant experiences an unchanging environment for a sufficient period to reach equilibrium (Knapp and Smith 1989). One of the best-known trade-offs within plants is the allocation of resources to roots versus the aboveground structures—the root:shoot ratio (Givnish 1986; Tilman 1988; Geiger and Servaites 1991; Mooney and Winner 1991). And given the argument that above- versus belowground resources should be about equally limiting at equilibrium, root:shoot ratios within a species are expected to be relatively constant under a range of conditions, although actual biomass may vary greatly. Clipping or grazing of foliage causes plants to allocate most productivity to regrowth (McNaughton 1983). Root growth also drops quickly in response to loss of aboveground tissue. Environmental conditions are also important because plants evolve in response to local conditions, often resulting in recognizable responses to high and low resource conditions (Chapin 1980; Chapin et al. 1987). Plants adapted to low-resource conditions are typically less responsive to an influx of nutrients than are plants adapted to high-resource levels.

This individual plant focus highlights trade-offs within the plant to alternative conditions; the plant cannot do everything well. As a result, expectations regarding which plant species should do best differ depending on environmental conditions, including the specific type of limiting resource, the level of disturbance, as well as the type and intensity of herbivory. Competitive outcomes are the result of the state of the environment, represent a dynamic and reversible response in a changing environment, and have populational and distributional effects.

Species-specific differences also exist that restrict the range of responses available to an individual plant in response to changing resource availabilities. A shortgrass species such as blue grama can never grow to the same height as big bluestem. If competing for light, big bluestem almost always grows taller and shades blue grama— if sufficient resources, especially water, exist. A useful and long-term goal is to predict competitive outcomes among plant species under specified conditions based on the physiological, morphological, and life history traits of the individual species coupled with the frequency of disruption from external sources.

Three predominant views currently exist attempting to explain how competitive interactions structure plant communities in unchanging environments—that is, where the system reaches a final equilibrium. These include (1) the resource-ratio model, (2) the competition–stress tolerator–ruderal model, and (3) the competitive hierarchy model. In changing environments, competition may act but not fully explain the final pattern. This provides an important fourth alternative for understanding natural patterns. I present these four views here. Empirical tests designed to determine which mechanism is likely operating are ongoing.

Resource-Ratio Model of Competition. In grasslands, light, water, or nitrogen is often most limiting, in a temperature-dependent fashion. Consumption or other restrictions to the limiting resource (e.g., shading) is the source of competitive ability so that species creating the lowest level of limiting resource (depicted by R*) will exclude other species. Since the ability to exploit soil nutrients versus light is generally negatively correlated (a trade-off), species will differ in the ratio of above- versus belowground competitive abilities. Based on a mechanistic model of belowground (water, nutrients) versus aboveground (light) resource consumption, Tilman (1982, 1988) predicts competitive outcomes of co-occurring grassland species (including coexistence). This is the resource-ratio model (Figure 6.4), based on the notion that each plant species has a ratio of light: soil mineral resource (especially nitrogen) at which it does better than any other species. If environmental conditions exist that match a particular light:mineral resource ratio for a species, and no disruptions occur, then that species wins any competitive encounters with other plant species in that habitat. Environmental characteristics external to the plants themselves typically determine the potential productivity of the habitat, usually as a function of nutrient and water availability. In turn, these determine the likelihood that light will ultimately be limiting.

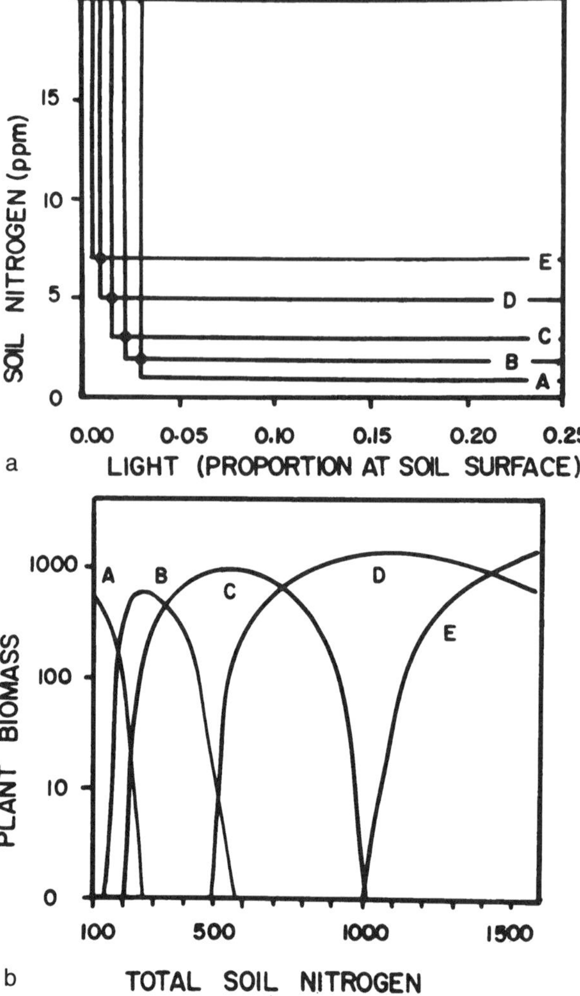

Figure 6.4. Model of resource competition applied to several hypothetical plant species: (a) Two plausible limiting resources (light and soil-N) are examined among several that might be important. Resource needs of five (A–E) hypothetical species illustrate how these species compare in resource needs when no other species are present. The lines indicate the minimum combination of each resource that is required for the plant to persist. As plants grow, they reduce resources (either through shading or by reducing soil-N in this example) to the minimum level that permits persistence. These final points represent the species equilibrium. In this example, species E requires more soil-N than species D to live, but less light. When two species appear together in this model, they both exploit the same resource base and affect each other in this manner. If both species can coexist, an equilibrium level of resources in the environment is attained. In this hypothetical example, these equilibria are represented by dots. (b) Using the conditions described by the resource isoclines, Tilman (1986) employed a computer simulation model to describe the distribution of these species along a soil-N: nutrient gradient. In the model, the density of each species is recorded after 50 years of simulated competition based on the relationships from the resource isoclines. The five species become separated along the

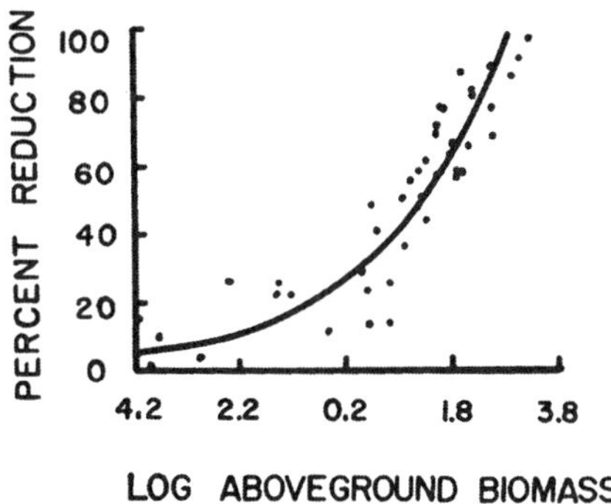

Figure 6.5. Gaudet and Keddy (1988) screened the competitive ability of 44 plant species by comparing growth ability of a standard competitor against each species when grown together. The higher the reduction in biomass, the greater the competitive ability. Plants on the left in this figure represent small evergreen species; plants on the right, large leafy species. (Redrawn from Keddy [1989], with permission of Chapman and Hall)

Species-specific growth rates depend on the levels at which specific plant species are limited by specific nutrients and at what level the nutrient exists in the environment; different nutrients may be limiting in different habitats so that competitive outcomes differ. At equilibrium (when nutrient supply rate just equals the uptake rate of nutrients by plants), the competitive encounter concludes, and the species that can exist at the lowest level of the limiting nutrient prevails. As an example, Tilman and Wedin (1991a) measured R^* for nitrogen among five perennial grasses. After growing monocultures of each grass species on a soil-fertility gradient (sand to black soil) coupled with low and high N-fertilizer treatments, they measured available soil nitrogen under the plants. Under low N-fertilizer conditions, significant differences in soil-N estimates of R^* resulted, while no differences in R^* resulted when grown under high N-fertilizer treatments.

Specific traits have been identified that are good predictors of R^*. For the grass species in Tilman and Wedin's experiment, root biomass explained 73% of the variance in belowground R^* (Figure 6.5), a result consistent with Gaudet and Keddy's (1988) finding that total biomass was the best overall predictor of plant competitive ability (Figure 6.5). Grime (1979) suggests that plant growth rate is a good estimator of competitive ability. In a similar fashion, tall plants have the best abilities to obtain light and shade shorter competitors.

Competition–Stress Tolerator–Ruderal (CSR) Model. Proposed by Grime (1979), this model (Figure 6.6) relies on plant life history responses to specific types of habitats coupled with mechanistic assumptions about how plants interact (Grime 1979; Grace 1990, 1991). According to Grime, the competitive ability of a plant is positively

---

resource gradients, much as species distribute themselves along natural gradients (Figure 6.1). Note that species A, with the lowest needs for soil-N, is found at the low end of the gradient, whereas species E, which requires the highest levels of soil-N, is along the upper end. (Redrawn from Tilman [1986], with permission from HarperCollins)

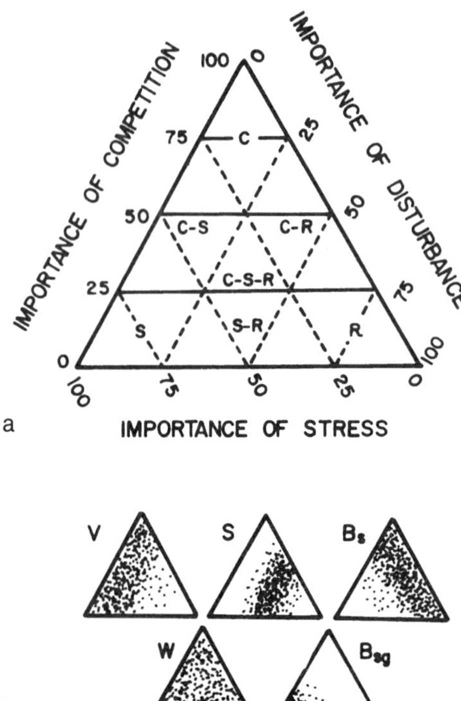

Figure 6.6. (a) Triangular model describing proposed hypothetical negative trade-offs existing among plants with needs to inhabit areas potentially exposed to three major environmental features: competition (C), stress (S), and disturbance (R). Values are %-units, ranging from 0% where the plant cannot withstand the conditions to 100% where the plant is specifically adapted to deal with this environmental feature. (b) Grime (1979) plots the likely range of conditions in which plants with the following types of life cycles will be found: V, combinations involving vegetative expansion; S, combinations involving seasonal regeneration; B$_s$, combinations involving a seasonal seedback; W, combinations involving numerous wind-dispersed seeds or spores; and B$_{sg}$ combinations involving persistent seedlings. (Redrawn from Grime [1979], with permission of Wiley)

correlated with the maximum relative growth rate of a plant under optimal resource conditions. Maximum growth rate leads to rapid development of resource-gathering organs and preemption of above- and belowground resources. These species are assumed to do well only when resources are abundant; when soil resources are not readily available, the plants are stressed.

However, Grime also proposes that evolutionary trade-offs exist such that not all species evolve traits that promote increased competitive ability. For species that regularly live in either low-resource environments or habitats that experience regular and extensive disturbance, life history characteristics evolve that permit populations to persist in these habitats. For example, in poor-quality habitats, *stress tolerators* are able to withstand lower levels of soil nutrients or higher levels of shading (stress tolerators) with relatively lower maximum growth rates. Better competitors cannot physiologically make it in these habitats independent of competitive interactions. Finally, when a habitat is routinely disturbed (partial or total destruction of plant biomass), competitors may never have much impact and plants that can exploit newly created, short-lived habitats will do best. Grime refers to this strategy as a *ruderal* type. A key point of this model is that three very different life histories have evolved that correspond to different types of habitats; alternative predictions account for species coexistence that follow from knowledge of the type of habitat under consideration, the differences based on the CSR strategy set. In each case, expected resource-based, stress responses are very different.

Competitive Hierarchies and the Centrifugal Community Organization Model. A third view argues that most plant species see the environment in a similar fashion rather than differently, as assumed in the first two views. Competitive hierarchies may exist that are centered on the same set of resource conditions (Wilson and Keddy 1986a, 1986b; Keddy 1989, 1990; Keddy and Shipley 1989). Competitive interactions among species are typically asymmetric, with the species exhibiting the same competitive relationships to other species over a range of conditions—the competitive hierarchy. A robust example for grasslands appears to be the competitive superiority associated with plant height or biomass (Clements 1933; Keddy 1990, 1991). Big and little bluestem always replaced ticklegrass along the entire soil-N continuum examined (Tilman and Wedin 1991a), suggesting that competitive hierarchies are often invariant. In addition, the intensity of competition varies from site to site, typically increasing toward more productive ends of environmental gradients.

The Centrifugal Community Organization model also assumes an underlying physiological basis that all species do best (when alone) under the same conditions, the so-called core habitat, which is preferred by all species (Figure 6.7). Interspecific competition is intense in this core area, and all species cannot coexist. However, each species also has a specialized peripheral habitat more or less unique to itself, a refugium from interspecific competition. Species that are poor competitors remain in the community by using the peripheral habitats almost exclusively. A key point is that

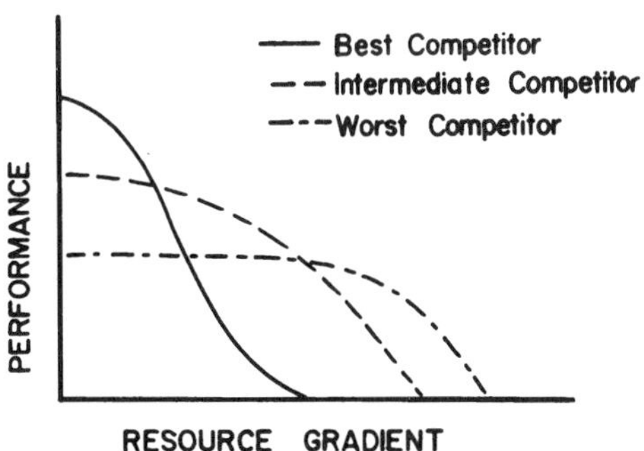

Figure 6.7. Relative performance along a resource gradient for three hypothetical species of differing competitive abilities. In each case, the plants do best for the same resource, even though the species differ in their relative abilities. The poorest competitors, however, can exist at resource levels where better competitors cannot, thus allowing them to persist in the community.

these same poor competitors do best (reach their highest biomass levels) in the core habitats if the dominant species are absent.

Community-Level Consequences of Plant Competition

Given these quite different models, is there a consensus about how plant communities operate? While most agree that interspecific competitive forces are critical for understanding plant assemblies, significant disagreement presently exists regarding the best framework for casting the problem. The biggest issue concerns the importance of alternative type of habitats, especially those lying along a productivity gradient (Grace 1990, 1991). The productivity gradient is largely determined by the availability of resources (mostly belowground) that promote plant growth. Each of these three primary contenders for explaining the dynamics of plant communities in grasslands includes some impact from competition.

1. The intensity of interspecific competition increases along a productivity gradient. Productive habitats support high growth rates and large biomass accumulation, resulting in preemption of light and resources. Species with high competitive abilities dominate productive habitats, while species with lower competitive ability are displaced to less productive habitats where competition is less intense. Grime's CSR model and Keddy's Centrifugal Community Organization model fall into this category.

2. The intensity of competition among species is equally intense along a productivity gradient, although the nature of limiting resources may change. In unproductive habitats, one expects intense competition for soil resources, while access to light is most important in productive environments. Species are specialized (physiologically and morphologically) in their ability to exploit particular ratios of soil resources and light, so species do best along that part of the productivity continuum where these conditions exist. Tilman's resource-ratio model falls into this category.

3. Environments are seldom unchanging, resulting in an additional view with perhaps the most explanatory power for actual communities (Collins and Glenn, Chapter 7). Perturbations, spatial and temporal heterogeneity, and interactions of plants with herbivores constantly beset strictly competitive interactions so that expected endpoints are seldom reached. Either the system is not in equilibrium, or potential equilibria due to competitive interactions are greatly influenced by other biotic interactions (e.g., herbivores) as well. As a result, community-level responses concerning plant species composition and coexistence must incorporate features other than only resource consumption. I explain this alternative in more detail.

Burrowing mammals, grazing ungulates, fire, or other natural participants in a natural grassland interrupt ongoing dynamic processes. Perturbations primarily act to limit competition in two ways, either by preventing competitive interactions from reaching completion or by changing the resource levels in the environment so that they are no longer limiting or so that another species is now favored. Disturbance from burrowing mammals, for example, often leads to highly diverse, forb-dominated patches that would otherwise be outcompeted by dominant grasses (Huntly and Inouye 1988; Whicker and Detling 1988). In this case, both disrupted competition and changed soil resources may be involved. The frequency of disruption will typically determine the importance of perturbation as an ecological force (Collins and Barber

1985; Collins and Glenn, Chapter 7). Fire often exhibits similar effects (Bragg, Chapter 4; Collins and Glenn, Chapter 7; Seastedt, Chapter 8).

Spatial and temporal heterogeneity per se may also alter expected competitive interactions (Shorrocks 1990; Shorrocks and Swingland 1990). A patchy environment with patches linked by dispersal may keep species in the system longer than is expected from competitive interactions. Competitive dynamics among species are operating largely independently within individual patches at a site. While local extinction of one or more species may take place within a patch, the larger unit of grassland still contains all species. As long as competitive dynamics among the patches are out of phase a bit, and some dispersal among patches takes place, more species can remain than expected (Atkinson and Shorrocks 1981; Holt 1984; Kareiva 1987). The determination of the community composition is a matter of scale; boundaries of the study may directly affect conclusions. Also, dispersal among patches means that a species that has gone extinct within a patch may reappear and coexist if conditions change in a suitable direction. Finally, the spatial mosaic interacts with forces responsible for creating new, suitable patches, and dispersal of appropriate propagules (usually seeds if thinking of plants) determines the initial conditions to start the competitive interactions over again.

A final mechanism that interacts with perceived competitive interactions is termed the "storage effect" (Chesson and Huntly 1988). Here, the environment fluctuates so that appropriate conditions for reproductive success occur only infrequently. However, in those few, exceptional years, reproductive output explodes and many seeds are produced. These seeds are then stockpiled in the seed bank and germinate over a period of years. While the stockpile of new individuals is continuously depleting until the next reproductive boom, the species is kept in the system, although it may become rare at times and may also be a poor competitor. For this mechanism to operate, the frequency of banner years must counter the "extinction curve" of recruits into the system. Other chapters (Hartnett and Keeler, Chapter 5; Collins and Glenn, Chapter 7) explore this issue in detail. This storage-effect mechanism will greatly influence conservation efforts for some species and deserves further study because the actual dynamic timing of the mechanism will be critical.

Debate continues regarding the relative importance of these mechanisms for structuring grassland plant communities. Important evidence both consistent with and counter to each possibility exists. The critical issue for our needs in this book is that detailed examination of interactions responsible for structuring grasslands is possible and ongoing. Resolution of such arguments is important both for the intrinsic understanding of grassland organization and for insights that will help direct conservation and rehabilitation efforts. And, depending on the species of concern (e.g., dominants versus interstitial species), different answers may arise (Grubb 1986).

HERBIVORY

To most observers, herbivores (particularly vertebrate grazers) are a dominant feature of grasslands. Large grazing ungulates are often conspicuous, as is the evidence for burrowing herbivores: gopher, badger, and ant mounds or prairie dog colonies. Still, herbivory from less obvious herbivorous invertebrates, particularly belowground root-feeding nematodes or insect larvae, may be as or more important.

Grasslands generally support large numbers of herbivores (Detling 1988). World-wide, native large mammalian herbivores and cattle remove, on average, 30 to 40% of the aboveground net primary production (ANPP) in grasslands, while insects remove another 5 to 15%. Belowground invertebrate consumers, primarily nematodes, consume an additional 6 to 40% of the belowground net primary production. Even these fairly high levels of consumption may be misleading, however, since these estimates greatly underrepresent the impact of seed predators or herbivores feeding on seedlings. Seed- and seedling-feeding herbivores destroy a stage required for rejuvenating populations while actually eating only a small amount of plant material. These estimates also assume that plant species are eaten uniformly. Even in years with generally low levels of overall herbivory, some plant species may be greatly impacted (e.g., more than 50% of ANPP eaten), while the rest of the species are largely uneaten (Joern 1988). It appears, then, that we must evaluate the impact of grassland herbivores in two ways: (1) herbivore impact on energy and material flow in aggregate (the total herbivore-impact levels), and (2) differential impact on specific life stages or species that affect population responses and interactions among species. The significance of the two may greatly differ, especially if a particularly influential species (keystone species) exists that exerts great impact but makes up a small part of the total net production.

In addition to removing plant tissue through feeding, herbivores often exert significant influence on grasslands in other ways, including (1) altering the physical environment and microtopography through burrowing; (2) changing patch structure of the community through intense, localized grazing (e.g., prairie dogs); (3) altering patterns and rates of soil development and nutrient availability, thus affecting nutrient cycling; (4) altering plant/soil–water relationships; (5) changing plant species composition; (6) altering gene frequencies with selection for prostrate growth forms in some cases; and (7) changing feeding-site selection by other herbivores (Detling 1988; Huntly and Inouye 1988; Whicker and Detling 1988). As an example, Table 6.1 lists ways that pocket gophers influence grasslands over a variety of spatial and temporal scales (Huntly and Inouye 1988). Each of the above changes the nature of the grassland, in effect altering the dynamics of species-level responses. My focus here concerns actual changes in the dynamics of species-level responses in response to herbivory.

Herbivores and Plant Population Dynamics

Since herbivores eat plants and can limit plant performance, it seems likely that herbivores play a significant role in directing plant population dynamics. This is a tricky relationship to untangle, however, because we must clearly distinguish the impact of herbivores on individual plants, say in reducing seed production, from its effect on population dynamics (Crawley 1988). Reduced seed production may be a direct response to the level of seed predation, but may not reduce recruitment to the plant population in the next generation, as in the case of the annual grama grass, *Bouteloua breviseta,* in Texas (Fowler 1986b). Specific sites may have resources to support only a limited number of new individuals, a number independent of the number of seeds

Table 6.1. Effects of ground-dwelling mammals on ecosystem properties over a variety of spatial and temporal scales

---

Small spatial and temporal scale (1m², 1 week)
   Increased light penetration
   Altered soil resources
   Decreased plant biomass
   Increased available resources
   New colonization sites

Intermediate spatial and temporal scale (100 m², 1 year)
   Increased resource heterogeneity
   Increased topographic heterogeneity
   Increased plant species richness
   Increased variability in plant biomass
   More microhabitats for consumers

Longer temporal scale (50 years)
   Altered soil fertility
   Altered rate of succession
   Altered path of succession
   Altered topography

Note: Events are assumed to operate in ecologically relevant rather than geologic or biogeographic domains, providing an upper limit on the scale considered (Huntley and Inouye 1988).

produced (as long as enough are produced). Consequences of mortality raise similar issues if population level responses to mortality are compensatory: Increased seedling mortality may not affect subsequent population dynamics if early mortality from herbivory simply reduces mortality suffered later on—say, from self-thinning. Still, even with the empirical difficulties of documenting the importance of herbivores to plant population dynamics in specific cases, herbivory is often important.

Strong evidence that herbivores have significant impacts on plant population dynamics comes from enclosure studies (Crawley 1988; Huntly and Inouye 1988; Whicker and Detling 1988). Herbivore exclusion almost always results in dramatic responses for individual species and species diversity. Such responses are clearly evident to anyone who has examined a heavily grazed pasture. Reduced plant growth in response to tissue loss from herbivores may open up resources to other plants and alter competitive interactions (Boryslawski and Bentley 1985).

Apparent Competition

Species sometimes respond to other species in a manner consistent with competition, but for other reasons. In "apparent competition," if the population density of one species increases, that of the other decreases, but the underlying response is driven by a shared predator. Here, the response is consistent with what is expected from competitive interactions, but it results from an unrelated mechanism. Herbivores may provide the driving mechanism behind apparent competition among plant species; the herbivore is shared by plant species, thus changing the resulting interactions in a way that superficially mimics competition between the plants (Louda et al. 1990). Rice

(1987) examined the importance of herbivory for maintaining the presence of two introduced annual forbs in the genus *Erodium* in California annual grasslands. As suggested earlier, persistence of these species in pastures grazed by sheep is enhanced by periodic soil disturbance from pocket gophers. However, when sheep are removed, *Erodium* is absent, even if soil disturbance occurs. This can possibly be explained by competition with grasses through shading of the rosette-forming *Erodium*. However, small seed-eating voles are common in the ungrazed areas, and Rice set up cages inside the sheep enclosures over artificially disturbed soil to examine the impact of these small herbivores on the establishment of *Erodium*. The annual forbs were able to persist in the ungrazed grassland in cages. It appears that the ungrazed areas afforded increased grass cover to serve as protection to the voles from avian predators. Then, the voles could readily forage beneath the grass cover, preventing the establishment of *Erodium*. The grass–*Erodium* interaction was indirect in its action rather than exhibiting a competitive result! The commonness of such indirect interactions is not known, but is likely to be much more important than suggested by known numbers of examples (Brown et al. 1986).

Herbivore–Herbivore Interactions

How do herbivores interact? An obvious type of interaction among herbivores includes direct and diffuse competition. While competition for limiting resources is certainly possible, this possibility has engendered much controversy (Strong et al. 1984). As a rule of thumb, many argue that folivorous herbivores probably do not compete for food under a wide range of normal conditions, but important exceptions exist. Some very clear evidence for competition exists among nectar/pollen feeders and seed eaters (Pleasants 1983; Waser 1983; Brown et al. 1986). And, as with the *Erodium* example from California annual grasslands, predator-mediated apparent competition is probably common, especially among insect herbivores, although there are few documented examples for grasslands. Important evidence is accumulating indicating that significant competition may occur among insect herbivores (Belovsky 1986a, 1986b; Karban 1986).

Above- and belowground herbivores each feed on the same plants, but on radically different parts. Because of the physiological linkage between roots and shoots, the two groups of herbivores interact and affect each other by their influence on the quantity and quality of plant tissue on tallgrass grassland in response to herbivory (Seastedt 1985; Seastedt et al. 1988). Plants respond physiologically to tissue loss from herbivores almost immediately. Severe defoliation results in nearly instantaneous cessation of root growth, further exacerbating any nutrient or water limitation suffered by the plant. Similarly, root-feeding invertebrates cause reductions in shoot growth rate (Ingham and Detling 1984). In addition to changes in the amount of available tissue, tissue quality may change; nutrients can be mobilized and translocated throughout the plant for regrowth or sequestered (Bloom et al. 1985). For example, soluble nitrogen in the leaf tissue may increase after grazing (Seastedt 1985), or the C:N ratio in roots may decrease after grazing as carbohydrates are mobilized for regrowth while nitrogen levels remain constant. Many soil invertebrates are greatly limited by the availability of nitrogen and any decrease in the C:N ratio results in rapid increases in growth rate (Seastedt 1985).

As a result of herbivore-mediated changes in plant quality, herbivores may have significant effects on other herbivores following feeding. Intermediate clipping levels (simulating moderate grazing) on blue grama from shortgrass prairie results in maximal numbers of nematodes (Stanton 1983). Similarly, belowground arthropod herbivores at Wind Cave National Park in South Dakota increased in abundance at intermediate levels of aboveground grazing (Seastedt et al. 1988). Biomass of white grubs (June beetle larvae, Scarabeidae), the dominant root-chewing insect in tallgrass prairie, increased significantly in mowed plots versus plots that were either burned/ungrazed or unburned/ungrazed, a result consistent with that of maximal densities of white grub with intermediate levels of sheep grazing (Seastedt 1985). Similar consequences exist for vertebrate grazers (Huntly and Inouye 1988; Whicker and Detling 1988). Under intense aboveground grazing, nematode or soil insect populations may decline relative to those in ungrazed areas.

Responses by root-feeding herbivores to aboveground herbivory may be quite involved and surprising (Seastedt et al. 1988). Root mass often decreased with loss of aboveground tissue, whereas the nutritional quality of remaining root tissue did not change. Plant resource availability (quantity and quality) was not sufficient to stimulate increased levels of belowground consumers. As a working hypothesis, large components of the belowground community seem implicated in the explanation. Since root tissues regularly grow and die in pulses, much root tissue in the soil recently died, producing nutrient-rich soils (Seastedt, Chapter 8). At senescence, roots are colonized by soil-dwelling microbes that find an adequate energy source in this root tissue but insufficient nitrogen for growth. To alleviate the nitrogen deficit, nitrogen is extracted from elsewhere (e.g., inorganic soil fraction) so that a portion of the diet of root-feeding consumers is directly or indirectly obtained from microbes as well. Intermediate levels of aboveground foliar tissue loss seem to provide the best conditions promoting such interactions and thus the highest densities of underground consumers (Seastedt et al. 1988).

Herbivores and Succession

Another excellent example of the dynamic interaction between above- and belowground herbivory concerns successional change in plant communities (Brown et al. 1988; Brown and Gange 1989a, 1989b, 1990; Gange and Brown 1989). A long-term, manipulative field experiment in pastures at Silwood Park, England, employed a range of sites at different successional stages and plant assemblies with different species composition to examine the relative roles of aboveground and belowground herbivory on the successional process. Key herbivores included sucking insects (Auchenorrhyncha) on grasses and chewing beetles on forbs.

Plant species richness continually increased over the course of the study in all plots, with the rate of species accumulation greatest when soil-residing herbivores were depressed. Ten (all forbs, mostly perennials) of the 98 total species that became established during these three years of colonization from bare ground were found only when soil insecticide was applied. Perennial grasses increased when foliar (aboveground) insecticides were applied, whereas perennial forbs increased when belowground herbivores were partially depressed. Two features of the dynamics of plant community development appear mediated by belowground herbivory:

1. Species recruitment is seemingly affected by belowground herbivory.
2. The competitive balance between plant species belonging to different life history groupings is modified (Brown and Gange 1990).

Results also indicate the degree to which trajectories of plant succession can be altered by insect herbivores. For example, removal of aboveground folivorous insect herbivores led to taller, denser stands of perennial grass with lowered species diversity (Brown et al. 1988).

PREDATOR–PREY INTERACTIONS

Predators consume prey, a simple dynamic interaction. While often conspicuous and dramatic, the act of predation is nearly instantaneous and its ultimate impact at population and community levels often obscure. For our interests in grassland communities, however, is not only who eats whom, but also the consequences of predation for population dynamics of both predator and prey species as well as any associated competitors. Experiments in the field are the only true means for untangling such causal relationships.

I raise the following issues in this section:

1. Can predators significantly alter prey populations?
2. What attributes of predator foraging behavior alter risk to individual prey?
3. What are the consequences of significant predation pressure at the community level?

I recommend Taylor (1984) and Stephens and Krebs (1986) for additional reading on this topic.

Predator Impact on Prey Populations

I begin with an example. Grasshoppers are very conspicuous insect herbivores in most grasslands and obvious candidates as prey for a number of predators and parasites/parasitoids. For example, in a single prairie, spiders, digger wasps, robber flies, lizards, turtles, ants, sarcophagid flies, scelionid wasps, and birds readily feed on various life stages of grasshoppers. Is predation pressure significantly strong to reduce grasshopper densities? Experiments are the most reliable way to assess this possibility, although few such studies exist for grassland situations.

Field experiments indicate that avian predation often significantly depresses grasshopper population densities from a variety of grasslands representing a wide range of conditions: Nebraska sandhills grassland (Joern 1986, 1992), North Dakota mixed grassland (Fowler et al. 1991), and Arizona arid grassland (Bock et al. 1992). Adult grasshopper densities at these sites were reduced by 25 to 35% in enclosures that prevented birds from foraging compared with similar areas that permitted avian predation over a six- to eight-week period. Predation pressure from birds to grasshoppers varied in time and space, however, both among patches and sites within years and among years at specific sites.

What about invertebrate predators? Robber flies (Diptera: Asilidae) may take large numbers of grasshoppers as prey in Nebraska sandhills grassland (Joern and Rudd 1982), at about the same level as birds in years when robber flies are common. Unlike birds, robber-fly populations fluctuate greatly in number and exert little impact on grasshopper populations in most years.

Interesting indirect interactions among predators also occur. While robber flies exerted periodic, strong effects on grasshoppers in Nebraska in some years, robber flies from Montana grasslands had a different effect (Rees and Onsager 1982, 1985). Dominant robber-fly species (*Efferia bicaudata, E. staminea,* and *Mallophorina guildiana*) feed primarily on Diptera (flies), including flies that are parasitoids of grasshoppers. Based on experiments, the presence of robber flies increased the survival rate of grasshoppers living in high-density patches, presumably by removing the important parasitoids. The reduction of robber flies by about 40% resulted in a doubling of parasitism rate to the grasshoppers. Parasitism rate at a site at which the robber flies were naturally absent was four-fold greater.

Frequency-Dependent and Higher-Order Interactions

Predation is often context dependent: Prey selection depends on what alternative prey types exist, the relative and absolute abundances of alternative prey types, and their spatial distribution relative to one another. Predators often have difficulty finding prey, largely because of remarkable blending of prey types with backgrounds (crypsis). Finally, prey may behaviorally respond to the risk imposed by predators and not use patches or habitats that they would otherwise use, with the result that other species interactions are changed.

Frequency-dependent predation occurs when prey are taken less than expected when rare and more than expected when common. In effect, risk to a prey individual to being eaten is dependent on what other prey are available and their relative densities. Many examples exist for frequency-dependent selection of prey by predators, especially by visually orienting predators such as birds. Frequency-dependent predation tends to stabilize the relative densities of prey because predators switch from one common prey type to another as relative densities shift in response to predation pressure (Joern 1988).

Behavioral responses of prey to avoid or elude predators may have as much impact on community-level responses as when actually eaten. Otherwise normal interactions may be disrupted, resulting in altered species composition or use of resources. Such disruptions are called higher-order interactions because the mechanisms go beyond the normal species—species interactions, as discussed earlier. For example, when prey sense that predators are nearby, movement and other behavioral patterns may change (Werner et al. 1983; Sih et al. 1988), or prey may seek refuge and not use specific habitats. For example, in aquatic habitats, the presence of bass causes small bluegill fishes to move from open water to seek refuge in weedy habitat. However, food reserves are often better in open water, and the bluegill grow more slowly. Slower growth ultimately affects future competitive interactions and changes the character of the community. It is easy to see how herbivores may alter habitat use in the presence of predators, as readily seen for many small mammalian herbivores such as voles. Rice's (1987) study of *Erodium* in California annual grassland (p. 116) illustrates that such higher-order interactions can have dramatic community-level effects.

INTERACTION WEBS

Natural communities are often diverse and the species-specific linkages complex. For example, when feeding interactions are arranged according to who eats what, food webs can be described. Plants capture energy from the sun and provide the base of the energy-transfer pyramid. Early ecologists such as Elton (1927) recognized the power that describing such relationships can bring to understanding community-level species relationships. With food webs and some careful quantitative measurements, energy flow in a community can be explicitly defined with such accounting methods. A great advantage of this approach is that community-level processes can be combined with ecosystem-level processes to investigate those forces that define ecosystem structure and function (McNaughton 1985; McNaughton et al. 1988; Pimm 1991; Hunter and Price 1992; Power 1992). Ecosystems, after all, are structurally organized as food webs within which energy is transmitted between trophic levels and dissipated into the environment (McNaughton et al. 1988).

Food webs by themselves do not sufficiently capture the interactive structure of an ecosystem, because they document only who eats whom. Other interactions, such as competition or the importance of indirect effects, are not explicitly included; interactive networks based exclusively on food webs can only partially explain the resulting structure. Pimm (1982, 1991) has argued, for example, that the trophic structure of a community observed in a food web results from the likelihood that all dynamic interactions among the species lead to a stable configuration, not just the feeding portion. In this section, I outline two related issues that indicate how important species interactions—as a network—are to understanding community organization.

Keystone Species

Some species exert such an incredible impact on communities that they are called keystone species—species that determine current structure in a community much as the keystone in an old stone arch maintains the integrity of the arch. In both cases, if the keystone is removed, the resulting structure quickly dissipates and a completely different state is achieved: rubble in the case of the arch, a different set of coexisting species in the case of a natural community—usually of lower diversity. Keystone species need not be dominant or conspicuous and may contribute little to energy and biomass flow in natural systems. However, based on available evidence, they contribute greatly through predator–prey, competitive, and mutualistic interactions with other species.

Keystone species can play a variety of roles. For example, a predator may directly control a dominant competitor and prevent it from causing local extinction of other poorer competitors. Likewise, a single competitor may exert such a strong effect on many species through a chain of indirect responses so that it acts mutualistically with most species; its demise results in a series of intense competitive interactions and local extinction of some former community members. In many ways, the response to the presence or absence of the keystone species results in a qualitative shift in how that particular community works, especially regarding species diversity and the actual identity of species in the community. In addition, physical disturbance by these species may be such that they disproportionately alter habitat structure, species composition, and biogeochemical processes (Brown and Heske 1990). From our standpoint, this issue is critical because we wish to keep keystone species populations intact in grassland communities. Unfortunately, no easy way presently exists to recognize which species are playing keystone roles. This is a vexing problem that must ultimately be confronted.

Keystone species readily illustrate that the worth of a species to community and ecosystem structure and function may be greater than immediate contributions to material and energy flow. In a transition zone between desert and arid grassland, vertebrates account for only a small fraction of biomass and energy flow. Brown and Heske (1990) showed that the long-term removal of a dominant seed-eating guild (three-species) of kangaroo rats near a transitional zone promoted a vegetational change from desert to arid grassland in 12 years. Tall perennial, annual grasses (exotic in this case) increased approximately three-fold in the absence of kangaroo rats at the expense of previous plant species. Dynamics of the shift seem largely caused by seed predation and soil disturbance. Selective seed predation on large seeded plants by the kangaroo rats was relieved on their removal. This then led to the initial shift from small-seeded to large-seeded winter annuals as a result of plant–plant competition (Brown et al. 1986). Elimination of physical disturbance of the soil and litter layers from foraging and burrowing kangaroo rats seemingly permitted the establishment and persistence of tall grasses. Cascading but otherwise indirect interactions with other trophic levels also resulted because of altered habitat structure. Other rodent species typical of arid grassland colonized plots with the new vegetation while avian foraging was reduced in plots where kangaroo rats were removed (Thompson et al. 1991). Because this study was performed near a natural transition from grassland to desert, it is unlikely that the specific results will be exactly duplicated whenever examined. However, a number of transitions among grassland types exist, and it is quite reasonable that biotic interactions instigated by as-yet-unrecognized keystone species may be playing a big role. The impact of prairie dogs or badgers discussed earlier is certainly a possibility here.

Trophic Cascades

Natural communities consist of multiple trophic levels, and all the interactions that I have discussed operate simultaneously. Do these simultaneously occurring processes influence one another, and how important is this issue? Do seemingly unrelated interactions affect one another? What are the relative roles of different ecological forces in determining population change and community structure? Answers to these questions are presently unfolding in a vigorous debate revolving around an attempt to determine how often and under what circumstances "top–down" forces, mediated through the impacts of predators, regulate community organization versus "bottom–up" forces, where participants in a community at each trophic level are largely food-limited so that competitive interactions dominate (Hunter and Price 1992; Hunter et al. 1992; Menge 1992; Power 1992; Strong 1992). Of course, the frequency and strength of any significant disturbances also determine community organization (Collins and Glenn, Chapter 7) so that the interactions that I discuss here, while significant, may never reach their concluding state. While experiments remain to be done in grasslands for the most part, results from aquatic systems indicate that strong, multiple-trophic-level events can greatly change the resulting community operations (Carpenter et al 1985; Vanni and Findlay 1990). Events happening at any level may "cascade" through the system, ultimately changing the impact of events at all other trophic levels, in both a dynamic population/community and an ecosystem/nutrient-cycling sense. For example, if predators of insect herbivores are removed or radically decreased in number (impact), population sizes of the insect herbivores may increase. The herbivore load on plants may then increase so that the amount of plant material that remains decreases or else certain plant species are more heavily eaten so that competitive outcomes among the plants shift. It is even possible that very different communities may result, depending on the strength and frequency of the cascade. An example provides the best introduction to this important topic.

Trophic cascades occur when population- and community-level consequences to species at a third trophic level vary because of interactions among species at two other trophic levels—the effects cascade through the system. For example, size-selective predation by fish on zooplankton in lakes often results in a zooplankton community consisting of smaller species (Carpenter et al. 1985). Often, the effect of predation at this level is to mediate the strong competitive effect of large zooplankton species on smaller species resulting in the shift. Zooplanktors eat algae (phytoplankton) in the lake, but the grazing rates of these herbivores is size-dependent. As a result, because of the trophic cascade, the fish predators can have a direct impact on plants (algae) without directly interacting with them. However, in some experiments, some predators have additional effects, notably the addition of nutrients to the system that another important invertebrate predator does not contribute. The phytoplankton react to nutrient additions with increased productivity (Vanni and Findlay 1990). In fact, Neill (1988) found that very different aquatic communities existed, depending on the nutrient regime found in the system; small shifts in the N:P nutrient ratios of the water column resulted in very different species membership in the community. As a result, both zooplankton grazing rates and nutrient recycling rates can drive community-level responses in these simple aquatic systems.

There is little doubt that multiple-trophic-level interactions must be understood to understand how communities are assembled. Yet plants and accompanying bottom—up forces occupy a fundamental position for understanding such relationships (Hunter and Price 1992)—removal of higher trophic levels leaves the plant level intact (if highly modified), while removal of the primary producers leaves no system at all. Primary producers provide energy to the system on which the rest of the community relies. This leads to a second (and highly debated) proposition. The degree of productivity in a system constrains the number of trophic levels (and sometimes species) that can stably coexist (Oksanen 1981; Fretwell 1987; Pimm 1991). Finally, the architecture of the food web, especially the degree of omnivory, may play an important role (Strong 1992) in determining the likelihood that top—down forces will have much impact, at least past one trophic level in either direction. Strong cascades will likely have a greater impact in low-species-diversity, ladderlike food webs, such as the aquatic situation described earlier. In many species-rich terrestrial communities, complex side-to-side and reticulated species relationships should act to diminish the full cascading effects among multiple trophic levels; there may simply be too much buffering within the system for it to be reactive to every perturbation at any level. Also, not all species will have equal impact: Keystone species alone may be able to redirect community structure in many situations.

Principles observed in aquatic systems undoubtedly apply to terrestrial grassland communities as well. As I have described, the impact of prairie dog colonies, pocket gophers, or large herds of grazing ungulates have sufficient impacts to create crosstrophic-level effects consistent with the cascade model for aquatic systems. The potential importance of such interactions to preservation or rehabilitation efforts is clear.

CONSEQUENCES

Clearly, a variety of species interactions exist in natural grassland communities. From these possibilities, we can devise a series of alternative views regarding the dynamic interactions that structure communities. The importance of contrasting competition models or the expected tempering effects of spatial heterogeneity, herbivory, or disturbance on competitive outcomes provide unresolved examples, despite clear predictions. More important, tests of the same hypotheses among the known variety of North American grassland systems have seldom been performed to see if the same rules are operating in all cases and, if operating, are of equal importance for determining the resulting community. This is an important gap for devising preservation and rehabilitation plans.

Are disputes such as the one regarding developing the correct form of competition model among coexisting plants to explain grassland communities of merely academic importance—of intrinsic interest but actually diverting attention away from important goals concerned with preserving biodiversity and restoring or rehabilitating grassland sites? How much species-specific detail is required to develop appropriate plans to maintain and recover North American prairie systems? Clearly, the number of potential specific interactions at one site, let alone many sites, are too numerous to carefully describe across continental gradients. If this amount of detail is required for successful application to real-world problems, the approach will not be of direct relevance, except for those rare occasions when the system is sufficiently simple to break down, or if we can identify keystone species. But if we are managing to promote a particular species diversity or the health of a few targeted species, we may have to untangle the specific species interactions.

Fortunately, only a few categories of major interactions exist that are replayed over and over. The confusing detail comes from (1) the multiple species in a community that interact directly, (2) the multiple indirect interactions in natural communities, (3) changes in the state of the system that lead to alternative outcomes for subtle reasons, often of a threshold nature, and (4) the dynamic properties of interacting units often behaving counterintuitively. The rate and order that processes proceed matter!

Even here, however, these indirect interactions are often compartmentalized into interacting units, which reduces the complexity significantly. In addition, key insights may be qualitative ones, responses that do not require the same amount of testing to uncover results critical to restoring or maintaining central features of grassland systems. Detecting and understanding the impact of keystone species is clearly a featured goal. Applied research should be directed at these issues to develop appropriate guidelines and methodologies so that prairie enthusiasts can apply them to local problems and campaigns. With care and focus, we can overcome these problems to investigate local problems. The United States has not yet declared an all-out assault on natural degradation with the same intensity that it addresses health-related issues. Now, too few individuals are expected to do too much with too few resources. When significant resources are directed at problems such as those addressed in this chapter, much as we examined important characters of cells to understand cancer, significant new insights with practical application will quickly arise.

Acknowledgments
My ideas have developed through frank discussions and interactions with many colleagues and students over the years. I thank Richard Alward, Spencer Behmer, Kathleen Keeler, and Yuelong Yang for comments on the manuscript. My research in this area has been supported by funds from the National Science Foundation, the USDA Competitive Grants Program, and the University of Nebraska.

LITERATURE CITED

Albertson, F. W. 1937. Ecology of mixed prairie in west-central Kansas. Ecological Monographs 7: 481–547.

Atkinson, W. D., and B. Shorrocks. 1981. Competition on a divided and ephemeral resource: a simulation model. Journal Animal Ecology 50: 461–471.

Barnes, P. W. 1985. Adaptation to water stress in the big bluestem–sand bluestem complex. Ecology 66:1908–1920.

Barnes, P. W., and A. T. Harrison. 1982. Species distributions and community organization in a Nebraska sandhills prairie as influenced by soil/water relationships. Oecologia 52:192–201.

Barnes, P. W., A. T. Harrison, and S. P. Heinisch. 1984. Vegetation patterns in relation to topography and edaphic variation in Nebraska sandhills prairie. Prairie Naturalist 16:145–158.

Begon, M., J. L. Harper, and C. R. Townsend. 1990. *Ecology: Individuals, Populations, Communities.* Blackwell, Boston.

Belovsky, G. E. 1986a. Generalist herbivore foraging and its role in competitive interactions. American Zoologist 26:51–69.

Belovsky, G. E. 1986b. Optimal foraging and community structure: implications for a guild of generalist grassland herbivores. Oecologia 70:35–52.

Bloom, A. J., F. S. Chapin III, and H. A. Mooney. 1985. Resource limitation in plants—an economic analogy. Annual Review Ecology and Systematics 16:363–392.

Bock, C. E., J. H. Bock, and M. C. Grant. 1992. Effects of bird predation on grasshopper densities in an Arizona grassland. Ecology 73:1706–1717.

Boryslawski, Z., and B. L. Bentley. 1985. The effect of nitrogen and clipping on interference between C_3 and C_4 grasses. Journal Ecology 73:113–121.

Brown, J. H., D. W. Davidson, J. C. Munger, and R. S. Inouye. 1986. Experimental community ecology: the desert granivore system. Pp. 41–61 in J. Diamond and T. J. Case (eds.). *Community Ecology.* Harper & Row, New York.

Brown, J. H., and E. J. Heske. 1990. Control of a desert–grassland transition by a keystone rodent guild. Science 250:1705–1708.

Brown, V. K., and A. C. Gange. 1989a. Differential effects of above- and belowground insect herbivory during early plant succession. Oikos 54:67–76.

Brown, V. K., and A. C. Gange. 1989b. Herbivory by soil-dwelling insects depresses plant species richness. Functional Ecology 3:667–671.

Brown, V. K., and A. C. Gange. 1990. Insect herbivory below ground. Advances Ecological Research 20:1–58.

Brown, V. K., M. Jepsen, and C.W.D. Gobson. 1988. Insect herbivory: effects on old field succession demonstrated by chemical exclusion methods. Oikos 52:292–302.

Carpenter, S. R., J. F. Kitchell, and J. R. Hodgson. 1985. Cascading trophic interactions and lake ecosystem production. Bioscience 35:635–639.

Chapin, F. S. III. 1980. The mineral nutrition of wild plants. Annual Review Ecology and Systematics 11:233–260.

Chapin, F. S. III, A. J. Bloom, C. B. Field, and R. H. Waring. 1987. Plant responses to multiple environmental factors. Bioscience 37:49–57.

Chesson, P. L., and N. Huntly. 1988. Consequences of life history traits in a variable environment. Annales Zoologica Fennici 25:5–16.

Clements, F. 1933. Competition in plant societies. News Service Bulletin, April 2. Carnegie Institution, Washington D.C.

Clements, F. E., J. E. Weaver, and H. C. Hanson. 1929. *Plant Competition: Analysis of Community*

Functions. Carnegie Institution, Washington, D.C.

Collins, S. L., and S. C. Barber. 1985. Effects of disturbance on diversity in mixed-grass prairie. Vegetatio 64:87–94.

Connell, J. H. 1990. Apparent versus "real" competition in plants. Pp. 9–26 in J. B. Grace and D. Tilman (eds.). *Perspectives on Plant Competition.* Academic Press, San Diego, Calif.

Crawley, M. J. 1988. Herbivores and plant population dynamics. Pp. 367–392 in A. J. Davy, M. J. Hutching, and A. R. Watkinson (eds.). *Plant Population Ecology.* Blackwell, Oxford.

Detling, J. K. 1988. Grasslands and savannas: regulation of energy flow and nutrient cycling by herbivores. Pp. 131–148 in L. R. Pomeroy and J. J. Alberts (eds.). *Concepts of Ecosystem Ecology: A Comparative View.* Springer-Verlag, Berlin.

Dix, R. L., and F. E. Smeins. 1967. The prairie meadow and marsh vegetation of Nelson Co., North Dakota. Canadian Journal Botany 45:21–58.

Elton, C. S. 1927. *Animal Ecology.* Macmillan, New York.

Fowler, A. C., R. L. Knight, T. L. George, and L. C. McEwan. 1991. Effects of avian predation on grasshopper populations in North Dakota grasslands. Ecology 72:1775–1781.

Fowler, N. L. 1986a. The role of competition in plant communities in arid and semiarid regions. Annual Review Ecology and Systematics 17:89–110.

Fowler, N. L. 1986b. Density-dependent population regulation in a Texas grassland. Ecology 67:545–554.

Fretwell, S. D. 1987. Food chain dynamics: the central theory to ecology? Oikos 50:291–301.

Gange, A. C., and V. K. Brown. 1989. Insect herbivory affects size variability in plant populations. Oikos 56:351–356.

Gaudet, C. L., and P. A. Keddy. 1988. A comparative approach to predicting competitive ability from plant traits. Nature 334:242–243.

Geiger, D. R., and J. C. Servaites. 1991. Carbon allocation and response to stress. Pp. 104–142 in H. A. Mooney, W. E. Winner, and E. J. Pell (eds.). *Response of Plants to Multiple Stresses.* Academic Press, San Diego, Calif.

Givnish, T. J. (ed.). 1986. *On the Economy of Plant Form and Function.* Cambridge University Press, Cambridge.

Goldberg, D. E. 1990. Components of resource competition in plant communities. Pp. 27–49 in J. B. Grace and D. Tilman (eds.). *Perspectives on Plant Competition.* Academic Press, San Diego, Calif.

Grace, J. B. 1990. On the relationship between plant traits and competitive ability. Pp. 51–66 in J. B. Grace and D. Tilman (eds.). *Perspectives on Plant Competition.* Academic Press, San Diego, Calif.

Grace, J. B. 1991. A clarification of the debate between Grime and Tilman. Functional Ecology 5: 583–587.

Grace, J. B., and D. Tilman (eds.). 1990. *Perspectives on Plant Competition.* Academic Press, San Diego, Calif.

Grime, J. P. 1979. *Plant Strategies and Vegetation Processes.* Wiley, Chichester.

Grubb, P. J. 1986. Problems posed by sparse and patchily distributed species in species-rich plant communities. Pp. 207–228 in J. Diamond and T. J. Case (eds.). *Community Ecology.* Harper & Row, New York.

Holt, R. D. 1984. Spatial heterogeneity, indirect interactions and the coexistence of prey species. American Naturalist 124:377–406.

Hunter, M. D., and P. W. Price. 1992. Playing chutes and ladders: heterogeneity and the relative forces of bottom–up and top–down forces in natural communities. Ecology 73:724–732.

Hunter, M. D., T. Ogushi, and P. W. Price. 1992. *Effects of Resource Distribution on Animal–Plant Interactions.* Academic Press, San Diego, Calif.

Huntly, N., and R. Inouye. 1988. Pocket gophers in ecosystems: patterns and mechanisms. Bioscience 38:786–793.

Ingham, R. E., and J. K. Detling. 1984. Plant–herbivore interaction in a North American mixed-grass prairie. III. Soil nematode populations and root biomass on *Cynomys ludovicianus* colonies and adjacent uncolonized areas. Oecologia 63: 307–313.

Joern, A. 1986. Experimental study of avian predation on coexisting grasshopper populations (Orthoptera: Acrididae) in a sandhills grassland. Oikos 46:243–249.

Joern, A. 1988. Foraging behavior and switching by the grasshopper sparrow, *Ammodramus savannarum,* searching for multiple prey in a heterogeneous environment. American Midland Naturalist 119:225–234.

Joern, A. 1992. Variable impact of avian predation on grasshopper assemblies. Oikos 64:458–463.

Joern, A., and N. T. Rudd. 1982. Impact of predation by the robber fly *Proctacanthus milbertii* on grasshopper populations. Oecologia 55:42–46.

Karban, R. 1986. Interspecific competition between folivorous insects on *Erigeron glauca.* Ecology 67:1063–1072.

Kareiva, P. 1987. Habitat fragmentation and the stability of predator–prey interactions. Nature 326: 388–390.

Kaspari, M. E. 1991. Central place foraging in grasshopper sparrows: opportunism or optimal foraging in a variable environment? Oikos 60: 307–312.

Keddy, P. A. 1989. *Competition.* Chapman and Hall, London.

Keddy, P. A. 1990. Competitive hierarchies and centrifugal organization in plant communities. Pp. 266–290 in J. B. Grace and D. Tilman (eds.). *Perspectives on Plant Competition.* Academic Press, San Diego, Calif.

Keddy, P. A. 1991. Plant competition and resources in old fields. Trends Ecology and Evolution 6: 235–237.

Keddy, P. A., and B. Shipley. 1989. Competitive hierarchies in herbaceous herb plant communities. Oikos 54:234–241.

Knapp, A. K. 1985a. Effect of fire and drought on the ecophysiology of *Andropogon gerardi* and *Panicum virgatum* in a tallgrass prairie. Ecology 66:1309–1320.

Knapp, A. K. 1985b. Water relations and growth of three grasses during wet and drought years in a tallgrass prairie. Oecologia 65:35–43.

Knapp, A. K., and W. K. Smith. 1989. Influence of growth form and water relations to stomatal and photosynthetic responses to variable sunlight in subalpine plants. Ecology 70:1069–1082.

Louda, S. M., K. H. Keeler, and R. D. Holt. 1990. Herbivore influence on competitive interactions among plants. Pp. 414–444 in J. B. Grace and D. Tilman (eds.). *Perspectives on Plant Competition.* Academic Press, San Diego, Calif.

Martin, T. 1986. Competition in breeding birds: on the importance of considering processes at the level of the individual. Current Ornithology 4: 181–210.

Martin, T. 1987. Food as a limit on breeding birds: a life-history perspective. Annual Review Ecology and Systematics 18:457–487.

McNaughton, S. 1983. Physiological and ecological implications of herbivory. Pp. 657–678 in O. L. Lange, P. M. Nobel, C. B. Osmond, and H. Ziegler (eds.). *Encyclopedia of Plant Physiology, New Series.* Springer-Verlag, Berlin.

McNaughton, S. J. 1985. Ecology of a grazing system: the Serengeti. Ecological Monographs 55: 259–294.

McNaughton, S. J. 1991. Dryland herbaceous perennials. Pp. 307–328 in H. A. Mooney, W. E. Winner, and E. J. Pell (eds.). *Response of Plants to Multiple Stresses.* Academic Press, San Diego, Calif.

McNaughton, S. J., R. W. Ruess, and S. W. Seagle. 1988. Large mammals and process dynamics in African ecosystems. Bioscience 38:794–800.

Menge, B. A. 1992. Community regulation: under what conditions are bottom–up factors important on rocky shores? Ecology 73:755–765.

Mooney, H. A., and W. E. Winner. 1991. Partitioning response of plants to stress. Pp. 129–142 in H. A. Mooney, W. E. Winner, and E. J. Pell (eds.). *Response of Plants to Multiple Stresses.* Academic Press, San Diego, Calf.

Neill, W. E. 1988. Community responses to experimental nutrient perturbations in oligotrophic lakes: the importance of bottlenecks in size-structured populations. Pp. 236–255 in B. Ebenman and L. Persson (eds.). *Size-Structured Populations.* Springer-Verlag, Berlin.

Oksanen, L. S., S. D. Fretwell, J. Arruda, and P. Niemela. 1981. Exploitation ecosystems in gradients of primary productivity. American Naturalist 118:240–261.

Paine, R. T. 1980. Food webs: linkage, interaction strength and community infrastructure. Journal Animal Ecology 49:667–685.

Pimm, S. L. 1982. *Food Webs.* Chapman and Hall, New York.

Pimm, S. L. 1991. *The Balance of Nature? Ecological Issues in the Conservation of Species and Communities.* University of Chicago Press, Chicago.

Pleasants, J. M. 1983. Structure of plant and pollinator communities. Pp. 375–393 in C. E. Jones and R. J. Little (eds.). *Handbook of Experimental Pollination Ecology.* S&AE (Van Nostrand), New York.

Power, M. E. 1992. Top–down and bottom–up forces in food webs: do plants have primacy? Ecology 73:733–746.

Rees, N. E., and J. A. Onsager. 1982. Influence of predators on the efficiency of the *Blaesoxipha* spp. parasites of the migratory grasshopper. Environmental Entomology 11:426–428.

Rees, N. E., and J. A. Onsager. 1985. Parasitism and survival among rangeland grasshoppers in response to suppression of robber fly (Diptera: Asilidae) predators. Environmental Entomology 14: 20–23.

Rice, K. J. 1987. Interaction of disturbance patch size and herbivory in *Erodium* colonization. Ecology 68:1113–1115.

Schimel, D. S., A. F. Kittel, A. K. Knapp, T. R. Seastedt, W. J. Parton, and V. B. Brown. 1991. Physiological interactions along resource gradients in a tallgrass prairie. Ecology 72:672–684.

Seastedt, T. R. 1985. Maximization of primary and secondary productivity by grazers. American Naturalist 126:559–564.

Seastedt, T. R., R. A. Ramundo, and D. C. Hayes. 1988. Maximization of densities of soil animals by foliage herbivory: empirical evidence, graphical and conceptual models. Oikos 51:243–248.

Shorrocks, B. 1990. Coexistence in a patchy environment. Pp. 91–106 in B. Shorrocks and I. R. Swingland (eds.). *Living in a Patchy Environment.* Oxford Science, Oxford.

Shorrocks, B., and I. R. Swingland (eds.). 1990. *Living in a Patchy Environment.* Oxford Science, Oxford.

Sih, A., J. Petranka, and L. B. Katz. 1988. The dynamics of prey refuge use: a model and tests with sunfish and salamander larvae. American Naturalist 132:463–483.

Smith, T., and M. Huston. 1989. A theory of the spatial and temporal dynamics of plant communities. Vegetatio 83:49–69.

Stanton, N. L. 1983. The effect of clipping and phytophagous nematodes on net primary production of blue grama, *Bouteloua gracilis.* Oikos 40:249–257.

Stephens, D. W., and J. R. Krebs. 1986. *Foraging Theory.* Princeton University Press, Princeton, N.J.

Strong, D. R. 1992. Are trophic cascades all wet? Differentiation and donor-control in speciose systems. Ecology 73:747–754.

Strong, D. R., J. H. Lawton, and T.R.E. Southwood. 1984. *Insects on Plants.* Harvard University Press, Cambridge, Mass.

Taylor, R. J. 1984. *Predation.* Chapman and Hall, New York.

Thompson, D. B., J. H. Brown, and W. D. Spencer. 1991. Indirect facilitation of graminivorous birds by desert rodents: experimental evidence from foraging patterns. Ecology 72:852–863.

Tilman, D. 1982. *Resource Competition and Community Structure.* Princeton University Press, Princeton, N.J.

Tilman, D. 1986. Evolution and differentiation in terrestrial plant communities: the importance of the soil resource: light gradient. Pp. 359–380 in J. Diamond and T. J. Case (eds). *Community Ecology.* Harper & Rev. New York.

Tilman, D. 1988. *Plant Strategies and the Dynamics and Structure of Plant Communities.* Princeton University Press, Princeton, N.J.

Tilman, D. 1990. Constraints and trade-offs: toward a predictive theory of competition succession. Oikos 58:3–15.

Tilman, D., and D. Wedin. 1991a. Plant traits and resource reduction for five grasses growing on a nitrogen gradient. Ecology 72:685–700.

Tilman, D., and D. Wedin. 1991b. Dynamics of nitrogen competition between successional grasses. Ecology 1038–1049.

Vanni, M. J., and D. L. Findlay. 1990. Trophic cascades and phytoplankton community structure. Ecology 71:921–937.

Waser, N. K. 1983. Competition for pollination and floral character differences among sympatric plant species: a review of evidence. Pp. 277–293 in C. E. Jones and R. J. Little (eds.). *Handbook of Experimental Pollination Ecology.* S&AE (Van Nostrand), New York.

Werner, E. E., J. F. Gilliam, D. J. Hall, and G. G. Mittelbach. 1983. An experimental test of the effects of predation risk on habitat use in fish. Ecology 61:233–242.

Whicker, A. D., and J. K. Detling. 1988. Ecological consequences of prairie dog disturbances. Bioscience 38:778–785.

Wiens, J. A. 1989. *The Ecology of Bird Communities.* 2 vols. Cambridge University Press, Cambridge.

Wilson, S. D., and P. A. Keddy. 1986a. Measuring diffuse competition along an environmental gradient: results from a shoreline plant community. American Naturalist 127:862–869.

Wilson, S. D., and P. A. Keddy. 1986b. Species competitive ability and position along a natural stress/disturbance gradient. Ecology 67:1236–1242.

Wilson, S. D., and D. Tilman. 1991. Components of plant competition along an experimental gradient of nitrogen availability. Ecology 72:1050–1065.

7

Grassland Ecosystem and Landscape Dynamics

SCOTT L. COLLINS
SUSAN M. GLENN

> With the day's field work done, we decided to take advantage of the chance
> of the moment and follow the calls to try to get a view of the wolves, and as
> the darkness fell we set off along one of the trials that led back into the forest.
> ... We knew that we could not predict exactly where the wolves would go or
> determine precisely from the calls just where the wolves were. Chance and
> uncertainty were inherent in our choices and in our understanding of the be-
> havior of the wolves in the wilderness. But that unpredictability added interest
> to the night. We did not find the wolves, but standing on the ridge in the
> darkness and listening to the fading sound of the calls, we enjoyed a deep sense
> of the wild that rewarded us as much as a view of the wolves.
>
> D. Botkin, *Discordant Harmonies*

THE CHANGING PRAIRIE

The primary focus of many ecologists during the first half of the twentieth century
was the description and classification of plant communities. Many believed that communities were distinct, precisely defined assemblages of species. A good community
description would allow the observer to name any plant community based on the
framework of an idealized vegetation classification. The focus was clearly on similarity. During the late 1940s and early 1950s, observations of variation, cycles, and
disturbance provided a spark that stimulated new views on plant community structure
and dynamics. Currently, ecological interests have shifted from seeking similarity for
the purpose of classification to understanding variability within the context of natural
disturbances. As a consequence, these general trends in community ecology certainly
reflect the focus of ecological research in prairies. For example, early grassland ecologists believed that fire was a catastrophic and unnatural component in prairies. To
some extent, this view developed during the drought of the 1930s, when fires were
damaging already overgrazed and eroded prairies. Today, fire is recognized as an

important aspect of the natural disturbance regime, especially in mesic tallgrass prairies.

Increased understanding of the importance of natural disturbances has led to the generalization that the structure of plant communities is variable in both space and time (Sousa 1984). *Structure* refers to the kinds and distribution of species in a community; *dynamics* refers to changes in structure over time. The dynamics of plant and animal communities, however, will vary with scale (Wiens 1989; Hoekstra et al. 1991). Delcourt et al. (1983) divided spatial and temporal scales into three components: the micro-, macro-, and mega-scales (Figure 7.1). Micro-scale phenomena occur in time frames of 1 to 5000 years and at a spatial scale from 1 to 10^6 square meters. Phenomena at this scale include disturbances such as mammal diggings, fire, and grazing, and the subsequent vegetation changes following disturbance (e.g., plant succession). Macro-scale events occur at intermediate levels ranging from 5000 to 10^6 years and at a scale of 10^6 to 10^{12} square meters. These phenomena encompass climatic changes and periods of glacial advances and retreats. The largest scale of resolution includes time periods greater than 10^6 years and larger than 10^{12} square meters. Events at this scale include evolutionary change as well as plate tectonics and continental drift. From an ecological perspective, many spatially "large-scale" phenomena, such as succession, are included under the micro-scale rubric defined by Delcourt et al. (1983). Nevertheless, we will use this approach because of its quantitative, albeit arbitrary, definition of scale.

Ecologists are becoming increasingly more enlightened about the role of hierarchy and scale in ecological research. Typically, macro-scale phenomena operate at higher hierarchical levels and have slower rates of change than micro-scale events (Allen and Starr 1982); for instance, postglacial migration of vegetation across the Great Plains takes longer than revegetation of an abandoned pasture. In addition, factors at higher hierarchical levels are said to "constrain" processes at lower levels (O'Neill 1989); for example, the number of species in a community is constrained to be a subset of the number of species in the region. Finally, macro-scale events can usually be explained by a small number of abiotic variables, whereas the study of micro-scale phenomena requires detailed information on a number of biotic and abiotic influences (Meentemeyer and Box 1987). Meentemeyer (1978) found that at a continental scale, rates of decomposition are correlated with precipitation, whereas local rates of decomposition are related to immediate conditions such as moisture, temperature, bacterial density, and nitrogen.

The current emphasis on experimental approaches in ecology has led to more detailed analyses of micro-scale mechanistic interactions. This is certainly true for research in grasslands (Collins and Glenn 1988). In response to this highly focused approach, it seems relevant to ask whether or not we are missing the grassland for the tillers? Many important and interesting events take place over large temporal and spatial scales. In this chapter, we discuss the dynamics of grassland communities at a variety of spatial scales, emphasizing the importance of large-scale or landscape-scale phenomena. First, we will briefly review the macro-scale developmental history of North American grasslands. Next, we will discuss the larger micro-scale patterns of community succession and the role of disturbance in grassland communities. Finally, we will briefly review and characterize the short-term dynamics of grasslands in the absence of disturbance.

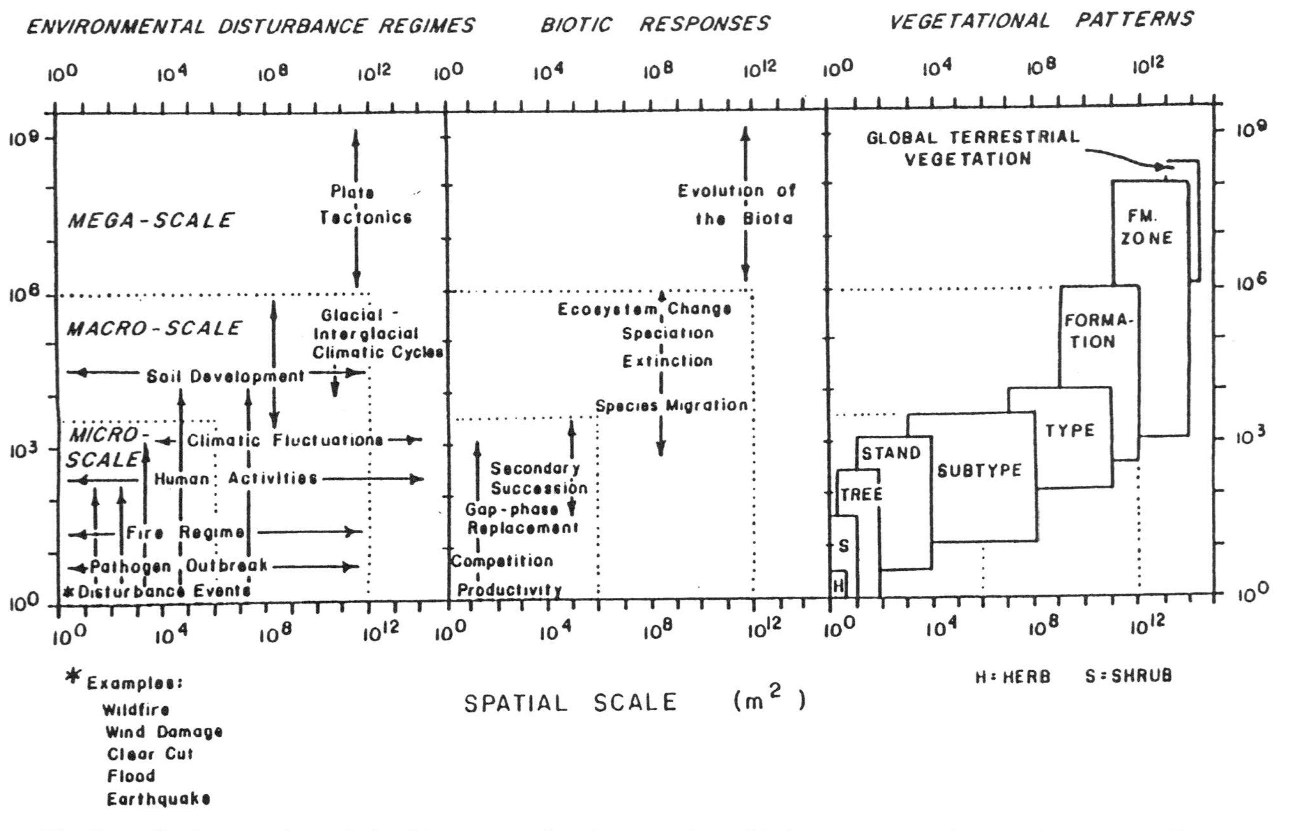

Figure 7.1. Generalized space–time relationships among disturbance regimes, biotic responses, and vegetation patterns. The models imply a general correlation between the spatial scale in which a variable occurs and the time scale over which it changes. (From Delcourt et al. [1983])

GLACIERS AND THE DEVELOPMENT OF NORTH AMERICAN GRASSLANDS

Large-scale dynamics are driven by latitudinal changes in temperature and seasonality produced by variation in the earth's orbit, and changes in the extent and height of continental glaciers (Prentice et al. 1991). More recent changes reflect the interaction of climate and man. Thus an understanding of the development of North American plant and animal communities may help explain current patterns of community structure and dynamics. A historical perspective may also help to predict future dynamics caused by global climate change and increasing levels of atmospheric CO_2.

Detailed information exists on vegetation history back to about 30,000 years before present (Y.B.P.). An understanding of historical change in plant communities comes primarily from analysis of fossil pollen, along with fossils of leaves, stems, wood, and flowers when available. Additional inferences are derived from knowledge of the biology of extant species that are related to species found in the fossil record. Description of paleoenvironments is based on the assumption that habitat requirements and tolerances of these species have not changed much over time (Delcourt and Delcourt 1991). Thus the presence of peccary bones in 20,000-year-old deposits in Kansas would imply that the climate there was frost free at that time because peccaries are currently found in the arid southwestern United States and Mexico.

General information on North American vegetation exists from the Tertiary (75–3 million Y.B.P). More detailed findings are available from the Quaternary (3 million Y.B.P.–present), especially for the eastern half of North America (Delcourt and Delcourt 1991). It is hypothesized that North America was covered by continental-scale geofloras during the Eocene epoch (60–40 million Y.B.P.) (Vankat 1979). These geofloras are species assemblages that covered wide latitudinal belts. The Arcto-Tertiary geoflora contained temperate forest species and extended across what is now Canada. The Neotropical-Tertiary geoflora, with many evergreen tropical species, occurred throughout much of southern North America. The Oligocene through the Pliocene epochs (40–2.5 million Y.B.P.) are marked by a gradual decrease in global temperature and a general southward shift of the geofloras. Large mountain ranges in western North America formed during this time. The Rocky Mountain uplift was particularly notable because it created a rain shadow to the east, which dramatically increased aridity in the central Great Plains.

Of primary relevance to the central Plains are climatic and vegetation changes that began during the Miocene (26 million Y.B.P.), when the grassland biome began to develop (Wells 1983; Axelrod 1985). Analysis of fossil floras and faunas suggests that during the Miocene much of the upland areas in the southern Great Plains was occupied by oak (*Quercus*) woodland with scattered patches of grasses. Throughout the Miocene, there was a gradual increase in grass pollen and a decrease in arboreal (tree) pollen other than oaks. As the drying trend continued through the Pliocene (13–2.5 million Y.B.P.), the uplands between river valleys became increasingly occupied by grasses, and trees were restricted to more mesic floodplain habitats (Axelrod 1985). An early Pliocene fossil site in Beaver County, Oklahoma, contained maples (*Acer*), hackberry (*Celtis*), ash (*Fraxinus*), and sycamore (*Platanus*). Several of these species currently have high population densities only many kilometers to the east (e.g., in Arkansas). Fossil remains of tortoise, rhinoceros, and peccary suggest that the habitat between rivers was open grassland or savanna (Chaney and Elias 1936). These animals indicate that the seasonal environment at this time was frost free. The maintenance of this open landscape is attributed to the preponderance of grazing animals and recurring fires. Thus grassland communities were well established by the end of the Pliocene.

The Pleistocene epoch is marked by dramatic changes in climate and vegetation. The Pleistocene lasted for only about 2.5 million years and represents a time period of global cooling that led to several sequences of glacial advance and retreat. Time spans of glaciation were on the order of 100,000 years, whereas interglacial cycles lasted about 20,000 years (Delcourt and Delcourt 1991). Approximately seven glacial–interglacial cycles occurred during the Pleistocene. Evidence of all but the most recent cycles is scanty because each subsequent glacier destroyed much of the evidence left by the previous ice sheet.

The last three glacial periods in North America—the Kansan, Illinoisan, and Wisconsinan (most recent)—are named for the southern extent of the ice. Accumulation and movement of the Wisconsin ice sheet was affected by a global temperature 7° to 8°C below current temperature (Daubenmire 1978). These temperature changes can alter the large-scale distribution of species; that is, species ranges appear to "migrate" as individuals die in the north while new individuals are established farther to the south. As ice accumulated, the upper Pleistocene plant communities were disrupted as species migrated southward ahead of the advancing ice. Also at this time, species in the Rocky Mountain region migrated down in elevation and expanded eastward across the plains (Kaul et al. 1988). At one time it was believed that plant communities were displaced as a unit; that is, what is now the tallgrass prairie shifted south as an intact community (Braun 1950). However, evidence from studies of fossil-pollen deposits in eastern North America shows that species were displaced to different degrees and that postglacial migration to the north and west occurred at varying rates among species (Davis 1976; Webb 1986). This resulted in a constant mixing of plant species in different communities.

Because few pollen-deposition zones occur in the arid central Plains, limited evidence exists concerning the distribution of vegetation in the Great Plains during the full extent of the Wisconsin ice sheet (Kaul et al. 1988). One source of information comes from packrat middens, because packrats gather and store seeds and other plant materials that can be used to infer what types of plants existed in the local region of the midden. Extrapolation from pollen cores and macrofossils in packrat middens from both Kansas and Texas suggests that during full glaciation, the southern and central Plains were occupied by an open forest of spruce and jack pine (Figure 7.2a), (Wells 1983; Axelrod 1985). The now extinct *Bison antiquus* and other large mammals, such as mammoth, were present in the southern Plains during the late Pleistocene (Axelrod 1985; Goetze 1989). Axelrod (1985) suggests that these large herbivores acted in concert with the changing climate to help eliminate tree species during the postglacial development of the grassland biome.

At the start of the Holocene about 18,000 to 12,000 Y.B.P., a gradual warming trend occurred and grassland species returned to the central Plains as forest species retreated northward. The forest–grassland ecotone (boundary) shifted to the east during the Hypsithermal period (8000–4000 Y.B.P.) (Figure 7.2b). A general cooling period followed the Hypsithermal, causing the ecotone to again shift to the west, but by this time the distribution of grassland vegetation was probably similar to the patterns we see today. Disjunct populations of sugar maple (*Acer saccharum*) in

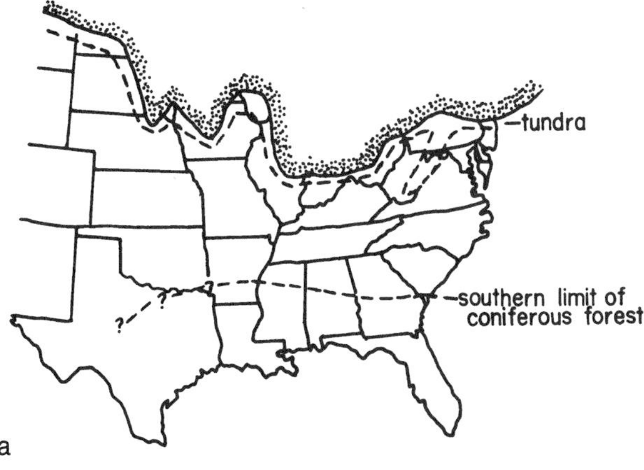

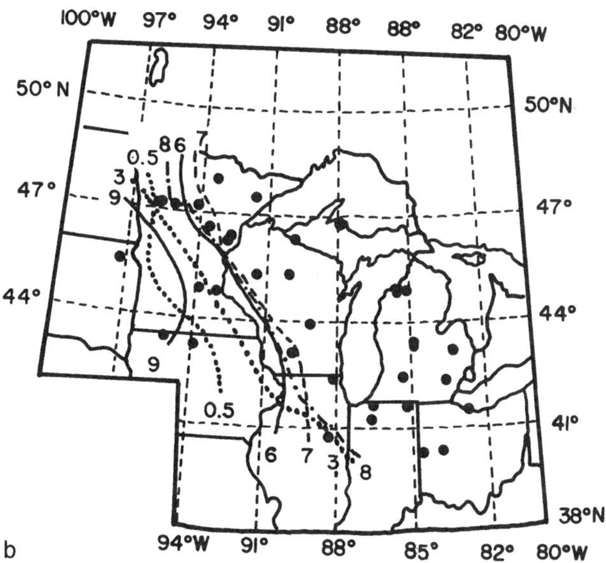

Figure 7.2. (a) Distribution of tundra and coniferous forest during the last full-glacial period. (From Woodcock and Wells [1990], with permission of Elsevier) (b) Changes in the forest–prairie boundary during the Holocene. (From Delcourt and Delcourt [1991]; Webb et al. [1983])

southwestern Oklahoma and paper birch (*Betula papyrifera*) on the northern edge of Nebraska (Kaul et al. 1988) may be relicts from range fluctuations occurring during the Hypsithermal.

The displacement of vegetation by glaciation leads to two central questions: Where were grassland plants and animals during the full glacial period, and how did these species rapidly migrate from refugia to repopulate the central Plains fol-

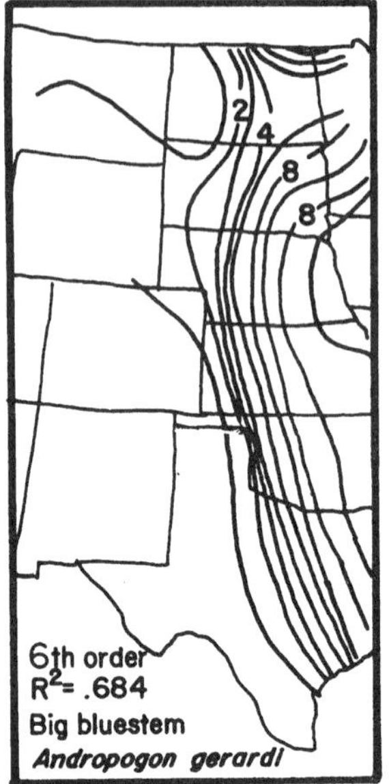

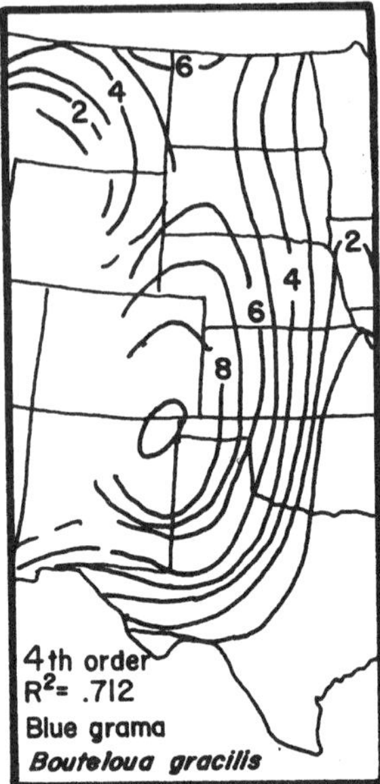

Figure 7.3. Isolines showing the dominance pattern of big bluestem and blue grama across the central Plains. The larger the value, the greater the relative dominance of each species where it occurs. These patterns suggest a glacial refugia in the southeast for big bluestem and in the southwest for blue grama. (From Brown and Gersmehl [1985], with permission of the Association of American Geographers)

lowing glacial retreat? The specific location of species during full glaciation is uncertain. Fossil pollen is not too helpful because most grass species have similar-looking pollen. Based on present-day geographic patterns of species richness, the center of origin for the genus *Andropogon* (includes big and little bluestem) appears to have been in the southeastern United States (Wells 1970). Some grassland species may have occurred in the spruce–jack pine savannas of the full glacial period. A map of current distribution patterns suggests that big bluestem could have migrated to its current range distribution as a wave from the southeast throughout the Great Plains (Figure 7.3), (Brown and Gersmehl 1985). Distribution of blue grama (*Bouteloua gracilis*) also suggests a wave movement pattern from the southwestern United States (Figure 7.3). However, current rates of seedling establishment of blue grama and big bluestem are low (Coffin and Lauenroth 1988; Glenn-Lewin et al. 1990; Hartnett and Keeler, Chapter 5), which may imply that conditions in the past were more favorable for establishment than are conditions today (Brown and Gersmehl 1985). The degree of local and regional genetic diversity in big bluestem populations would imply that new individuals of big bluestem were established frequently in tallgrass plant communities (Keeler 1990; Hartnett and Keeler, Chapter 5), and regular establishment may still occur. More research on the population dynamics of grassland species is needed.

The grassland biome is considered to be relatively recent in origin because it lacks endemic species (Wells 1970) and a unique climate (Axelrod 1985; Bragg,

Chapter 4). Indeed, some climate-based vegetation models predict the presence of woodlands in the tallgrass prairie region. Many grassland genera and species occur throughout eastern, northern, and western North America. It would also appear that the ranges of many grassland species are still capable of expanding. Thus distributions of prairie plants are not in equilibrium with climate (Wells 1983; Axelrod 1985; Brown and Gersmehl 1985). We believe the highly dynamic nature of grassland communities currently operating at several spatial and temporal scales is consistent with the relatively recent development of the grassland biome, annual climatic variability in the region, nonequilibrium distribution of species in relation to climate, and complex disturbance regime imposed by large grazing mammals and frequent fires.

SUCCESSIONAL CHANGE: NATURAL RESTORATION OF THE PRAIRIE

Succession is defined as a directional change in species composition at a site over time (McIntosh 1980). Directional implies the notion that species composition changes in a generally predictable way; for instance, if cropland is left fallow in northern Oklahoma, then an orderly progression of plant communities starting with annual weeds and ending in tallgrass prairie may occupy the abandoned field. Plant succession occurs following both natural and man-made events, such as abandonment of agricultural land (old-field succession), natural disturbances (postfire succession), or the creation of new land surfaces—for example, by volcanoes (primary succession). Old fields have been the cornerstone of successional research in North America (Pickett et al. 1987). One of the first theoretical models of old-field succession was developed by Frederick Clements while at the University of Nebraska. Clements believed that the development of vegetation was similar to the development of an organism; that is, the association of species "arises, grows, matures, and dies" (Clements 1916). Succession was like a healing process when some external factor damaged part of the natural climax vegetation. Clements's viewpoint of a highly structured and organized plant community developed partly from life in the prairie, where vast tracts of seemingly similar vegetation were evident (Allen and Hoekstra 1992). Clements proposed that the variation seen in the landscape was simply different stages of succession following abandonment of agricultural land. The Clementsian scheme of highly integrated species associations has been carried to an extreme by many ecologists, and as a result, his contributions are often unfairly criticized. Although very little evidence exists for the community-as-organism analogy, much of Clements's writing contains valuable insights into the structure, development, and dynamics of plant communities.

Because successional change may occur over decades to centuries, succession is most commonly studied using a space-for-time substitution method called a *chronosequence* (Pickett 1989). Essentially, a number of sites of different age since abandonment are sampled at one time within a region. From this information, it is possible to piece together the general pattern of successional change in the region. The technique assumes that all the sites are similar in soil type, starting conditions, and climate. Although this assumption is rarely true, it has allowed ecologists to study some patterns of species replacement over time. Specific mechanisms of replacement and processes during succession, though, are best studied at a single site over time. Such studies provide detailed information on the mechanisms that allow one species to replace another during succession. Most of such research has focused on the earliest and most rapid stages of succession. In this section, we will briefly review some general successional patterns in tall grass prairie. Then we will summarize a few studies that have dealt with the mechanisms of successional change in grasslands.

Booth (1941) conducted one of the most comprehensive surveys of species replacement patterns during old-field succession in tallgrass prairie. Booth sampled a chronosequence of 106 abandoned fields in Oklahoma and Kansas, ranging from 1 to 31 years since abandonment. From these data, Booth identified four distinct successional stages: annual-weed, annual-grass, bunchgrass, and mature prairie. Particularly curious was the rapid disappearance of the annual-weed stage (within 1–3 years), and the duration of the annual-grass stage, which replaced it (9–13 years and occasionally up to 30 years [Rice 1976]). The persistence of the annual-grass stage was intriguing because it was dominated by a seemingly diminutive annual, three-awn grass (*Aristida oligantha*).

The four successional stages defined by Booth (1941) served as the cornerstone for detailed mechanistic analyses of species replacement and persistence by Rice and his students (Rice 1984) at the University of Oklahoma. One of the primary mechanisms influencing the rate of old-field succession in Oklahoma is *allelopathy,* biochemical interactions (inhibitory and stimulatory) between plants and microorganisms (Rice 1984). Wilson and Rice (1968) demonstrated that at least one member of the annual-weed stage, annual sunflower (*Helianthus annuus*), is autotoxic. This means that sunflowers produce chemical by-products that inhibit the growth of other sunflowers. This can certainly account for the rapid disappearance of the annual-weed stage. Rice and colleagues also demonstrated that three-awn grass, the dominant of the second successional stage, produces chemicals that reduce germination and growth of competing species, thus allowing three-awn grass to rapidly achieve and retain dominance of a site for many years. Finally, laboratory experiments indicated that three-awn grows best on soils containing nitrate nitrogen (NO_3–N), whereas later successional species prefer ammonia (NH_4–N) (Smith and Rice 1983). Three-awn grass, and other early successional species, produce toxins that inhibit the conversion of nitrate to ammonia by soil bacteria (Rice 1964). Nutrients with a positive charge, such as ammonia, remain longer in the root zone of soils, which increases soil fertility, whereas negatively charged molecules, such as nitrate, are easily leached from the soil by rain. As a result, soils with a high ratio of nitrate to ammonia will generally have less nitrogen available for plants. Additional research showed that three-awn grass requires less soil nitrogen than does little bluestem, the dominant of the next successional stage (Rice et al. 1960).

The studies by Rice (1984) represent an important link between ecosystem nutrient dynamics and population dynamics in the study of successional causes. Because not all successional pathways in grasslands follow the pattern proposed by Booth (1941), the mechanistic detail of allelopathy cannot be generalized to explain succession in all prairies. For example, Collins and Adams (1983) analyzed succession over 32 years at a single site in Oklahoma. Neither the annual-grass stage nor the bunchgrass stage could be identified during succession on these plots (Collins and Adams

1983). Rather, succession proceeded quickly, and different patterns of vegetation change occurred among adjacent study areas.

Succession may also occur following other changes in land use. Overgrazing may dramatically reduce the abundance of perennial grasses and lead to an increase in annual weed species. Glenn-Lewin (1980) reported that revegetation of a previously overgrazed area in Iowa exhibited no clear species replacement pattern at all. Instead, succession occurred by the encroachment of "more-or-less whole prairie vegetation" from adjacent undisturbed prairie. At this rate, Glenn-Lewin calculated that succession would require 450 to 900 years to completely revegetate the previously overgrazed area. Thus the mechanistic approach yields detailed information on species replacement patterns at a site, but does not necessarily provide a generalizable model to describe succession in grasslands.

Another example of pattern and mechanism during grassland succession is provided by Inouye et al. (1987), who sampled vegetation and soils at 22 sites in a 56-year-long chronosequence in the sand-plain grassland of southern Minnesota. Again, nitrogen availability was the focus. Soil nitrogen increased during succession. Area occupied by annual species decreased, and cover of perennial species increased during succession. Cover of grasses was positively correlated with nitrogen, whereas cover of forbs was unrelated to soil nitrogen. Finally, cover of true prairie species was positively correlated with soil nitrogen.

At the University of Minnesota, Tilman (1982, 1985) proposed a simple mechanism by which it is possible to predict differences in competitive abilities between species, and thus predict patterns of species replacement during succession. According to Tilman's model, if one species can reduce the level of a limiting resource below the tolerance level of a second species, then the second species cannot survive and will be eliminated from a site. This represents a micro-scale mechanism of competitive interaction between neighboring plants. Tilman considered light and nitrogen to be the two most critical resources that drive succession. Light is readily available early in succession, but decreases at the soil surface as plant cover increases during succession. Nitrogen availability is initially low, but increases during succession. Competitive ability in plants represents a trade-off system (Joern, Chapter 6). Species that are good competitors for light are often poor competitors for nitrogen and vice versa. Using field experiments, Tilman and Wedin (1991) demonstrated that patterns of dominance during a 56-year-long chronosequence could be explained by this trade-off in competition for light versus nitrogen. Both big bluestem and little bluestem, dominant species in the late stages of succession, could reduce soil nitrogen to levels below which early successional species could survive.

How well does this process-oriented model explain the patterns of species replacement during succession in grasslands? Experimental results suggest that the nutrient-versus-light-competition trade-off can explain successional dynamics once trees are established. However, the first 40 years of succession are more a function of the trade-off between good colonizing ability and a weak ability to compete for limiting resources (Tilman 1990). Since grassland succession can take less than 30 years on some sites (Collins and Adams 1983), the resource–trade-off model probably does not apply to grasslands. It does appear, however, that successional replacement is a function of the interaction between competitive ability and population growth rates in response to nutrient levels (Huston 1979). If propagules are available for colonization, then enhancing soil fertility in grasslands will decrease the time required to restore degraded tallgrass prairie by increasing the growth rates of mature prairie grasses (Rice 1976; Tilman and Wedin 1991).

SHORT-TERM CHANGE: BISON, GOPHERS, AND FIRE

The previous sections described patterns and mechanisms of vegetation development over relatively long (by human standards), but still micro-scale, time periods. Much of the dynamics in grassland vegetation occurs at very short time spans and over small patches of prairie. For example, a gopher may create a small mound of soil that covers and kills a small patch of plants. Revegetation of this soil patch occurs very quickly as species like big bluestem send new tillers into the bare soil, while other species colonize the patch by dispersal of seeds. In this section, we will discuss the short-term spatial and temporal dynamics of grasslands in response to grazing and burning. We will then expand our spatial resolution somewhat and review regional patterns of species distribution and abundance. Local patch dynamics are related to regional dynamics of distribution and abundance (Ricklefs 1987). Regional dynamics are often a function of spatial variation in disturbance within a region, in conjunction with annual variations in climate. In combination, these factors define the natural disturbance regime in grasslands (Collins and Glenn 1988).

The historic occurrence of fire in the tallgrass prairie must have been spectacular. Good precipitation years bring high growth rates of grasses, in particular. The drying winds, low humidity, and warm summer temperatures set the stage for dramatic fires, with little in the landscape to prevent them from spreading. At times, enormous areas must have been burned by a single fire. Fire is now one of the most commonly studied natural disturbances in mesic prairie (Daubenmire 1968; Vogl 1974; Wright and Bailey 1982), and is commonly used as a management tool in restoration and preservation of tallgrass prairies (Cottam 1987; Kline and Howell 1987). Fires increase the growth of grasses and prevent woody vegetation from invading tallgrass prairies (Hulbert 1969, 1988; Bragg and Hulbert 1976; Bragg, Chapter 4). In other words, periodic fires arrest succession in mesic grasslands. Often, however, ecological research on fire has been done in the absence of grazing, and much fire research is restricted to burning in the spring. April may be the best time to control and use fire to increase productivity of forage grasses like big and little bluestem, but early spring represents only a small portion of the time span in the natural fire regime of grasslands (Bragg 1982; Ewing and Engle 1988).

In the absence of fire, litter accumulates on the soil surface in tallgrass prairie (Knapp and Seastedt 1986; Seastedt, Chapter 8). The litter slowly decomposes, contributing to a buildup of soil organic matter. If a grassland is burned after several years without fire or grazing, the litter layer is removed, exposing the mineral soil. Solar radiation increases the soil temperature following fire. This, along with more favorable moisture conditions, increases soil decomposition rates and, subsequently, soil fertility (Ojima et al. 1990). Increased soil fertility yields high growth rates of the C_4 grasses, such as big bluestem (Knapp 1985). Annual fires, on the contrary, eventually lead to a chronic shortage of soil nitrogen. The nutrient pulse following the initial burn is temporary. The stimulation of productivity following fire eventually reduces soil nitrogen supplies. Thus over years of annual burning, the grasses produce leaf tissue on less and less nitrogen. The C:N ratio in the litter increases, and soon the soil bacteria outcompete the grasses for nitrogen released during decomposition. This creates low soil fertility, so only those species tolerant of low soil nitrogen can survive.

Effects of spring fires on prairie plant and animal communities have been the focus of a long-term research program at Konza Prairie Research Natural Area in northeastern Kansas. At Konza, the landscape has been divided into replicated fire management units burned at 1-, 2-, 4-, 10-, and 20-year intervals. These fire regimes have a dramatic effect on plant species composition (Figure 7.4a). Unburned sites are undergoing a successional change from grass- to shrub-dominated vegetation (Gibson and Hulbert 1987). Composition of annually burned sites has been relatively stable over time. Sites burned once every four years are considered to mimic presettlement fire frequencies (Gibson 1988), but also appear to be progressively changing (Figure 7.4a). Once woody species are established, they will not be eliminated by an occasional (e.g., once every four years) fire (Adams et al. 1982). Fire suppression and land-use changes during the first half of the twentieth century allowed the spread of woody species into grasslands, with lasting consequences. Seed sources are now widely available, and colonization of grasslands by bird-dispersed species (e.g., eastern red cedar) is rapid. Unburned sites contain more species than burned sites (Figure 7.4b), indicating that fire reduces the number of species in this tallgrass prairie. In addition, fire also reduces local spatial variation (Briggs and Nellis 1991; Collins 1992); that is, prairies become more ''uniform'' in that a small number of species dominate the landscape (Figure 7.4c). In the absence of fire, prairies appear more ''patchy'' because a great variety of species is scattered across the landscape.

The decrease in the number of species may result, in part, from the increase in productivity following fire. Numerous studies have documented a negative correlation between the number of plant species and biomass, which may be estimated by cutting and weighing the vegetation. Sites with high biomass generally indicate greater soil fertility than sites with low biomass (e.g., Grime 1973; Tilman 1982). This relationship has been found in a variety of different types of plant communities. It is interesting to note that in a single-community type of tallgrass prairie, there are fewer species on sites with high biomass compared with sites with low biomass (Figure 7.5), the implication being that prairies on nutrient-rich soils will have fewer species than prairies developing on nutrient-poor soils. Indeed, management efforts on species-rich chalk grasslands of western Europe have focused on decreasing and maintaining low soil fertility.

Fire also affects animal abundances in tallgrass prairies. These changes may reflect a secondary response to fire more than direct mortality from burning. Fire alters the kinds and amounts of soil resources (e.g., nitrogen, organic matter content); the distribution, abundance, and nutrient quality of plant species; and habitat structure (e.g., amount of plant litter, height of vegetation) in prairies. James (1988) revealed that native prairie earthworm populations increased with burning in response to post-fire increases in soil organic matter. Likewise, vegetation composition and structure affect abundance and distribution of plant-feeding insects like grasshoppers (Joern 1979, 1982; Kemp et al. 1990). Evans (1984) reported that the greatest species di-

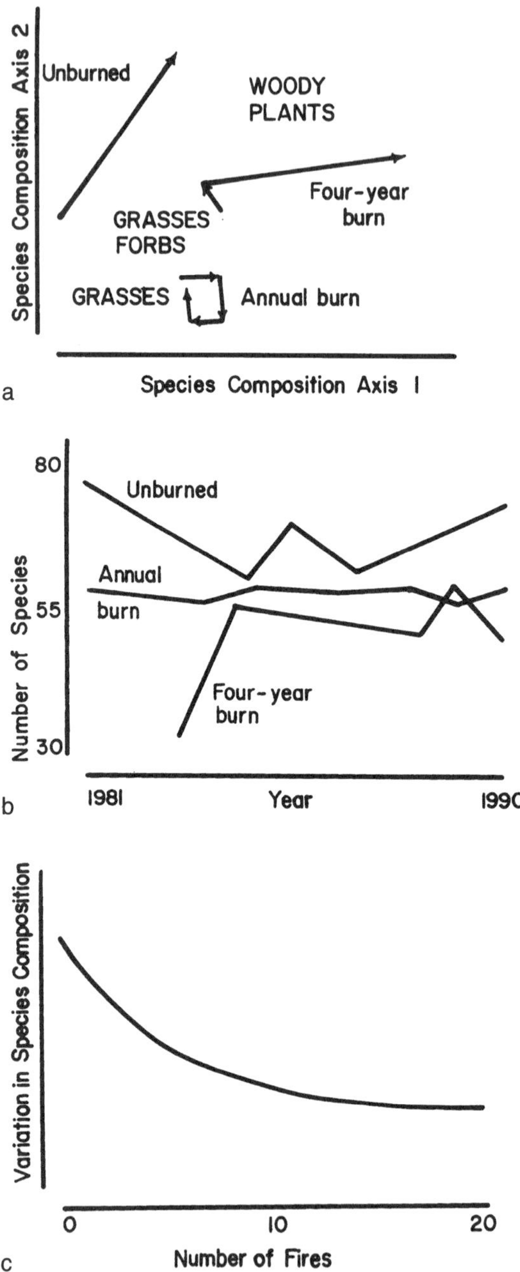

Figure 7.4. Changes in communities over nine years in three upland management units with different burning regimes at Konza Prairie, Kansas. Points in the figure near one another indicate samples with similar vegetation. The distribution of lines indicates that vegetation on each of the management units is different. The vegetation on the unburned and four-year burn sites is changing in a somewhat linear fashion, whereas the vegetation on the annually burned prairie is fluctuating round a stable community-type. (b) Plant species richness on the three manage-

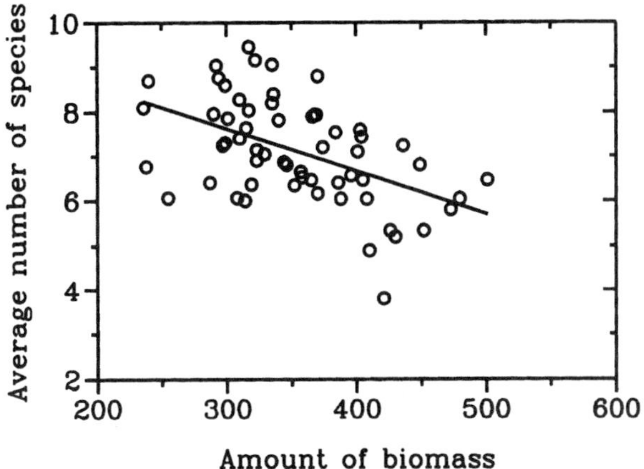

Figure 7.5. Relationship between the amount of biomass in a prairie and the number of plant species. Values are averages over seven to nine years for 15 different sites at Konza Prairie, Kansas. (Collins and Briggs, unpublished data)

versity and richness of early instars of grasshoppers occurred on tallgrass prairies burned once every four years. No differences were found for later instars. In annually burned grassland, forb-feeding species had higher mortality rates following burning than grass-feeding species. Burning reduced the abundance of forbs and favored the subset of grasshoppers that feeds on grasses.

Anderson et al. (1989) found that overall insect abundance was reduced on burned prairie in Illinois in the year following a spring burn. Differences between burned and unburned prairies were no longer evident the second year after burning. Thus fire effects on insect abundances appear to be temporary. Evans (1988) reported similar short-term dynamics for grasshoppers at Konza Prairie over a four-year burn cycle. The relative abundance of grass feeders increased during the year of the burn and decreased each year thereafter. Long-term community structure essentially fluctuated around a mean; however, changes in composition and abundance from one year to the next were only weakly predictable within this four-year burn cycle.

Kaufman et al. (1990) reported that small mammal species were little affected by fire, but as a whole, composition of mammal assemblages differed between burned and unburned prairies. Species could be classified as fire-positive, fire-negative, or fire-neutral with regard to short-term responses to burning. In particular, abundances of deer mice and thirteen-lined ground squirrels were greater on burned compared with unburned prairies. In contrast, Schramm and Willcutts (1983) reported that small mammal diversity was higher on burned grasslands shortly after burning. As with

---

ment units over time. Droughts occurred in 1985 and 1988. (c) The effects of fire frequency on variability in grassland communities. The sites burned most often since 1972 have low variability in species composition. That is, annual burning in the spring in the absence of grazing tends to make prairies more homogeneous. (From Collins [1992], with permission of the Ecological Society of America)

insects, it appears that many of the animal responses are short-term and result from the temporary alteration of habitat structure by fire.

Of greater importance for understanding grassland community structure is the interaction between grazing and burning (Bragg, Chapter 4; Wallace and Dyer, Chapter 9). For example, Hobbs et al. (1991) recently demonstrated that grazing by cattle offset the losses of soil nitrogen caused by burning: Combustion losses of nitrogen from ungrazed prairie were 1.8 g m^2 yr^{-1}, twice that of grazed prairie. Because grazing reduced the standing stock of biomass, burned prairie had less fuel, less volatilization of nitrogen during combustion, and cooler fire temperatures than unburned prairie. Because bison are naturally attracted to burned areas (Campbell and Hinkes 1983; Shaw and Carter 1990; Vinton 1990), fire enhances grazing on a site, and the animals continue to return to forage at previously grazed areas. This tends to reduce the fuel load and overall nitrogen losses during combustion. Grazing may also influence fire frequency. It is hypothesized that grasslands burned once every three to five years during presettlement times. This estimate, based on current rates of woody species colonization of unburned areas, does not necessarily account for fire frequencies when fuel loads were affected by huge herds of bison roaming the Great Plains or for the effects of increased seed availability of woody species in response to fire suppression during the first half of this century. Therefore, fires may have been less common in areas frequented by bison.

Some ecologists have suggested that species diversity will be highest in communities that are subjected to a combination of natural disturbances that reflect, to some extent, the presettlement disturbance regime. Periodic disturbances over space and time allow the coexistence of species that are tolerant of disturbance along with those species that are sensitive to disturbance. Unfortunately, no prairie currently exists that has not been altered by human activities sometime during this century. Species diversity in mixed-grass prairie was, indeed, highest on sites with a combination of natural disturbances, including fire, grazing, and wallowing by bison (Figure 7.6a). Diversity was also enhanced by the interaction of fire and grazing in tallgrass prairie; diversity was highest on burned and grazed grassland and lowest on burned but not grazed treatments (Figure 7.6b). An increase in the number of forb species rather than grasses best explains this response to disturbance (Figure 7.6c). These results support the notions of academic ecological theory that a combination of natural disturbances promotes species diversity in grassland communities. Alternatively, this increase could simply reflect the expanse of weedy species now well established in the region as a result of poor management practices until the 1940s, along with increasing urbanization and human intrusion on prairie communities.

Patterns of species diversity and community variability in response to disturbance differ with scale (Figure 7.7). Variability in species composition (which is positively correlated with diversity) within several square meters in undisturbed sites was greater than in grazed and burned sites (Glenn et al. 1992). At the regional scale, however, the opposite was true. Several grazed and burned grasslands were very dissimilar, and several undisturbed grasslands were quite similar. This may result from the scale of grazed patches within burned and unburned grasslands. At Konza Prairie, bison occur three times more often on burned management units than would be expected if the animals moved at random among management units (Vinton 1990). Patches of grazed vegetation were larger on burned compared with unburned sites. Thus local spatial

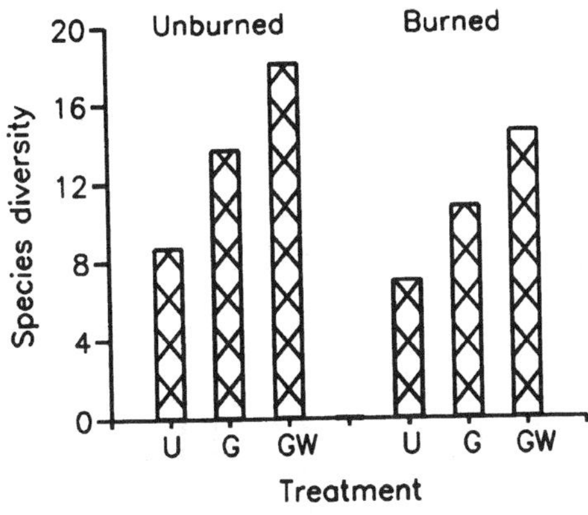

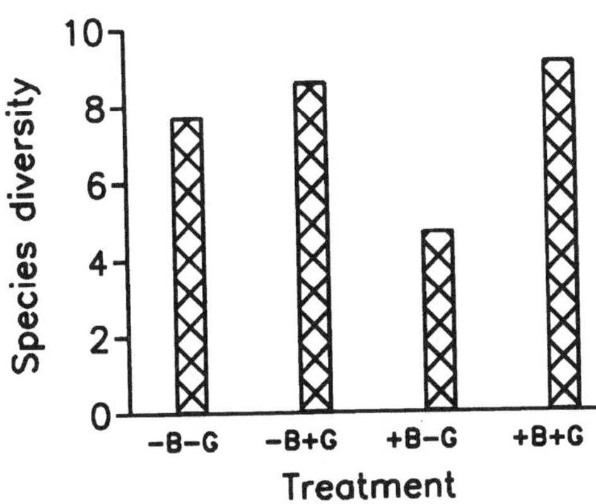

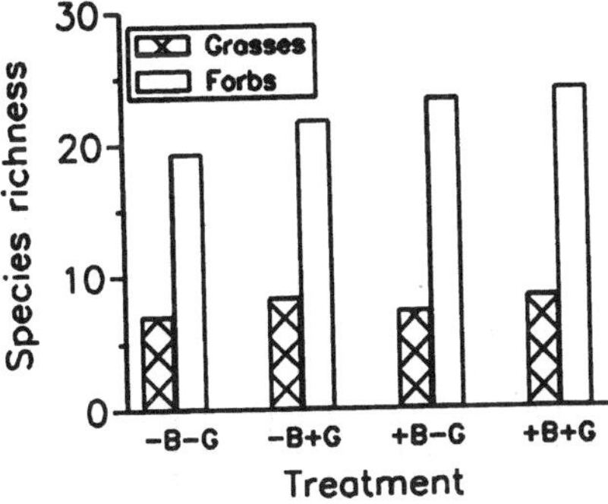

Figure 7.6. Patterns of plant species richness under different regimes in (a) mixed-grass prairie in the Wichita Mountains Wildlife Refuge, southwestern Oklahoma (Collins and Barber 1985), and (b) tallgrass prairie at the U.S. Department of Agriculture Range Research Station in El Reno, Oklahoma (Collins 1987). (c) Contributions of grasses and forbs to species richness (Collins 1987). U, ungrazed; G, grazed; B, burned; W, buffalo wallows. (Reproduced with permission of the Ecological Society of America)

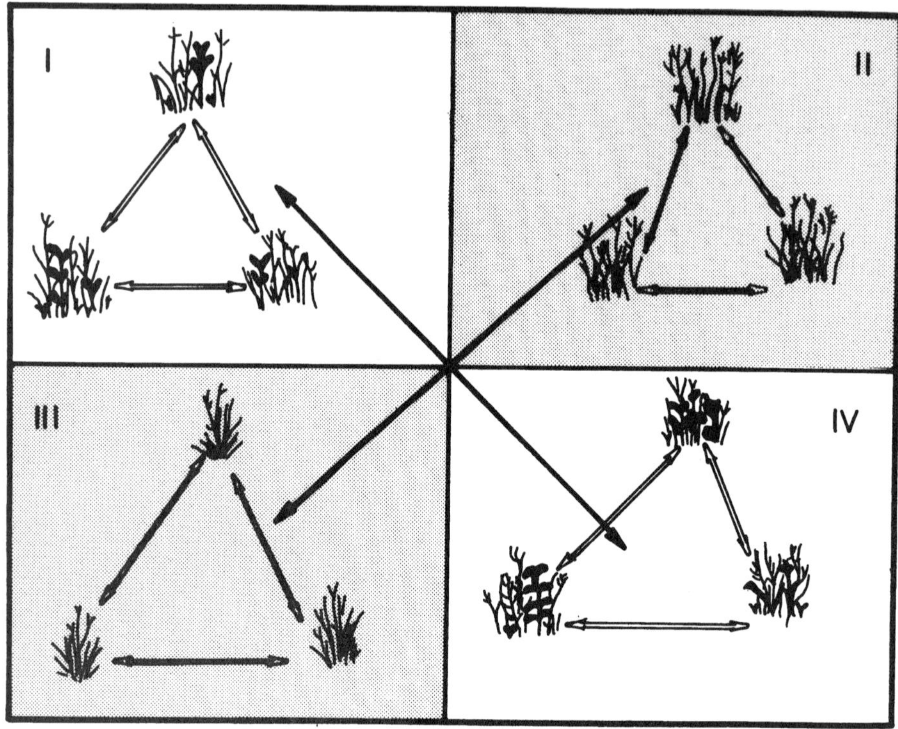

Figure 7.7. Differences in measurements between local and regional variability. The "region" in this case consists of four local sites (I–IV), each of which is sampled at three points. Although the sites are contiguous in this example, they do not have to be. Local or within-site variation is measured as the amount of variation among the three samples within a site (double lines connecting patches). Regional variation is determined by averaging local samples (X) and comparing the variation among sites in a region (solid arrows).

variation may be reduced in burned and grazed grasslands because burned grasslands are more homogeneous in the large patches of "grazing lawns." A grazing lawn is a dense mat of prostrate grass tillers formed by intensive grazing (McNaughton 1984). A similar phenomenon develops in suburban lawns from the use of lawn mowers. At larger scales, variability results from initial differences in community structure among sites, in soils and topography (White and Glenn-Lewin 1984; Gibson and Hulbert 1987), and in surrounding land use, as well as from comparing samples from both grazed and ungrazed patches. In contrast, the short-grass prairie in eastern Colorado shows little change in composition and structure between grazed and ungrazed vegetation (Milchunas et al. 1990). Fire is much less important here than in the eastern prairies because the limited amount of fuel available for burning tends to prevent fires from spreading across the landscape. In these systems, plant community change is limited by environmental extremes, including shortages of moisture and soil nitrogen (Milchunas et al. 1988).

So although fire is an important management tool in mesic prairies, it must be used in moderation and, if possible, in conjunction with grazing or mowing. An ideal fire-management scheme might include two to three fires every five to six years (e.g., Allen and Wyleto 1983), depending on fuel loads and precipitation patterns (Steuter et al. 1990), with burning occurring at different times during the growing season. Such a management scheme has been proposed for the Nature Conservancy's Niobrara Preserve in the Nebraska Sand Hills (Steuter et al. 1990) and Tallgrass Prairie Preserve in northern Oklahoma.

CHANGES IN ABUNDANCE AND DISTRIBUTION: DOES CHAOS REIGN ON THE RANGE?

To the casual observer, grasslands appear to be homogeneous and stable communities. Nothing could be further from the truth. Indeed, one of the most intriguing aspects of grassland communities is the seemingly random change in species composition among sites in a region and within a site over time. For example, Penfound (1964) described a pattern of pulse-phase dominance whereby local populations suddenly and unexpectedly increased in abundance in one growing season only to become rare again the following year. Similar population pulses were observed over 30 years in grazed and ungrazed sandsage prairie (Collins et al. 1988). The temporal variation in short-grass prairie in western Kansas resembled ''Brownian noise,'' a term from physics describing the rapid movement of particles from one state to another (Collins et al. unpublished data). For prairies, this implies that annual changes in community composition were rapid and unpredictable, constrained from one year to the next only by the presence of a few long-lived dominant species. This was true for vegetation in both grazed and ungrazed shortgrass prairie.

The observed small-scale random dynamics of undisturbed tallgrass prairie is especially intriguing. Much current research in ecology focuses on ''patch dynamics.'' A *patch* is a relatively discrete spatial pattern formed by a collection of plant species (Pickett and White 1985). *Patch dynamics* is the study of the change in shape, size, and number of patches in an area. While most research on patch dynamics of plant communities focuses on succession following disturbance, our analyses indicated that small patches of tallgrass prairie vegetation were highly variable in space and time *in the absence of disturbance* (Glenn and Collins 1990). In addition, patch structure was the same as that produced if species were distributed at random among samples. Because most of the space in tallgrass prairie was occupied by the dominant perennial grasses, such as big and little bluestem (Figure 7.8), these patches were more subtle, being defined by the numerous forbs that occupied the spaces between clumps of the dominant grasses. Even removing clumps of one of the dominants, little bluestem, did not change patch structure, although the number of species increased in plots where little bluestem was removed.

Changes in species composition over time are really a function of two simple variables: immigration and local extinction. MacArthur and Wilson (1967) used these variables to develop a simple theoretical model to explain the number of species occurring on islands. Essentially, at equilibrium, the number of species on an island

Figure 7.8. Distribution of clumps of little bluestem in a 10 meter $\times$ 10 meter area at the U.S. Department of Agriculture Range Research Station, El Reno, Oklahoma. About 50% of the area is occupied by little bluestem. (From Glenn and Collins [1990], with permission of *Oikos*)

is stable when the rate of immigration of new species to the island equals the rate of extinction on the island. The species composition may be changing, however, resulting in a "dynamic equilibrium." Although this model was developed for true oceanic islands, the fundamental aspects of the model apply to all communities, as well as to patches within communities. Unfortunately, rates of local immigration and extinction are difficult to measure in plant communities. We calculated rates of immigration and extinction of species over 9 years in eastern Kansas tallgrass prairie and over 18 years in western Kansas shortgrass prairie (Glenn and Collins 1992). Here, *immigration* refers to the appearance of living aboveground vegetation, and *extinction* refers to the local disappearance of aboveground vegetation of a particular species. As such, extinction does not necessarily imply that the species has died. Instead, individuals may be dormant, such that they are not participating in the interactions among species (e.g., competition) that define a community. Extinction rates did not differ between burned and unburned grasslands in the tallgrass prairie, but immigration rates were somewhat higher on unburned grasslands (Figure 7.9a), implying that more species are being added to unburned sites each year compared with burned sites. In shortgrass prairie, immigration rates were similar on grazed and ungrazed treatments, but extinction rates were slightly higher on grazed prairie (Figure 7.9b).

It is important to recognize that there is considerable variation around the immigration and extinction lines in Figures 7.9a and 7.9b. If confidence intervals were calculated for these lines, then where the lines intersect is defined as a region in which species richness fluctuates randomly over time (Figure 7.9c). In fact, this conclusion meshes well with theoretical and empirical models of the regional distribution and abundance of species that predict such random fluctuations (Hanski 1982, 1991). Data from long-term permanent plots indicate that the distribution of many plant species

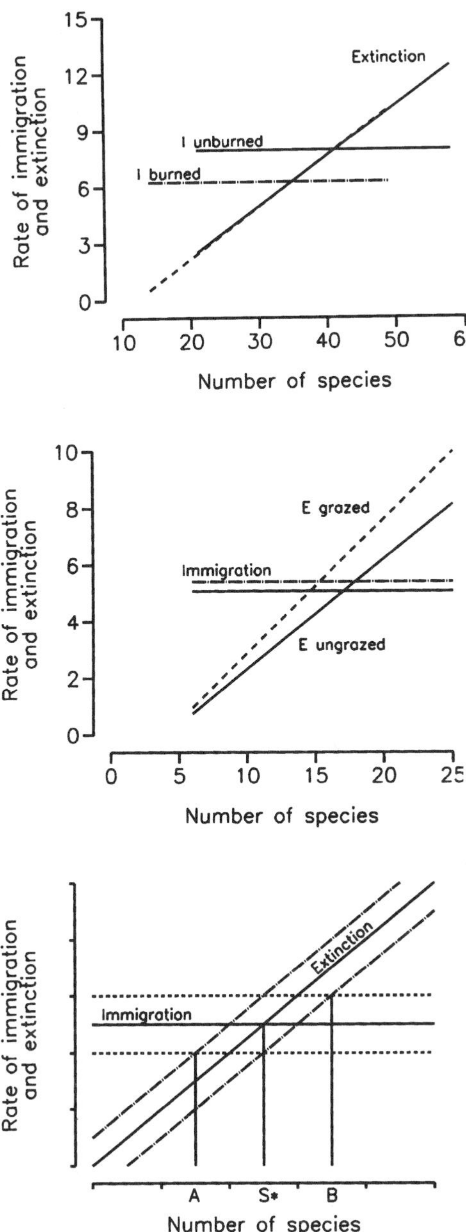

Figure 7.9. (a) Rates of immigration (I) and extinction (E) in burned and unburned tallgrass prairie at Konza Prairie, Kansas, and (b) in grazed and ungrazed shortgrass prairie in southwestern Kansas. (c) Conceptual model of immigration and extinction showing the natural variation in species richness due to random processes of species immigration and extinction. Essentially, species richness in a prairie will randomly fluctuate between A and B based on regional species richness, disturbance regimes, isolation of prairie fragments and the local species pool (From Glenn and Collins [1992], with permission of *Oikos*)

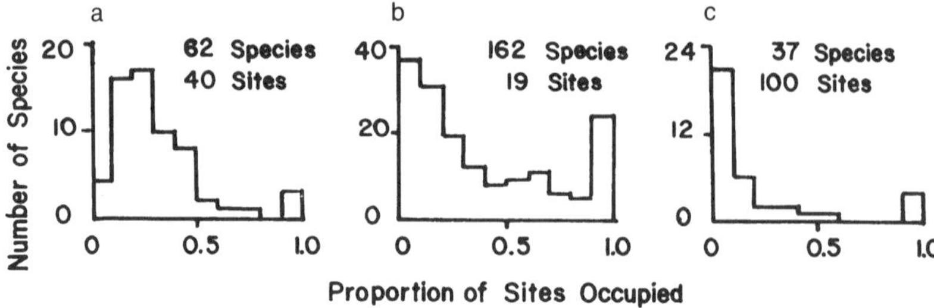

Figure 7.10. Pattern of distribution of plant species (a) among grasslands in the central Plains; (b) among 19 sites in 12 management units at Konza Prairie; and (c) among 100 1-square-meter adjacent sample units at El Reno, Oklahoma. Each figure shows that at all spatial scales most species are regionally rare occurring at less than 10% of the sample areas, but there is a "core" group of species that occurs in all samples. ([a] data from Diamond and Smeins [1988]; [b, c] from Collins and Glenn [1990], with permission of the University of Chicago Press)

among sites in grasslands changes randomly over time (Collins and Glenn 1990, 1991).

Random changes in distribution of species among sites in a region and the abundance of species over time conform to a model called the core–satellite species hypothesis (Hanski 1982, 1991). This model assumes that there is a positive correlation between the number of places at which a species occurs (i.e., present or absent) and the average abundance (e.g., biomass) of that species (Hanski 1982). That is, if a species is abundant in a given prairie, we might guess that it will also be abundant in other prairies subject to similar environmental conditions. However, the distribution and abundance of a species change over time. Species can immigrate to new sites and go locally extinct at some sites. If immigration and extinction are a function of the number of sites occupied, then the core–satellite model predicts that species will be distributed among sites in a region such that most species are either very common or very rare. By definition then, "core" species occur at over 90% of the sites in a region and are typically dominants, whereas "satellite" species occur at less than 10% of the sites in a region and are typically subordinates. Data from Konza Prairie fit this model; core and satellite species can be identified for plants, birds, grasshoppers, and small mammals (Figures 7.10 and 7.11). The implication of this pattern is that the number of sites at which some species occur changes dramatically from year to year (Collins and Glenn 1991).

For plants, this core–satellite pattern of species distribution (Figure 7.10) among sites appears at a variety of spatial scales (Collins and Glenn 1990). Core species can be found among 100 square meters of prairie, among sites within 36 square kilometers of prairie, as well as across the prairie region as a whole. The time frame for these changes will increase with spatial scale. Changes occur annually at the smallest scale and may take millennia at the largest scales, as discussed for postglacial migration. Overall, this implies that there may be general patterns of community structure that translate across spatial scales. In the parlance of fractals and chaos (Gleick 1989), this implies that community structure exhibits "self-similarity" in which the large-scale entity is composed of smaller units of similar structure. Much further research is needed to determine if this pattern is repeatable in other systems.

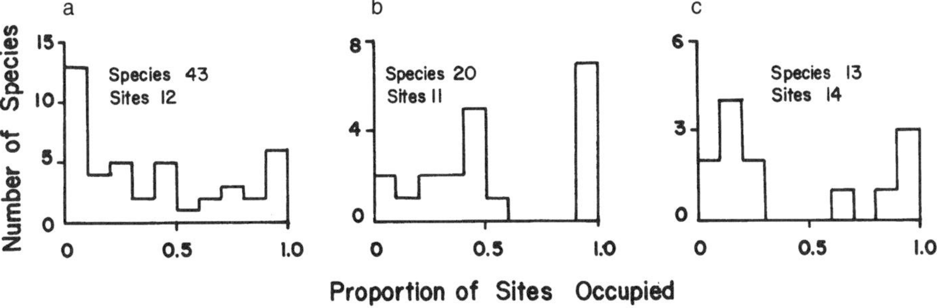

Figure 7.11. Patterns of distributions of (a) birds, (b) grasshoppers, and (c) small mammals among sites at Konza Prairie. Similar to plant species distributions (Figure 7.8), different animal groups contain "core" species which can be found at greater than 90% of the sample areas in a region.

Our analyses of distribution and abundance at Konza Prairie are based on studies of plants and animals within a continuous habitat of grassland vegetation. Fragmentation of grasslands by agriculture and development may have severe consequences on these dynamics and, therefore, the patterns of species distribution. As grasslands are fragmented, the ability of species to disperse among sites decreases. In some models, local extinction can be compensated for by the "rescue effect" (Levins 1969; Gotelli 1991), in which a prairie fragment may be recolonized by immigrants from extant populations on other prairie fragments in the region. The ability of the rescue effect to maintain a high probability of colonization is a function of size of local populations and distance between sites. If habitat fragmentation increases the distance between sites, then the potential for immigration and colonization will be reduced (Opdam 1991; Saunders et al. 1991). Also, if fragmentation reduces the size of habitat patches, then smaller populations occur within patches and the probability of emigration from small populations is reduced and the chance of local extinction increases.

LANDSCAPE CHANGE: GLOBAL CLIMATE AND HUMAN INTRUSION

A *landscape* is traditionally defined as a cluster of interacting patches or ecosystems repeated in similar form (Forman and Godron 1986). A landscape contains many elements, such as isolated patches of prairie, wind breaks, cropland, housing lots, and urban areas. Landscape ecology represents a newly emerging ecological perspective that deals both with the number and kinds of elements in the landscape and their interactions, such as the exchange of materials between landscape elements, and with changes in the number and kinds of landscape elements over time (Risser et al. 1984; Urban et al. 1987). As a discipline, landscape ecology has explicitly included the impacts of humans on landscape structure and function at large spatial scales.

One of the most fundamental anthropogenic impacts presently occurring is the gradual increase of carbon dioxide (CO_2) and other trace gases in the earth's atmosphere. The increase in CO_2 began with the industrial revolution in the late nineteenth century, and if current trends continue, there will be a doubling of atmospheric CO_2

by the middle of the twenty-first century. Several global circulation models (GCMs) have been devised to assess the potential impact of this relatively rapid increase in CO_2 on global climate. The increase in atmospheric CO_2 may be favorable for plant growth by providing more of a carbon resource for photosynthesis, which may also yield an improved internal water balance in grassland plants. However, on the down side, GCMs generally predict a 2° to 4°C increase in mean annual temperature for the central Great Plains. Temperature increases in northern latitudes are expected to be even higher, especially during winter. The effects of global change on regional precipitation is uncertain. It appears that there may be a 10% increase in precipitation in the southern Great Plains, and a 10% decrease in precipitation in the north (Burke et al. 1991). The increase in atmospheric CO_2 and temperature and the altered precipitation regimes are likely to have significant impacts on prairie communities. For example, increased temperature accompanied by decreased growing-season precipitation could tip the balance in favor of C_4 species, such as big bluestem, in northern regions where C_3 species, like the wheatgrasses (*Agropyron* spp.), are currently abundant, producing dramatic changes in species composition.

Efforts are now under way to develop models of atmospheric–biospheric interactions at the landscape scale (micro- to mega-scale) to further our understanding of the potential impacts of global warming on ecosystem function in grasslands (Hall et al. 1988; Pielke and Avissar 1990; Burke et al. 1991; Hunt et al. 1991). Because of the uncertain consequences of global change on precipitation in the Great Plains, predictions from landscape-level models are variable. Compared with that in western grasslands, moisture in eastern or tallgrass prairie is often less of a limiting factor on grassland productivity than is nitrogen (Sala et al. 1988; Seastedt, Chapter 8). Nitrogen also appears to be more limiting than moisture on local production in the shortgrass prairie (Dodd and Lauenroth 1979). Regionally, however, annual net primary production is highly correlated with precipitation (Burke et al. 1991; Hunt et al. 1991; Bragg, Chapter 4).

If moisture regimes remain favorable, landscape-level models indicate that the increase in temperature will produce increased mineralization rates of soil organic matter. The resulting burst in soil fertility will be followed by an increase in grassland productivity. As high growth rates continue, the fertility pulse will decline as stored soil organic matter is mineralized at a rate greater than it is accumulated. Thus the rate of nutrient removal will not keep pace with nutrient supplies. As a consequence, soil nitrogen levels will eventually decline; however, the increased atmospheric CO_2 will continue to produce a favorable carbon balance in the plant. Expected increases in the C:N ratio in leaf tissue will further decrease the decomposition rates of leaf litter, reducing soil nutrient quality (Hunt et al. 1991), and possibly have adverse effects on the number of species in a community.

By their nature and scaling, regional and global climate models do not explicitly incorporate estimates of species diversity and composition. In many models, production is simply "averaged" over a stand dominated by a single species whose physiology is well known (e.g., blue grama). Nevertheless, increasing global temperatures and the subsequent effects on ecosystem function will have important impacts on local and regional biodiversity (Woodward and Rochefort 1991). It is interesting to note that the alterations in ecosystem dynamics in response to climate change closely mirror the changes in nitrogen availability with annual burning of ungrazed tallgrass prairie.

Based on this analogy, we can assume then that climate change will have a significant effect on local species richness and diversity in grasslands. Remember, there is a negative relationship between productivity and species richness (Figure 7.5). A sustained level of increased productivity in response to global warming may decrease species diversity in prairies. As soil fertility decreases and productivity declines, the number of plant species will likely remain low because the warmer, drier environment will make establishment of immigrants less likely. The reduction in regional diversity will decrease rates of immigration from site to site. Also, rates of extinction within sites will increase as population sizes decrease. In other words, the natural patterns and dynamics of species distribution and abundance will be severely altered. It is important to recall that these climatic changes are superimposed on the already fragmented prairie landscape. The reduction in species diversity could be even greater because fire will still be an important disturbance in prairies. Fire temperature is positively related to fuel availability. The increase in productivity will provide a greater fuel load, leading to hotter, more severe fires. The combination of higher fire intensity and reduced precipitation could further reduce species diversity.

Changes in plant community composition and structure could lead to changes in animal diversity. Those species that feed on grasses, for instance, may at first be favored because of the initial burst in productivity. However, the long-term consequences of such changes are difficult to predict. An interesting and relevant experiment was performed by Fajer and colleagues at Harvard (Fajer et al. 1992). In one growth chamber, they grew English plantain at ambient CO_2 levels; in another growth chamber, the same species was grown in an atmosphere with twice the CO_2 as ambient. They then allowed the caterpillars of the buckeye butterfly to forage on these plants. Growth rates of the caterpillars were reduced and death rates were higher on the high CO_2-grown plants because these plants offered a lower quality food than plants grown at ambient CO_2. Essentially, the nitrogen in the leaves of plants from the CO_2-enriched environments was diluted by excess carbon. So, what was a high-quality resource under current levels of atmospheric CO_2 became insect junk food at higher levels of CO_2.

Davis (1989) has warned that the changes in regional climate in response to global warming are predicted to occur at a rate much greater than climatic changes during the Pleistocene. She suggests that species will not be able to migrate rapidly enough in some regions to compensate for these dramatic changes, if the predictions hold true. The repercussions may be large-scale decreases in prairie biodiversity. The only species that may be able to migrate rapidly enough to keep pace with climate change might be the weedy, invasive species commonly associated with human disturbance. With regard to grasslands, if the nutrient pulse in response to warmer temperatures leads to higher productivity and reduced diversity, it is unlikely that diversity will subsequently increase when productivity eventually declines. Remnant prairies are already highly fragmented, with limited migration corridors, so local extinctions are unlikely to be compensated by the "rescue effect." Fortunately, it may appear that the rate of climate change has not been as rapid as models predicted (Karl et al. 1991). Mitigating factors include increasing particulate matter in the atmosphere along with other trace gases that might lead to reduced temperatures. Nevertheless, the data still seem to indicate that spring and summer minimum temperatures in the Great

Plains have increased by about 1°C in the past century (Karl et al. 1991). Clearly, conservation efforts must be designed to incorporate the potential consequences of global change.

CONSERVATION: PRESERVING CHANGE ON THE RANGE

The highly dynamic nature of grasslands requires a protection strategy that allows the community to change. Therefore, to minimize loss of components of grassland biodiversity, a series of preserves needs to be established across the range of grassland environments (Glenn 1990). In designing these preserves, sources of immigrating species must be present in, or adjacent to, the preserve. This may be accomplished by placing the preserves in areas with variable topography and soils (Noss 1987). A prairie preserve with a range of elevations—cool deep canyons, north-facing hillsides, warmer south-facing hillsides, and shaded forests along larger perennial streams— may be preferable to a flat, homogeneous ''sea of grass.'' Such a design provides habitats for a variety of plant and animal species, and may provide a refuge of habitats in the event of global change. A large initial pool of species is necessary to maintain high rates of immigration to balance rates of local extinction. Otherwise, the preserve risks becoming dominated by a few highly competitive species, and no longer supports the dynamics characteristic of natural grasslands.

The area covered by native species in and around the preserve should be as large as possible: The number of species increases naturally as area increases, and large areas generally incorporate more habitat types. Also, some animals, such as bison and badgers, require large areas in order to maintain healthy populations. Gopher mounds, badger and coyote dens, anthills, buffalo wallows, and grazing are some of the natural disturbances that add to the variety of habitats in grasslands. Populations of animals are necessary to maintain these habitats, and thus contribute to the natural dynamics of grassland communities (Joern, Chapter 6). The area requirements of these species must be met to preserve a viable grassland ecosystem. In some cases, small prairie remnants may be linked by corridors of restored prairies, creating an immigration pathway for grassland plants and animals.

A large preserve makes it easier to manage for a natural abiotic disturbance regime. Some areas can be burned, and other areas left as refuges for fire-sensitive species. Incorporation of entire watersheds allows for a natural water regime, including periodic floods. Small, grassy prairie drainageways to streams with deep pools and shallow riffles, and a closed canopy of gallery forest, are all part of the tallgrass prairie ecosystem. Flooding may help maintain the gallery forests along prairie streams, adding to regional habitat structure. These areas may also serve as natural fire breaks. Flooding may also scour streams, leaving deep pools for fish and invertebrates in times of drought. Protecting entire watersheds is necessary to ensure that the water is free from sediments and pollutants that would threaten the aquatic communities.

The image of a vast expanse of big bluestem so tall it could hide a horse is an illusion of a dramatically altered disturbance regime. Such a picture could most likely develop only in the eastern tallgrass prairie following a spring fire in a year with abundant precipitation, and in the absence of many other natural disturbances. To us, such an image of the tallgrass prairie is no more interesting than a monoculture of corn or wheat. Therefore, instead of designing for a homogeneous grassland preserve, no more variable than cropland, grassland preserves should be designed to maintain the rapid dynamics and diversity typical of grassland ecosystems. This requires a large, diverse area with many habitat patches. Several such preserves should be established across the variety of grasslands to incorporate the range of biodiversity characteristic of prairies. A natural landscape context surrounding each preserve, with minimal human disturbance, will reduce invasion by exotic species (Janzen 1983; Glenn and Nudds 1989), while providing a large pool of potential natural immigrants to the preserve.

LITERATURE CITED

Adams, D. E., R. C. Anderson, and S. L. Collins. 1982. Differential response of woody and herbaceous species to summer and winter burning in an Oklahoma grassland. Southwestern Naturalist 27:55–61.

Allen, T.F.H., and T. W. Hoekstra. 1992. *Towards a Unified Ecology*. Columbia University Press, New York.

Allen, T.F.H., and T. B. Starr. 1982. *Hierarchy: Perspectives for Ecological Complexity*. University of Chicago Press, Chicago.

Allen, T.F.H., and E. P. Wyleto. 1983. A hierarchical model for the complexity of plant communities. Journal Theoretical Biology 101:529–540.

Anderson, R. C., T. Leahy, and S. Dhillion. 1989. Numbers and biomass of selected insect groups on burned and unburned sand prairie. American Midland Naturalist 122:151–162.

Axelrod, D. I. 1985. Rise of the grassland biome, central North America. Botanical Review 51:163–202.

Booth, W. E. 1941. Revegetation of abandoned fields in Kansas and Oklahoma. American Journal Botany 28:415–422.

Bragg, T. B. 1982. Seasonal variations in fuel and fuel consumption by fires in a bluestem prairie. Ecology 63:7–11.

Bragg, T. B., and L. C. Hulbert. 1976. Woody plant invasion of unburned Kansas bluestem prairie. Journal Range Management 29:19–24.

Braun, E. L. 1950. *Deciduous Forests of Eastern North America*. Blakiston, Philadelphia.

Briggs, J. M., and D. Nellis. 1991. Seasonal variation of heterogeneity in the tallgrass prairie: a quantitative measure using remote sensing. Photometry: Engineering Remote Sensing 57:407–411.

Brown, D. A., and P. J. Gersmehl. 1985. Migration models for grasses in the American midcontinent. Annals Association American Geographers 75:383–394.

Burke, I. C., T.G.F. Kittel, W. K. Lauenroth, P. Snook, C. M. Yonker, and W. J. Parton. 1991. Regional analysis of the Central Great Plains. BioScience 41:685–692.

Campbell, B. H., and M. Hinkes. 1983. Winter diets and habitat use of Alaska bison after wildfire. Wildlife Society Bulletin 11:16–21.

Chaney, R. W., and M. K. Elias. 1936. *Late Tertiary Floras from the High Plains*. Publication No. 476. Carnegie Institution, Washington, D.C.

Clements, F. E. 1916. *Plant Succession: An Analysis of the Development of Vegetation*. Publication No. 242. Carnegie Institution, Washington, D.C.

Coffin, D. P., and W. K. Lauenroth. 1988. The effects of disturbance size and frequency on a shortgrass plant community. Ecology 69:1609–1617.

Collins, S. L. 1987. Interactions of disturbance in tallgrass prairie: a field experiment. Ecology 68:1243–1250.

Collins, S. L. 1992. Fire frequency and community heterogeneity in tallgrass prairie vegetation. Ecology 73:2001–2006.

Collins, S. L., and D. E. Adams. 1983. Succession in grasslands: thirty-two years of change in a central Oklahoma tallgrass prairie. Vegetatio 51:181–190.

Collins, S. L., and S. C. Barber. 1985. Effects of disturbance on diversity in mixed-grass prairie. Vegetatio 64:87–94.

Collins, S. L., J. A. Bradford, and P. L. Sims. 1988. Succession and fluctuation in *Artemisia*-dominated grassland. Vegetatio 73:89–99.

Collins, S. L., and S. M. Glenn. 1988. Disturbance and community structure in North American prairies. Pp. 131–143 in H. J. During, M.J.A. Werger, and J. H. Willems (eds.). *Diversity and Pattern in Plant Communities*. SBP Academic, The Hague.

Collins, S. L., and S. M. Glenn. 1990. A hierarchical analysis of species abundance patterns in grassland vegetation. American Naturalist 135:633–648.

Collins, S. L., and S. M. Glenn. 1991. Importance of spatial and temporal dynamics in species regional abundance and distribution. Ecology 72: 654–664.

Cottam, G. 1987. Community dynamics on an artificial prairie. Pp. 257–270 in W. R. Jordan, M. E. Gilpin, and J. D. Aber (eds.). *Restoration Ecology*. Cambridge University Press, Cambridge.

Daubenmire, R. 1968. Ecology of fire in grasslands. Advances Ecological Research 5:209–266.

Daubenmire, R. 1978. *Plant Geography with Special Reference to North America*. Academic Press, New York.

Davis, M. B. 1976. Pleistocene biogeography of temperate deciduous forests. GeoScience and Man 13:13–26.

Davis, M. B. 1989. Insights from paleoecology on global change. Bulletin Ecological Society America 70:222–228.

Delcourt, H. R., and P. A. Delcourt. 1991. *Quaternary Ecology: A Paleoecological Perspective*. Chapman and Hall, New York.

Delcourt, H. R., P. A. Delcourt, and T. Webb III. 1983. Dynamic plant ecology: the spectrum of vegetational change in space and time. Quaternary Science Review 1:153–175.

Diamond, D. D., and F. E. Smeins. 1988. Gradient analysis of remnant true and upper coastal prairie grasslands of North America. Canadian Journal Botany 66:2152–2161.

Dodd, J. L., and W. K. Lauenroth. 1979. Analysis of the response to a grassland ecosystem to stress. Pp. 43–58 in N. R. French (ed.). *Perspectives in Grassland Ecology*. Springer-Verlag, New York.

Evans, E. W. 1984. Fire as a natural disturbance to grasshopper assemblages of tallgrass prairie. Oikos 43:9–16.

Evans, E. W. 1988. Community dynamics of prairie grasshoppers subjected to periodic fire: predictable trajectories or random walks in time? Oikos 52:283–292.

Ewing, A. L., and D. M. Engle. 1988. Effects of late summer fire on tallgrass prairie microclimate and community composition. American Midland Naturalist 120:214–223.

Fajer, E. D., M. D. Bowers, and F. A. Bazzaz. 1992. The effects of nutrients and enriched CO_2 environments on production of carbon-based alleleochemicals in *Plantago*: a test of the carbon/nutrient balance hypothesis. American Naturalist 140:707–723.

Forman, R.T.T., and M. Godron. 1986. *Landscape Ecology*. Wiley, New York.

Gibson, D. J. 1988. Regeneration and fluctuation in tallgrass prairie vegetation in response to burning frequency. Bulletin Torrey Botanical Club 115: 1–12.

Gibson, D. J., and L. C. Hulbert. 1987. Effects of fire, topography and year-to-year climatic variation on species composition in tallgrass prairie. Vegetatio 72:175–185.

Gleick, J. 1979. *Chaos: Making a New Science*. Viking, New York.

Glenn, S. M. 1990. Regional analysis of mammal distributions among Canadian parks: implications for park planning. Canadian Journal Zoology 68: 2457–2464.

Glenn, S. M., and S. L. Collins. 1990. Patch structure in tallgrass prairie: dynamics of satellite species. Oikos 57:229–236.

Glenn, S. M., and S. L. Collins. 1992. Effects of scale and disturbance on rates of immigration and extinction of species in prairies. Oikos 63:273–280.

Glenn, S. M., S. L. Collins, and D. J. Gibson. 1992. Disturbances in tallgrass prairie: local versus regional effects on heterogeneity. Landscape Ecology 7:243–251.

Glenn, S. M., and T. D. Nudds. 1989. Insular biogeography of mammals in Canadian parks. Journal Biogeography 16:261–268.

Glenn-Lewin, D. C. 1980. The individualistic nature of plant community development. Vegetatio 43: 141–146.

Glenn-Lewin, D. C., L. A. Johnson, T. W. Jurik, A. Akey, M. Leoschke, and T. Rosburg. 1990. Fire in central North American grasslands: vegetative reproduction, seed germination and seedling establishment. Pp. 28–45 in S. L. Collins and L. L. Wallace (eds.). *Fire in North American Tallgrass Prairies*. University of Oklahoma Press, Norman.

Goetze, J. R. 1989. Mammalian faunas of the late Pleistocene–Holocene terrace of the Red River, Tillman County, Oklahoma. Texas Journal Science 41:205–209.

Gotelli, N. J. 1991. Metapopulation models: the rescue effect, the propagule rain, and the core satellite hypothesis. American Naturalist 138: 768–776.

Grime, J. P. 1973. Control of species diversity in herbaceous vegetation. Journal Environmental Management 1:151–167.

Hanski, I. 1982. Dynamics of regional distribution: the core–satellite species hypothesis. Oikos 38: 210–221.

Hanski, I. 1991. Single-species metapopulation dynamics: concepts, models and observations. Biological Journal Linnaean Society 42:17–38.

Hobbs, N. T., D. S. Schimel, C. E. Owensby, and D. S. Ojima. 1991. Fire and grazing in the tallgrass prairie: contingent effects on nitrogen budgets. Ecology 72:1374–1382.

Hoekstra, T. W., T.F.H. Allen, and C. H. Flather. 1991. Implicit scaling in ecological research. BioScience 41:148–154.

Hulbert, L. C. 1969. Fire and litter effects in undisturbed bluestem prairie in Kansas. Ecology 50: 874–877.

Hulbert, L. C. 1988. Causes of fire effects in tallgrass prairie. Ecology 69:46–58.

Hunt, H. W., M. J. Trlica, E. F. Redente, J. C. Moore, J. K. Detling, T.G.F. Kittel, D. E. Walter, M. C. Fowler, D. A. Klein, and E. T. Elliott. 1991. Simulation model for the effects of climate change on temperature grassland ecosystems. Ecological Modeling 53:205–246.

Huston, M. 1979. A general hypothesis of species diversity. American Naturalist 113:81–101.

Inouye, R. S., N. J. Huntly, D. Tilman, J. R. Tester, M. Stillwell, and K. C. Zinnel. 1987. Old-field succession in a Minnesota sandplain. Ecology 68:12–26.

James, S. W. 1988. The post-fire environment and earthworm populations in tallgrass prairie. Ecology 69:476–483.

Janzen, D. H. 1983. No park is an island: increase in interference from outside as park size decreases. Oikos 41:402–410.

Joern, A. 1979. Resource utilization and community structure in assemblages of arid grassland grasshoppers (Orthoptera: Acrididae). Transactions American Entomological Society 105:253–300.

Joern, A. 1982. Vegetation structure and microhabitat selection in grasshoppers (Orthoptera: Acrididae). Southwestern Naturalist 27:197–209.

Kaufman, D. W., E. J. Finck, and G. A. Kaufman. 1990. Small mammals and grassland fires. Pp. 46–80 in S. L. Collins and L. L. Wallace (eds.). *Fire in North American Tallgrass Prairies*. University of Oklahoma Press, Norman.

Karl, T. R., R. H. Heim, Jr., and R. G. Quayle. 1991. The greenhouse effect in central North America: if not now, when? Science 251:1058–1061.

Kaul, R. B., G. E. Kantak, and S. P. Churchill. 1988. The Niobrara River Valley, a postglacial migration corridor and refugium of forest plants and animals in the grasslands of central North America. Botanical Review 54:44–81.

Keeler, K. H. 1990. Distribution of polyploid variation in big bluestem (*Andropogon gerardii*: Poaceae) across the tallgrass prairie region. Genome 33:95–100.

Kemp, W. P., S. J. Harvey, and K. M. O'Neill. 1990. Patterns of vegetation and grasshopper community composition. Oecologia 83:299–308.

Kline, V. M., and E. A. Howell. 1987. Prairies. Pp. 75–84 in W. R. Jordan, M. E. Gilpin, and J. D. Aber (eds.). *Restoration Ecology*. Cambridge University Press, Cambridge.

Knapp, A. K. 1985. Effect of fire and drought on the ecophysiology of *Andropogon gerardii* and *Panicum virgatum* in a tallgrass prairie. Ecology 66:1309–1320.

Knapp, A. K., and T. R. Seastedt. 1986. Detritus accumulation limits productivity of tallgrass prairie. BioScience 36:662–668.

Levins, R. A. 1969. Some demographic and genetic consequences of environmental heterogeneity for biological control. Bulletin Entomological Society America 15:237–240.

MacArthur, R. H., and E. O. Wilson. 1967. *The Theory of Island Biogeography*. Princeton University Press, Princeton, N.J.

McIntosh, R. P. 1980. The relationship between succession and the recovery process in ecosystems. Pp. 11–62 in J. Cairns (ed.). *The Recovery Process in Damaged Ecosystems*. Ann Arbor Science, Ann Arbor, Mich.

McNaughton, S. J. 1984. Grazing lawns: animals in herds, plant form, and coevolution. American Naturalist 124:863–886.

Meetemeyer, V. 1978. Macroclimate and lignin control of litter decomposition rates. Ecology 59:465–472.

Meentemeyer, V., and E. Box. 1987. Scale effects in landscape studies. Pp. 15–34 in M. G. Turner (ed.). *Landscape Heterogeneity and Disturbance*. Springer-Verlag, New York.

Milchunas, D. G., O. E. Sala, and W. K. Lauenroth. 1988. A generalized model of the effects of grazing by large herbivores on grassland community structure. American Naturalist 132:87–106.

Milchunas, D. G., W. K. Laurenroth, P. L. Chapman, and M. K. Kazempour. 1990. Community attributes along a perturbation gradient in a shortgrass steppe. Journal Vegetation Science 1:375–384.

Noss, R. F. 1987. From plant communities to landscapes in conservation inventories: a look at The Nature Conservancy (U.S.A.). Biological Conservation 41:11–37.

Ojima, D. S., W. J. Parton, D. S. Schimel, and C. E. Owensby. 1990. Simulated impacts of annual burning on prairie ecosystems. Pp. 118–132 in S. L. Collins and L. L. Wallace (eds.). *Fire in North American Tallgrass Prairies*. University of Oklahoma Press, Norman.

O'Neill, R. V. 1989. Perspectives in hierarchy and scale. Pp. 140–156 in J. Roughgarden, R. M. May, and S. A. Levin (eds.). *Perspectives in Ecological Theory*. Princeton University Press, Princeton, N.J.

Opdam, P. 1991. Metapopulation theory and habitat fragmentation: a review of holarctic breeding bird studies. Landscape Ecology 5:93–106.

Penfound, W. T. 1964. The relation of grazing to plant succession in tallgrass prairie. Journal Range Management 17:256–260.

Pickett, S.T.A. 1989. Space-for-time substitution as an alternative to long-term studies. Pp. 110–135 in G. E. Likens (ed.). *Long-term Studies in Ecology: Approaches and Alternatives*. Springer-Verlag, New York.

Pickett, S.T.A., S. L. Collins, and J. J. Armesto. 1987. Models, mechanisms and pathways of succession. Botanical Review 53:335–371.

Pickett, S.T.A., and P. S. White. 1985. *The Ecology of Natural Disturbance and Patch Dynamics*. Academic Press, New York.

Prentice, J. C., P. J. Bartlein, and T. Webb III, 1991. Vegetation and climate change in eastern North America since the last glacial maximum. Ecology 72:2038–2046.

Rice, E. L. 1964. Inhibition of nitrogen-fixing and nitrifying bacteria by seed plants. I. Ecology 45:824–837.

Rice, E. L. 1976. Allelopathy and grassland improvement. Pp. 90–111 in J. R. Estes and R. J. Tyrl (eds.). *The Grasses and Grasslands of Oklahoma*. S. R. Noble Foundation, Ardmore, Okla.

Rice, E. L. 1984. *Allelopathy*, 2nd ed. Academic Press, New York.

Rice, E. L., W. T. Penfound, and L. M. Rohrbaugh. 1960. Seed dispersal and mineral nutrition in succession in abandoned fields in central Oklahoma. Ecology 41:224–228.

Ricklefs, R. E. 1987. Community diversity: relative roles of local and regional processes. Science 235:167–171.

Risser, P. G., J. R. Karr, and R.T.T. Forman. 1984. *Landscape Ecology: Directions and Approaches.* Illinois Natural History Museum Special Publication No. 2. Illinois Natural History Museum, Champaign.

Sala, O. E., W. J. Parton, L. A. Joyce, and W. K. Lauenroth. 1988. Primary production of the central grassland region of the United States. Ecology 69:40–45.

Saunders, D. A., R. J. Hobbs, and C. R. Margules. 1991. Biological consequences of ecosystem fragmentation: a review. Conservation Biology 5:18–32.

Schramm, P., and B. J. Willcutts. 1983. Habitat selection of small mammals in burned and unburned tallgrass prairie. Pp. 49–55 in R. Brewer (ed.). *Proceedings of the Eighth North American Prairie Conference.* University of Western Michigan Press, Kalamazoo.

Shaw, J. H., and T. S. Carter. 1990. Bison movements in relation to fire and seasonality. Wildlife Society Bulletin 18:426–430.

Smith, J. L., and E. L. Rice. 1983. Differences in nitrate reductase activity between species of different stages of old-field succession. Oecologia 57:43–48.

Sousa, W. P. 1984. The role of disturbance in natural communities. Annual Review Ecology and Systematics 15:353–392.

Steuter, A. A., C. E. Grygiel, and M. E. Biondini. 1990. A synthetic approach to research and management planning: the conceptual development and implementation. Natural Areas Journal 10:61–68.

Tilman, D. 1982. *Resource Competition and Community Structure.* Princeton University Press, Princeton, N.J.

Tilman, D. 1985. The resource-ratio hypothesis of plant succession. American Naturalist 125:827–852.

Tilman, D. 1990. Constraints and tradeoffs: toward a predictive theory of competition and succession. Oikos 58:3–15.

Tilman, D., and D. Wedin. 1991. Plant traits and resource reduction for five grasses growing on a nitrogen gradient. Ecology 72:685–700.

Urban, D. L., R. V. O'Neill, and H. H. Shugart, Jr. 1987. Landscape ecology. BioScience 37:119–127.

Vankat, J. L. 1979. *The Natural Vegetation of North America.* Wiley, New York.

Vinton, M. A. 1990. Bison grazing patterns and plant responses on Kansas tallgrass prairie. M.S. Thesis, Division of Biology, Kansas State University, Manhattan.

Vogl, R. J. 1974. Effects of fire on grasslands. Pp. 139–194 in T. T. Kozlowski and C. E. Ahlgren (eds.). *Fire and Ecosystems.* Academic Press, New York.

Webb, T., III. 1986. Is vegetation in equilibrium with climate? How to interpret late-Quaternary pollen data. Vegetatio 67:75–92.

Webb, T., III, E. J. Cushing, and H. E. Wright, Jr. 1983. Holocene changes in the vegetation of the Midwest. Pp. 142–165 in H. E. Wright, Jr. (ed.). *Late Quaternary Environments of the United States.* Vol. 2: *The Holocene.* University of Minnesota Press, Minneapolis.

Wells, P. V. 1970. Historical factors controlling vegetation patterns and floristic distributions in the Central Plains region of North America. Pp. 211–221 in W. Dort, Jr., and J. K. Jones, Jr. (eds.). *Pleistocene and Recent Environments of the Central Great Plains.* University Press of Kansas, Lawrence.

Wells, P. V. 1983. Late Quaternary vegetation of the Great Plains. Transactions Nebraska Academy of Sciences 11 (Special Issue):83–89.

White, J. A., and D. C. Glenn-Lewin. 1984. Regional and local variation in tallgrass prairie remnants of Iowa and eastern Nebraska. Vegetatio 57:65–78.

Wiens, J. A. 1989. Spatial scaling in ecology. Functional Ecology 3:385–397.

Wilson, R. E., and E. L. Rice. 1968. Allelopathy as expressed by *Helianthus annuus* and its role in old-field succession. Bulletin Torrey Botanical Club 95:432–448.

Woodcock, D. W., and P. V. Wells. 1990. Full-glacial summer temperatures in eastern North America as inferred from Wisconsinan vegetational zonation. Palaeography, Palaeoclimatology and Palaeoecology 79:305–312.

Woodward, F. I., and L. Rochefort. 1991. Sensitivity analysis of vegetation diversity to environmental change. Global Ecology Biogeography Letters 1:7–23.

Wright, H. A., and A. W. Bailey. 1982. *Fire Ecology.* Wiley, New York.

8

Soil Systems and Nutrient Cycles of the North American Prairie

TIMOTHY R. SEASTEDT

> For every atom lost to the sea, the prairie pulls another out of the decaying rock. The only truth is that its creatures must suck hard, live fast, and die often, lest its losses exceed its gains.
>
> A. Leopold, *Sand County Almanac*

Beneath the sea of grass is a much larger sea of roots, soil animals, and microbes. The Great Plains still contain some great soils, something worth boasting about. And while Aldo Leopold in his essay ''Odyssey'' correctly depicted the cycling of nutrients within and through the prairie system, he underemphasized the power of the prairie to accumulate and store both nutrients and organic matter. Over the past 10,000 years, the prairie has accrued a vast store of soil wealth; its annual losses of nutrients and organic matter seldom exceeded its gains in these materials. Agricultural practices have destroyed much of this accumulated reserve, but restoration, albeit slow, is possible.

Plants require water, nutrients, and light energy for growth; soils provide all but the last resource. We therefore need to understand those processes associated with creation, maintenance, and limitations of soil fertility if we are to understand the prairie ecosystem. Jenny (1941, 1980) showed us how to do this; the interactions of climate, biota, parent material, and time since last disturbance explain patterns within and between sites. My goal here is first to identify the unique features of the North American grassland soils and then to show how these unique features both influence and have been influenced by the prairie biota. I emphasize four points. First, the rich soils we associate with the mesic regions of the North American grasslands are as new or newer than the biotic components of the biome, and these soils are not static. Second, comparisons with adjacent soils beneath forests suggest that the grasses, themselves, have strongly modified soil features. The extent to which biota control the physical and chemical characteristics of the soil is large, and patterns are observed in grasslands that are not observed in other biomes. Third, just as fire is essential in maintaining the prairie biota, fire is essential in maintaining prairie soils as well. The interaction of fire, climate, and biota produces the characteristics we associate with soil fertility. These same characteristics result in these soils being "overachievers" in terms of carbon storage. Fourth, present soil characteristics tell us about previous land use. The soil characteristics of undisturbed mesic grassland soils suggest that grazing by bison was probably infrequent and certainly not a chronic activity, at least not over the past 1000 years.

PALEOHISTORY OF SOILS

The Pleistocene glaciers in North America functioned as ecosystem and soil erasers. At the height of the last glacial period, circa 20,000 years before present (Y.B.P.), present-day prairie regions not buried under ice were part of a boreal coniferous forest region that extended well into Kansas (Axelrod 1985). The building of the prairie soils, called Mollisols, did not begin until the retreat of the Wisconsin Glacier and the demise of the forests. As there are no areas in North America too arid for native trees (Sauer, 1950), frequent fires interacted with drought to move forest boundaries to the east and north. Fire return frequencies did not have to be of short duration to maintain the prairie once the sources of seeds for woody plant species had been removed. Indeed, this fact is probably responsible for an underappreciation of the importance of fire by early prairie ecologists. However, the combination of the use of fire by Native Americans and lightning strikes resulted in very high return frequencies across the prairie (Pyne 1982).

Droughts and recurrent fires removed the forest, but the forest soils and zones recently freed from ice would not have any resemblance to the current prairie soils. Many portions of the glaciated regions would have been scoured free of weathered surface materials. Even if patches of soil survived under the icepack, the soil biota did not. Those soils formed beneath boreal forests would likely resemble similar modern-day, northern coniferous forest soils. Unlike the rich, dark loams we now associate with prairies, such soils would have a bleached appearance beneath a layer of partially decomposed plant-litter residues. Nutrient-rich limestone occurred in many areas, but nitrogen, an element not supplied by the bedrock, was present in amounts much smaller than those found today. The soil needed a new climate and a new flora and fauna to develop its full potential.

Many soils formed after the retreat of the last glaciers. Perhaps associated with retreating glaciers, or perhaps associated with extensive droughts during the Hypsithermal (ca. 7000 Y.B.P.), winds redeposited sand and silt from glacial debris over large areas. The heavy sand remained relatively close to the moraines and riverbeds that provided the source of this material. Dunes formed by this activity marched across the land until finally stabilized by vegetation. Lighter silts and clays were carried farther downwind from the source of glacial debris. These accumulated to remarkable depths, forming fertile loess caps that subsequently contributed to high plant productivity. These wind-deposited soils were highly erodible, yet they are still present in areas such as eastern Nebraska and Iowa that have substantial erosion potential. This suggests that vegetation rapidly stabilized and protected these soils. The presence of paleosols (buried "fossil" soils) in upland areas attest to the likelihood that the late Pleistocene dust storms rivaled or exceeded the human-caused dust storms of the 1930s.

In at least the western grasslands, the presence of layers of organic material separated by layers of sand indicates a more complex history. The dunes were stabilized by vegetation. However, extended drought, overgrazing, or some combination of these two factors resulted in the destabilization of the soil surface and subequent burial of the grassland. The multiple organic layers indicates that this happened several times since the retreat of the glaciers. Desertification of dune grasslands is a recurring activity and, as our planet warms, is something to be anticipated.

THE IMPORTANCE OF SOIL TEXTURE

Prairie soils were derived from multiple sources: (1) in situ redevelopment in regions previously occupied by forests, (2) new soils that developed in ice-scoured areas, or (3) imported soils, brought in by wind and deposited as sand or silt loess caps. The relative amounts of sand, silt, and clay in a soil is called soil texture, and this characteristic strongly influences the ability to retain water for plants and to hold plant-available nutrients (Figure 8.1). Sandy soils are composed of mostly large particles; silts are of intermediate size; and clays are very small. Because of their small size, the surface area to volume ratio of the clays is very large, and these surfaces have the ability to attract and loosely hold plant nutrients such as ammonium (NH_4), calcium (Ca), potassium (K), and magnesium (Mg). Sand and silts lack this ability; hence, the presence of some clay in the soil is desirable.

Solids occupy only about 50 to 60% of the soil volume; the remaining 40 to 50% is pore space. These pores can be filled with water, air, or some mixture of the two. The last is most desirable over the long term. Pore spaces filled with water result in oxygen starvation of roots and anaerobic conditions. These anaerobic conditions not only are harmful to most soil biota, but also result in the production of anaerobic gases such as methane (CH_4), a greenhouse gas associated with global warming. On the other extreme, soil pores filled with air provide no water for plant evapotranspiration or for nutrient uptake by roots.

Large pores (macropores) at the soil surface are needed to allow rapid water infiltration into the soils during rains so that the water is not lost as surface runoff. Both plants and soil animals such as earthworms are important in creating macropores (Figure 8.1). Most prairie soils with intact plant canopies above them seldom experience runoff except when the soils are frozen. Leaves and litter intercept raindrops and prevents the force of the raindrops from destroying the macropores. Overgrazing by ungulates can remove the canopy and compress the soil surface, thereby directly and indirectly destroying the macropores and enhancing runoff. Chronic grazing therefore reduces a soil's potential water-storage capacity.

Soil texture strongly affects pore size. Sandy soils have such large pores that rain infiltrates easily. However, these same soils hold water poorly against the pull of gravity. In contrast, clay soils have very small pores. Clay soils may have very slow infiltration rates and therefore be more susceptible to surface runoff. These soils hold their water very tightly, not only against gravity but also against the forces of root uptake. Thus the extremes in soil texture tend to be less efficient in providing water for plant roots. Loam soils, formed from equal mixtures of sand-, silt-, and clay-size particles, have the greatest ability to provide plants with nutrients and water.

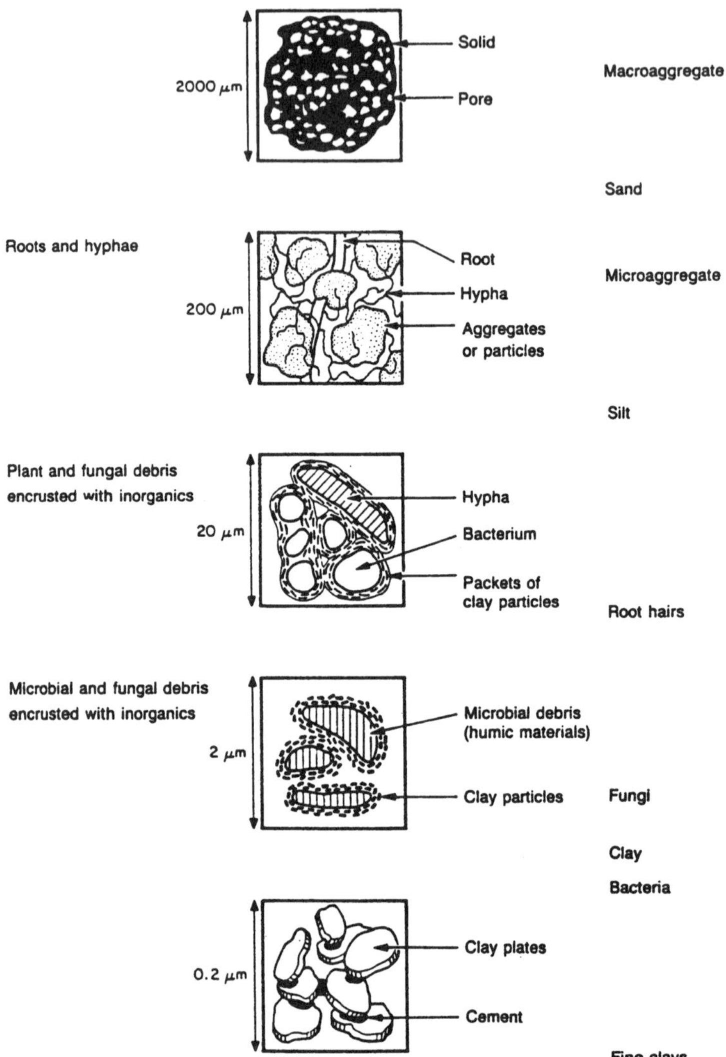

Figure 8.1. Composition of soil constituents. Large roots and earthworms are too big to be illustrated. Also not shown are soil arthropods, nematodes, and a variety of invertebrates that range from several centimeters to clay-sized particles. (From Tisdall and Oades [1982], as drawn by Paul and Clark [1989], with permission of Academic Press)

Soil texture strongly interacts with climate to provide the regional patterns of plant productivity *and* the year-to-year variability in plant productivity (Parton et al. 1987; Sala et al. 1988). For example, productivity on sandy soils is constrained in wet years due to lack of sufficient nutrients, but in drought years the sandy textured soils have lower water losses to surface evaporation than do more finely textured soils (Figure 8.2). Plant production on sandy soils during drought years can therefore be relatively greater than on the more inherently fertile, but more water-limited, clay soils. The net result of this phenomenon is that the overall var-

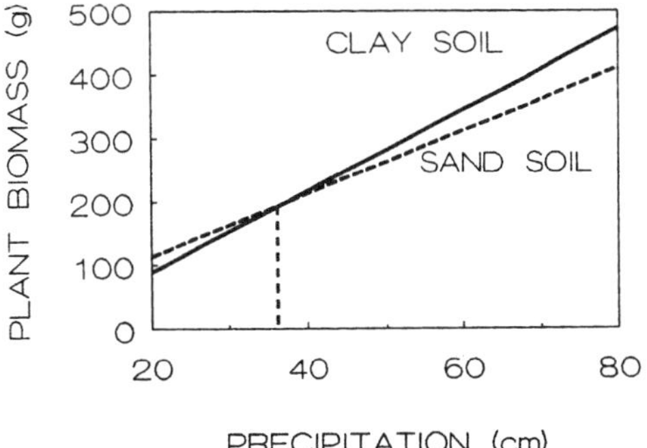

Figure 8.2. Example of a soil–climate interaction. Sandy soils are less productive than clay soils under high rainfall amounts, but more productive under low rainfall conditions. At 37 centimeters of rainfall (dashed vertical line), soil texture has no effect on plant production. (Data from Sala et al. [1988])

iability in productivity on coarse soils is much less than that of fine-textured soils. Productivity of grasslands on coarse soils may therefore be less responsive to climatic shifts than grasslands on fine-textured sites. Figure 8.2 also predicts that at locations with an average of 37 centimeters (about 15 inches) of rain, soil texture is unimportant to grassland production. Species diversity and successional patterns in disturbed areas may also be related to the inherent variation in both water and nutrient retention status of the various prairie soils (Collins and Glenn, Chapter 7). The nutrient and water status of a soil is also correlated with the species diversity of that area. This variation in resource abundance is also under biotic influence, as discussed later. Over all but the longest time scales, biota cannot affect soil texture; however, the biota can influence the forms and availability of nutrients for a given texture.

The chemical and physical properties of parent material from which soils are derived appear to be more important to nutrient-cycling patterns and carbon storage than these are to plant productivity. Sandy soils can be moderately productive, but they cannot store much carbon or nitrogen. Clay soils can also be moderately productive, but they can represent some of the largest storage reservoirs for organic carbon found in temperate, terrestrial ecosystems. With the exception of the Sand Hills of Nebraska, most of the eastern to midwestern prairie soils fall into the latter category.

HOW CLIMATE AFFECTS SOILS

The North American grasslands are composed of temperature and precipitation gradients (Figure 8.3). The importance of other variables, such as nutrient availability, is not constant across this gradient. The soil texture–water interaction illustrated in Figure 8.2 is a good example of an interaction between the relative importance of nutrients and water. Soil nutrient status is important to the prairie plants only if sufficient water is present to allow for plant growth. If water is most critical, then nutrient status is probably unimportant, at least during that portion of the growing season.

Given a particular parent material as a source for soil solids, the realized fertility

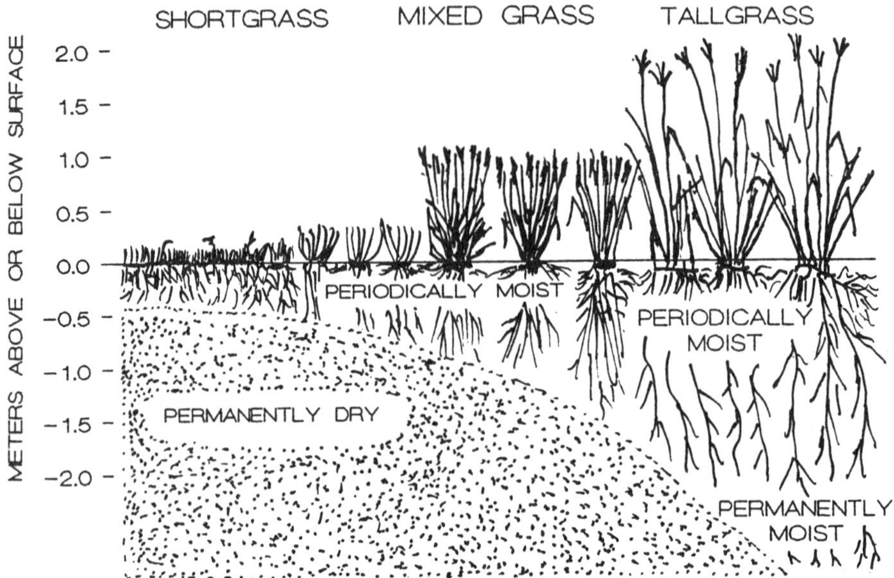

Figure 8.3. The North American grassland varies in soil characteristics. A major difference results from the east–west moisture gradient. This gradient affects rooting depth and root production, which then affects soil organic-matter accumulation and a number of soil chemical reactions. Due to the absence of rainfall in the West, soils tend to accumulate salts. In contrast, eastern soils are leached of nutrients if soils cannot retain precipitation (Shantz 1923).

of this soil is strongly influenced by the interaction of precipitation and solar energy. The amount of sunlight and amount of precipitation, operating over long periods of time, influence the physical and chemical composition of the soil solids, which, in turn, has a major impact on the ability of a soil to hold nutrients.

The energy environment of the prairie is determined by the fate of incoming solar radiation. A portion of this energy is reflected back to the atmosphere, unchanged and unused. Plants are somewhat better reflectors than are dark soils, but do not reflect as much sunlight as do light, sandy soils. Thus if plants are removed by grazing or by agricultural procedures, the amount of solar energy retained by the ecosystem can either increase or decrease. This fact is very important to our understanding of how land-management practices can either warm up or cool down the earth's surface (Pielke and Avissar 1990).

If the sunlight is not reflected, most of it is used to evaporate water from surfaces, creating what is called latent heat. Most of the remaining energy is absorbed by other molecules, increasing their energy content. This increase in energy content (which also occurs to water molecules) is termed sensible heat, and air temperature is related to this energy term. Only a very small fraction (usually well under 4% during the growing season) of solar radiation is converted by plants into stored chemical energy. If rainfall is abundant, most of the energy is converted to latent heat; during drought or when vegetation is senescent, most of this energy is converted to sensible heat. Drought intervals are hotter not because the absolute amount of energy entering or leaving the ecosystem has increased, but because most of the energy exchange is through conversion of sunlight to sensible heat. The absence of surface or plant water for evaporation results in a warmer midday surface environment.

If too much rainfall occurs such that solar energy is insufficient to evaporate this water, then that water not held by the soil is removed by gravity. This water can leach the soil of nutrients, salts, and soluble organic chemicals. The combination of precipitation and organic acids created by decomposing plant materials tends to reduce the potential fertility of the soil as well by dissolving the clays that hold plant nutrients. While small amounts of leaching are beneficial to prevent salts from accumulating in soils, well-leached soils tend to be infertile. Even when large nutrient sources are added, these weathered soils cannot hold these nutrients for any length of time. Thus soil fertility is difficult to maintain under high-rainfall conditions. Under the opposite extreme, too little rainfall, soils tend to accumulate salts deposited by dust and rain, and even during times of adequate moisture these salts interfere with plant metabolism.

Between the negative endpoints of excess leaching and insufficient precipitation lies the amount of rainfall equal to the amount of water that could be evaporated by solar energy. This amount is greater in the southern prairies, which have more intense solar energy inputs. In most mesic, temperate areas, precipitation exceeds the ability of the system to evaporate water during the nongrowing season, and is less than the potential evaporative demand in the summer. However, if the soils can store this precipitation during the nongrowing season until it is removed and transpired by plants, then this balance can be obtained (Risser et al. 1981). Such sites, where solar radiation inputs are roughly equivalent to the energy needed to evaporate annual precipitation, have been hypothesized to produce the optimal growing environment for plants (Budyko 1974). Such environments can occur in cold climates (where little rainfall occurs with little solar radiation) and in hot climates (where much rainfall is needed to balance solar inputs). While grassland biomes are not the exclusive locations of such environments, many of the world's grasslands and savannas encompass or are adjacent to these "balanced" combinations of temperature and precipitation.

A number of researchers have plotted global patterns in precipitation and evaporation potential (solar radiation) and examined the characteristics of biomes possessing similar climates (Post et al. 1982; Meentemeyer et al. 1985). Grasslands are found in climates also occupied by moist-to-dry forests and shrublands. While there exists substantial variation within a single biotic community type, amounts of soil organic matter found in the nongrassland systems are clearly lower than those found in grasslands (Figure 8.4). Thus the type of biotic community, in addition to the climate, appears to be important in creating the soil characteristics.

HOW THE GRASSES AND SOIL BIOTA BUILT THE SOILS

Grasses tend to be less productive than forests; that is, the annual amount of carbon dioxide removed from the atmosphere by plants and converted into organic carbon forms is less per unit of area than in an equivalent area of forest (Figure 8.4). While relatively more of this carbon is allocated to roots by the grasses, the abso-

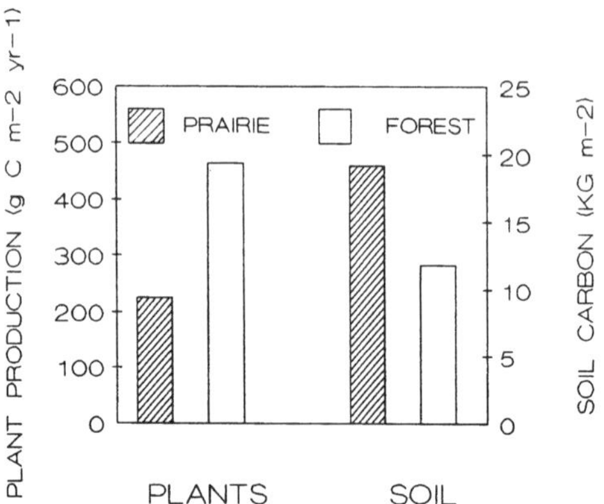

Figure 8.4. The amount of carbon found in annual plant production and soils of temperate grasslands and temperate forests. When compared with forests, prairies are underachievers in terms of plant production but overachievers in terms of soil carbon storage. Note that the amount of soil carbon is almost 100 times greater than the amount of carbon in plant production. (Production data from Whittaker [1975]; soil carbon data from Schlesinger [1977])

lute amounts of organic matter placed on and into the soil as a result of plant productivity is less in grasslands than in forests. In spite of this, Schlesinger (1991) reports that the average organic-matter content in temperate grasslands is almost double the amount found in temperate forests. The difference between the two systems must therefore be attributed to the conservation of this organic matter. Prairies retain more of their fixed carbon instead of returning it to the atmosphere as carbon dioxide. This retained soil carbon is in a variety of forms in various stages of decomposition; about half is in the form of a very complex, amorphous group of chemicals collectively called humus. This material is important for several reasons. Humus helps hold soil moisture against gravitational forces and increases potential soil fertility; plant-available nutrients such as calcium (Ca), potassium (K), and ammonium (NH_4^+) are weakly attracted to the surface of humus molecules, just as these are to clays. Because of this attraction, these nutrients are retained rather than leached from the soil. Finally, the humus molecules contain substantial amounts of organic nitrogen. Almost 1 atom of nitrogen is retained with every 10 atoms of carbon. Since soil carbon amounts often exceed 10,000 grams per square meter, over 1,000 grams of nitrogen per square meter can be held by the soil in organic forms. This amount is well over 50 times the amount of nitrogen found in prairie plants at any one point in time (Risser and Parton 1982).

Life within the prairie soil would seem less precarious than that of a surface dweller. Temperature and moisture extremes are ameliorated by the soil environment. Soil temperatures below a few centimeters seldom increase more than a few degrees Celsius in response to fires. Accordingly, a square meter of prairie soil is home to millions of animals, billions of microbes, and miles of fungal hyphae. Earthworms, including species that followed the retreating glaciers northward and species introduced from other continents during the past two centuries by man, are in great abundance in the more mesic prairies. Their biomass rivals that of cattle in terms of mass of animal per unit area of prairie. About 1 to 10% of the sand, silt, and clay that composes most of the soil mass has been passed through the guts of these animals every year (James 1991). The burrows of earthworms provide macropores for rainwater to move into the soil where it can be stored, rather than lost as surface runoff. Yet the impacts of the earthworms may be less than that of the millions of small unsegmented worms, the nematodes, that feed on everything from live and dead roots, bacteria, fungi, and other animals, to one another. The nematodes have been identified as the most important plant-root feeders of the prairie (Stanton 1988). The surface horizons of the prairie soils are, in actuality, invertebrate fecal material, intertwined with living organisms of every trophic status. (Clearly, Mom was right when she told us to go wash our hands following any soil explorations!)

Surprisingly, soil organisms appear to be somewhat inefficient at consuming plant foliage and root litter in grasslands; that is, plant material decays slowly, and a high percentage of dead plant litter is converted into humus. Humus is difficult for microbes to consume. The molecules are too big to be absorbed internally by microbes; the organisms must excrete enzymes into the soil to attempt to break down these molecules. This process is slow at best; the humus molecules are often physically protected by inorganic clay particles. Hence, decomposition rates of humus are very slow. Thus while prairie soils house an abundance of microbes and associated invertebrates, these groups do not appear to be able to completely break down and convert all organic compounds back to their inorganic constituents. What chemical, physical, or biological factors have prevented the soil biota from using this resource at a more rapid rate?

Jenny (1930) analyzed the patterns in abundance of soil carbon and nitrogen in forest and prairie soils in the central United States. He was among the first scientists to detail the large differences between prairies and forests in terms of their soil organic-matter holdings. He also reported on the north–south and east–west patterns in soil organic matter and soil nitrogen in the prairies (Figure 8.5). Soil carbon and nitrogen amounts are highest in the moist, cool regions of the northeastern grasslands and least in the hot, dry regions of the southwestern grasslands. The amounts of soil carbon and nitrogen in soil illustrated in Figure 8.5 are the result of two conflicting processes: (1) how fast plants can make roots, and (2) how fast these roots die and are consumed by soil animals and microbes. The resulting pattern is complex and is generated by an east–west decline in root production, and a south–north decline in soil decay rates. Given the presence of frequent fires, the source of this organic matter was largely of belowground origin.

The amount of soil organic matter in the soil is therefore the result of two processes: inputs of root detritus from plants, and the decay of this detritus by microbes and soil invertebrates. The difference between these two amounts represents the net gain or net loss of soil organic matter to the ecosystem each year. Accumulation is the result of the difference between production and decay and not necessarily due to large amounts of plant production. For example, tropical forests have substantially more plant production, but even higher rates of decomposition. The net result is reduced amounts of soil organic matter when compared with that in the prairie. If decay always equalled production, regardless of amounts per year, no accumulation would result.

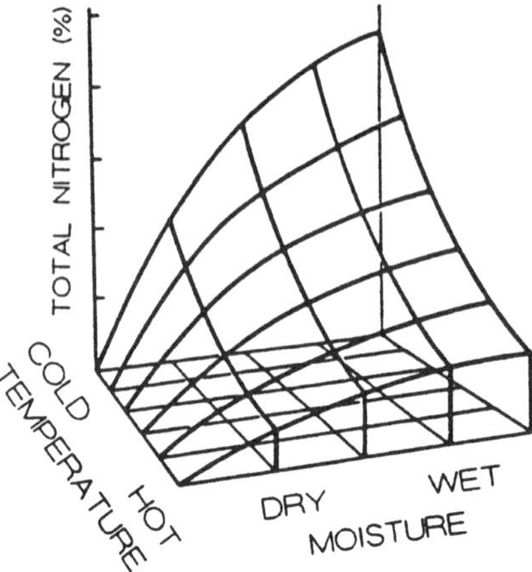

Figure 8.5. Three-dimensional projection of the relationship among prairie soil nitrogen concentrations and annual average temperature and precipitation values found across the North American grasslands. The same pattern exists for the amounts of soil carbon, nitrogen, and most other nutrients as well (Redrawn from Jenny [1980], with permission of Springer-Verlag)

The amount of organic matter present in the soil at any point in time is equal to the initial amount present at some previous time, plus the input of new organic material from plants and animals, minus the losses of soil organic matter to decomposition, erosion, and leaching. The total amount of soil organic matter will therefore vary from year to year, but often this year-to-year variation is small, and amounts approach an average or equilibrium value. Expressed mathematically, this occurs when the fraction of litter decay per year multiplied by the amount of litter present equals plant production; that is, [(% decay/100)*(total detritus amount) = annual plant litter production].

Jenny (1980) and other authors have noted that decay in cold environments tends to lag behind production. This is why the northern boreal forest and tundra soils contain so much organic matter. Unlike that in the prairies, however, much of that organic matter is in the form of partially decayed plant residues, not as humus. In the prairies, the combination of microbes and soil invertebrates appear to adequately process plant litter to the humus stage. The absence of leaching and the cooler temperatures found in the deeper soil profiles, however, prevent this latter group of chemicals from being further processed at a rapid rate. Surprisingly, the ratio of microbial biomass to soil organic matter in the prairie zones is among the lowest observed in North America (Insam 1990). This ratio may be more a consequence of the high humus accumulation than an indication of constraints on microbial growth. Nonetheless, ongoing research shows that additions of either water or nitrogen to the soil can stimulate microbial growth and increase microbial biomass (C. W. Rice, Kansas State University, personal communication). Microbes appear to suffer from resource limitations that are created, in part, by plant activities.

Physical limitations to microbial activity are the result of imposition of climatic factors that switch in terms of their importance over the growing season. In the fall, winter, and spring, activities are limited by temperature. By the time the soil warms to near optimal rates for microbial decomposition, the soil is often too dry for optimal growth by these organisms (Wildung et al. 1975). Plants are of no help to the microbes here; the canopy shades and cools the soil early in the season, and the plant roots extract water so that a midsummer water limitation in soil microbial processes is likely.

The presence of both living and dead plant materials above the soil surface acts as a soil insulator. Energy exchange occurs in the plant canopy rather than on the soil surface, and shaded areas are correspondingly cooler. We can use Jenny's diagram (Figure 8.5) to predict what happens to soil carbon and nitrogen when a site is warmed. Canopy removal, by either grazing or fallow agricultural techniques, creates a warmer and drier environment. This is the equivalent of moving the site to the south and west, and we predict less soil organic-matter accumulations under these conditions *if* plant productivity were unaffected. If less plant production occurs because of canopy removal, even less soil accumulation would be predicted. The use of Jenny's figure as a predictive model has limitations. Since forest soils are generally wetter and cooler than prairie soils because of canopy shading, the model would predict more carbon and nitrogen in these soils. In fact, forest soils have less (Figure 8.4), suggesting that the model is limited to grasslands only. At the same time, however, this observation again emphasizes that the type of vegetation is at least as important as the soil climate in generating patterns of soil carbon and nitrogen.

The production of large amounts of plant roots in a climate that does not strongly leach the soils created the prairie Mollisols, rich in soil organic matter. And, once the soils were formed, the positive benefits to the plants from the humus as a water and nutrient reservoir stimulated further plant growth. This feedback loop generates additional humus: Accumulation results if plants continue to overproduce and/or if decomposer organisms continue to underconsume. Plants, however, are not known to waste their resources. Plants adopt growth characteristics that match up with resource availability; excessive growth in the absence of nutrients or excessive nutrient uptake in the absence of adequate energy is, theoretically at least, not adaptive (Chapin 1980, 1991). If, however, the physical and chemical environments keep changing, then plants must adopt a more opportunistic growth and nutrient-uptake strategy. Frequent fires perpetuated change in microclimatic conditions and resource availability, and, in doing so, resulted in the perpetuation of this organic-matter accumulation.

FIRE: THE MANIPULATOR OF PLANT AND SOIL BIOTA CHARACTERISTICS

Fire results in substantial losses of nitrogen from the prairie; perhaps twice as much nitrogen is lost in a single fire as enters the system yearly in rainfall or by nitrogen-fixing organisms (Seastedt 1988; Ojima et al. 1990; Hobbs et al. 1991). Annual fires are therefore believed to mine the prairie of its nitrogen reserve. Less frequent fires, however, remove a smaller fraction of the annual inputs. The removal of vegetation and plant surface litter results in an exposed, blackened soil surface. This surface is both warmer and drier than unburned prairie. This change in microclimate probably has a negative or at most a neutral effect in the western prairies, where increased water loss caused by enhanced surface evaporation would negate the benefits of soil

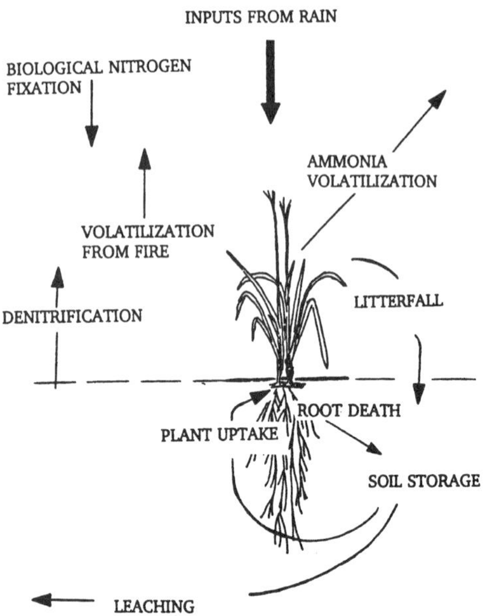

Figure 8.6. Movements of nitrogen in and through the prairie ecosystem. Nitrogen enters either as nitrate and ammonium in rain or by biological nitrogen fixation. Losses occur to volatilization (both from fire and by the conversion of plant and soil nitrogen to ammonia gas), to denitrification (a form of microbial respiration in the absence of oxygen), and by leaching. Leaching losses can involve all forms of nitrogen, but nitrate is by far the most mobile form of dissolved nitrogen. Nitrogen contained in plant detritus or soil would also be lost if surface erosion occurred. Surface erosion, however, is very rare in unplowed prairie.

warming. However, given adequate water, these conditions generally stimulate plant growth (Knapp and Seastedt 1986; Hulbert 1988). These same conditions also create the potential for nitrogen and water rather than solar energy to act as factors limiting plant growth. Fire frequency is very important here. Losses of nitrogen in a fire followed by losses of water due to increased surface evaporation result in both of these resources becoming less available. The plants further exacerbate this limitation by their enhanced growth. Available nitrogen becomes locked away in plant tissues, while higher rates of photosynthesis place strong demands on soil water. Plants respond to these limitations by allocating relatively more photosynthate to the roots in order to maintain adequate water and nitrogen. This input of new roots to the prairie soil was critical to the accumulation of soil organic matter and humus; frequent fires perpetuated the nitrogen and water demands and therefore perpetuated the inputs of roots.

Nitrogen is a very mobile element that occurs in a variety of gaseous and particulate forms (Figure 8.6). Total nitrogen amounts in the ecosystem are dictated in a manner similar to carbon accumulation. The sum of inputs minus outputs, over the history of the site, determines nitrogen amounts. Inputs tend to be low, often under 1 gram of nitrogen per square meter per year, and are in the form of nitrate (NO_3^-) or ammonium (NH_4^+) in rainwater. A much smaller amount of nitrogen is made available when inert nitrogen gas (N_2) from the atmosphere is converted to a usable form for plants by specialized microbes. As previously stated, fire can oxidize organic or in organic particulate nitrogen into a gaseous form. However, the absence of fire does not ensure that nitrogen will be available for plant growth. Nitrogen can be easily leached from the soil if it occurs as nitrate. Soils lacking plant roots can lose substantial amounts of nitrate to groundwaters if leached, and nitrate enrichment is a serious water-quality problem in many of the former prairie states. Chemical reactions associated with the breakdown of organic forms of nitrogen in bison or cattle urea will result in a fraction of the nitrogen leaving as ammonia gas (NH_3). If soils remain too wet and lack sufficient oxygen, a group of opportunistic microbes will use nitrate nitrogen in place of oxygen for respiration. The net result is that this nitrogen emerges from the soil as either inert nitrogen gas (N_2) or nitrous oxide (N_2O), a greenhouse gas associated with atmospheric warming. These potential losses of nitrogen from the soil suggest why, even in the absence of frequent fire, unlimited nitrogen availability to plants was not a guaranteed condition. The presence of large amounts of organic nitrogen in humus provided a source of stability for this otherwise foot-loose element.

Frequent fire stimulated plant growth and root production, but why should fire diminish the ability of the soil biota to consume this production? The same limitations to plant growth, nitrogen, and water also appear to constrain the complete oxidation of dead plant materials to carbon dioxide and water. Both above- and belowground plant litter are very low in nitrogen content, and this limitation likely slows the initial rate of decay. The absence of adequate moisture, caused in part by root uptake, would also slow decomposition during those periods when the soil temperature was optimal for decomposition processes (Buyanovsky et al. 1987).

Frequent fire would consume much of the aboveground plant litter; this material does not compose a major fraction of the soil biota's diet, nor does it contribute as much carbon to the soil as do roots. The annual amount of root death has been estimated at about one-third of belowground root mass in mesic grasslands (Dahlman and Kucera 1965; Hayes and Seastedt 1987). Perhaps only one-eighth of the root mass dies per year in the dry, shortgrass prairie (Milchunas and Lauenroth 1992). As microbes and soil invertebrates consume this material, those carbon molecules not broken down by the biota are slowly converted to more complex, less usable forms such as fulvic and humic acids that have even slower decay rates. The fact that much of the grass litter is deposited in the soil rather than on the surface appears to be a factor in humus formation; the relatively high carbon allocation to roots by grasses grown under frequent fire conditions is significant to soil organic-matter accumulation and results from the location of the inputs as well as the absolute amount of inputs.

Microbes and soil animals attempting to use either live or senescent roots as an energy source are faced with a dilemma. Their bodies often contain as much as 20% nitrogen by dry weight, but they are consuming foods that contains about 0.5% nitrogen. To grow and retain a 20% nitrogen content, the biota must (1) "waste" or respire much of the carbon in their food and (2) conserve or immobilize all available sources of nitrogen. Their activities, instead of liberating nitrogen from the decaying plant material, appear to decrease its availability. Some of these microbes have the ability to remove inorganic nitrogen from the soil, and actually compete with the plants for this resource. Their feeding activities help perpetuate the nitrogen-limitation problem faced by the plants. Only the death of the microbes and animals liberates this nitrogen, and then the plants must compete with the next generation of soil organisms for this resource. Meanwhile, a portion of the plant litter continues to be converted to humic substances, thereby accruing both nitrogen and carbon into an ever-increasing pool of soil organic matter. The plants must feel the frustration of the Ancient Mariner: "Nitrogen, nitrogen everywhere, but not in a form to uptake."

Wedin and Tilman (1990) have provided some new thoughts regarding nitrogen limitation on the prairie. They pointed out that the dominant C_4 grasses tended to have the lowest nitrogen requirements per gram of plant or the highest nitrogen use efficiencies (i.e., grams of carbon fixed per gram of nitrogen in the plant) of all the species they tested. Soils beneath these plants also exhibited low nitrogen availability, because all available nitrogen was tied up in either plant or microbial biomass. The plants are able to retain their dominance because of their ability to withstand and persist under low-nitrogen conditions. These species demonstrate enhanced growth when given additional nitrogen, but after a few years, these species appear to lose their dominance to other plants (Seastedt et al. 1991). In other words, plants such as big bluestem are most dominant on the prairie under environmental conditions suboptimal for their own growth and reproduction. These persist under low-nitrogen conditions, but under chronic nitrogen enrichment they are outcompeted by other species. These species are abundant not just because of their ability to survive fire, but because they can withstand, thrive, and even help create the nitrogen-limited conditions that are maintained by frequent fires.

Other authors have noted that plants living in zones where resource availability is moderate to low have the highest nutrient-use efficiencies (Chapin et al. 1986). Plants growing in areas where nutrients are never abundant lack the ability to exploit sudden additions of nutrients. Conversely, species that grow where nutrients are always abundant are under no selection pressures to develop conservation mechanisms. Species with high nutrient use efficiencies tend to withdraw nutrients back into their living tissues before senescence. Such species will produce large amounts of nutrient-poor litter. If these plant communities are persistent (due to fire or other mechanisms), then the plant species themselves help maintain the high-carbon and low-nutrient conditions of the soils.

Wedin and Tilman's work suggests why biological nitrogen fixation, the process whereby plants with specialized symbiotic microbes convert N_2 to a usable form of nitrogen, is not common in prairies. Vitousek and Howarth (1991) note that nitrogen fixation is a very costly process, in terms of both energy requirements and needs for high concentrations of other nutrients such as phosphorus (P). If the dominant grasses in the prairie are capable of growth using low concentrations of nutrients such as nitrogen, then the ability to fix atmospheric nitrogen does not necessarily give such species the advantage. Only if nitrogen became essentially unavailable (e.g., through many, many years of annual fires) would nitrogen-fixing species become dominant.

THE ENDPOINT TO ACCUMULATION

Even prairie soils cannot accumulate carbon and nutrients indefinitely. Sooner or later, sufficient humus reserves accumulate such that enough nitrogen would be liberated per year by decay of this substrate to meet the needs of the dominant plant species, even in the face of frequent fires and large nitrogen losses to the atmosphere. This nitrogen would be ''dilute''—that is, in low concentrations per unit of soil or soil water. But added up over the depth of the soil column, the amount available would meet minimal plant needs in at least some years. Water or sunlight would limit annual plant production in those years when sufficient nitrogen was available. A reasonable question to ask, however, is whether these soils, most under 10,000 years old, have finished acquiring nutrients and carbon. The answer for those few remaining undisturbed soils is probably yes. If a square meter of soil could store away about 10 grams of carbon per year as humus, it would also store about 1 gram of nitrogen. Ten thousand years later, we would have 10,000 grams of nitrogen in that soil, about 5 to 10 times more than that found in the most fertile of North American grasslands. Given current rates of plant production and decay, we should assume about 1000 years are required for prairie soils to approach equilibrium values (Parton et al. 1987).

GRAZERS AS MANIPULATORS OF THE NITROGEN CYCLE AND SOIL CARBON STORAGE

Removal of foliage by grazing changes a plant's view on life. If such a plant was previously growing on a nitrogen-limited or water-limited site, the plant would now find itself lacking sufficient leaves to fix enough energy to feed its roots. The net result is that plants use both new and stored carbohydrate reserves to replace the foliage that was lost to the grazers (Detling et al. 1979). Instead of pumping fixed carbon from foliage into the root system to find new sources of nitrogen and water, the plant uses those resources aboveground to construct new leaf tissue. Nitrogen is also required for this process, and both plant reserves and root uptake of nitrogen go into this new foliage. If the grazing is chronic, plants have little choice but to replenish aboveground foliage at the expense of both new and old roots, and a reduction in root mass occurs.

Weaver (1950, 1958) provided substantial information about the effects of chronic grazing on the prairie roots. Roots of grazed plants were always shorter and less branched and had dry weights less than those of ungrazed plants. Grazing therefore had a twofold effect on soil processes:

1. The source of new carbon for humus development was reduced.
2. The chronic nitrogen limitation of frequently burned prairie, created in part by fire and in part by the plant roots themselves, would also be reduced.

Grazed sites burn poorly, if at all (Hobbs et al. 1991). Since root production is diminished, the demand for nitrogen from microbes and soil invertebrates feeding on those roots is also diminished and soil nitrogen availability tends to increase (Holland and Detling 1990). The short-term effect of chronic grazing is therefore a more rapid nitrogen cycle, which allows a diminished root mass to provide sufficient nitrogen to maintain foliage production. In the western portions of the prairie, this system may be the rule, with the dominant species well adapted to grazing. In the more easterly grasslands, however, the propensity for dominant species to be outcompeted under high-nitrogen conditions suggests that these species could not persist under chronic grazing. The prairie could not have been dominated by ''tallgrass'' species under chronic grazing. The massive amount of carbon accumulation in these soils is simply inconsistent with the concept that such areas were chronically grazed. Historical accounts (e.g., Mattes 1969) support this hypothesis; bison were present but infrequent in the eastern grasslands.

A hypothesis just emerging is that infrequent grazing, like infrequent fire, tends to cause a transient pulse in plant productivity (Seastedt and Knapp 1993; Turner et al. 1993). The conversion of unburned prairie to burned prairie results in a shift in the relative importance of solar energy, water, and nutrients as limiting resources. Since nitrogen and water can be stored when self-shading by the plant canopy is limiting plant growth, a fire produces a period of time when water, nitrogen, and usable solar energy are more abundant than average, and a pulse of production results. In similar fashion, infrequent grazing on a water- or nitrogen-limited system causes a change to an energy limitation on plant growth because of the absence of leaf area. Plants respond by using carbohydrate storage to replace lost foliage. This storage might otherwise not be used at all that year, or might have been used for seed production, which does not contribute much to the biomass of the plant. In essence, consumers force the plants to take accrued resources out of a noninterest-earning activity and invest them in the energy-capture process. However, since reserves are limited, the production response can only be ephemeral. Unlike fire, however, the long-term benefits of such grazing-induced pulses to soil fertility are few, if any, because these grazed plants reduce their carbon allocation to new roots.

ECOSYSTEM REHABILITATION AND GLOBAL "RELEAF"

Prairie "restoration" is an illusive concept. While we can certainly reestablish a number of prairie species to their former habitats, we do not know the presettlement species composition of the prairie or how this composition varied over time. Moreover, the large number of species introductions and extinctions over the past 150 years makes true restoration essentially impossible. What we can do is practice ecosystem rehabilitation and create a biotic system that embodies those traits beneficial not only to desirable species, but to the biome and biosphere as a whole. The desirable ecosystem traits exhibited by the prairie include the conservation of nitrogen and carbon in relatively inert soil forms, which causes a concurrent reduction in streamwater and groundwater contamination by excess nutrients, and maximization of midsummer cooling by evapotranspiration processes. Trees may make for a cooler surface microclimate, but most species would not time their maximum water use with the period of maximum solar inputs. These same trees would put more carbon into plant materials on an annual basis, but unless this plant material is protected, it will be rereleased to the atmosphere.

Soils subjected to conventional agricultural methods for the past century have lost about half of their soil organic matter (Allison 1973). The speed at which these soils can be rebuilt is limited. Carbon can be accumulated rapidly, but long-term storage as humus is probably limited by nitrogen inputs. Even in areas, such as the eastern prairie regions, where the atmosphere has been nitrogen-enriched by human activities, the amount of nitrogen deposited on the soil surface is probably under 2 grams per square meter per year. Ironically, extra additions of nitrogen will not speed this process and may retard it. This is because (1) the dominant species will limit root production if excess nitrogen is available, thereby slowing carbon deposition, and (2) weedy species, which do not produce as much root biomass, tend to outcompete the native species under conditions of high nitrogen availability.

The prairie was created by the constant switching of energy, water, and nitrogen limitations, with changes in these limitations induced by both fire and grazers as well as by climatic variability. As such, the system never established a true equilibrium with its climate, and the absence of this equilibrium was responsible for the prairies being "overachievers" in terms of their storage potential of soil carbon and nitrogen. Good combinations of climate and fire benefited the plants, and the plants responded by hording both nitrogen and water from the soil-decomposer organisms. Hence, plant and soil carbon accumulated in good years. In drought years, both the plants and the microbes suffered from water limitations; thus both production and decomposition were limited in such years. Interludes of foliage or root grazing would disrupt the accumulation pattern, but the relatively large amount of carbon currently found in the soils suggest that these interludes did not dominate over the past 1000 years of prairie history.

Acknowledgments
Discussions with many colleagues, but especially Alan Knapp, Dennis Ojima, David Schimel, and William Parton, helped shaped my views about the prairie ecosystem. I apologize for the eastern prairie biases, but, if water is generally limiting as it is in the western prairies, then the importance of the rich interactions between energy and nitrogen is diminished, as is soil fertility. The Konza Prairie near Manhattan, Kansas, was the home for most of my research, and these studies were supported by the National Science Foundation, NASA, and the Department of Energy, Global Environmental Change programs.

LITERATURE CITED

Allison, F. E. 1973. *Soil Organic Matter and Its Role in Crop Production.* Elsevier, Amsterdam.

Axelrod, D. I. 1985. Rise of the grassland biome, central North America. Botanical Review 51:163–202.

Budyko, M. I. 1974. *Climate and Life.* International Geophysics Series 18. Academic Press, New York.

Buyanovsky, G. A., C. L. Kucera, and G. H. Wagner. 1987. Comparative analysis of carbon dynamics in a native and cultivated ecosystems. Ecology 68:2023–2031.

Chapin, F. S., III. 1980. The mineral nutrition of wild plants. Annual Review Ecology and Systematics 11:233–260.

Chapin, F. S., III. 1991. Integrated responses of plants to stress. BioScience 41:29–36.

Chapin, F. S., P. M. Vitousek, and K. Van Cleve. 1986. The nature of nutrient limitation in plant communities. American Naturalist 127:48–58.

Dahlman, R. C., and C. L. Kucera. 1965. Root productivity and turnover in native prairie. Ecology 46:84–89.

Detling, J. K., M. I. Dyer, and D. Winn. 1979. Net photosynthesis, root respiration, and regrowth of *Bouteloua gracilis* following simulated grazing. Oecologia 41:127–134.

Hayes, D. C., and T. R. Seastedt. 1987. Root dynamics of tallgrass prairie in wet and dry years. Canadian Journal Botany 65:787–791.

Hobbs, N. T., D. S. Schimel, C. E. Owensby, and D. J. Ojima. 1991. Fire and grazing in the tallgrass prairie: contingent effects on nitrogen budgets. Ecology 72:1374–1382.

Holland, E. A., and J. K. Detling. 1990. Plant response to herbivory and belowground nitrogen cycling. Ecology 71:1040–1049.

Hulbert, L. C. 1988. Causes of fire effects in tallgrass prairie. Ecology 69:46–58.

Insam, H. 1990. Are the soil microbial biomass and basal respiration governed by the climatic regime? Soil Biology and Biochemistry 22:525–532.

James, S. W. 1991. Soil nitrogen, phosphorus and organic mater processing by earthworms in tallgrass prairie. Ecology 72:2101–2110.

Jenny, H. 1930. *A Study on the Influence of Climate upon the Nitrogen and Organic Matter Content of the Soil.* University of Missouri Agricultural

Experiment Station Research Bulletin No. 152. Columbia.

Jenny, H. 1941. *Factors of Soil Formation.* McGraw-Hill, New York.

Jenny, H. 1980. *The Soil Resource.* Springer-Verlag, New York.

Knapp, A. K., and T. R. Seastedt. 1986. Detritus accumulation limits productivity of tallgrass prairie. BioScience 36:662–668.

Leopold, A. 1949. *A Sand County Almanac.* Oxford University Press, New York.

Mattes, M. J. 1969. *The Great Platte River Road,* vol. 25. Nebraska State Historical Society, Lincoln.

Meentemeyer, V., J. Gardiner, and E. O. Box. 1985. World patterns and amounts of detrital soil carbon. Earth Surface Processes and Landforms 10: 557–567.

Milchunas, D. G., and W. K. Lauenroth. 1992. Carbon dynamics and estimates of primary production by harvest, ^{14}C dilution, and ^{14}C turnover. Ecology 73:593–607.

Ojima, D. S., W. J. Parton, D. S. Schimel, and C. E. Owensby. 1990. Simulated impacts of annual burning on prairie ecosystems. Pp. 118–132 in S. L. Collins and L. L. Wallace (eds.). *Fire in North American Tallgrass Prairies.* University of Oklahoma Press, Norman.

Paul, E. A., and F. E. Clark. 1989. *Soil Microbiology and Biochemistry.* Academic Press, San Diego, Calif.

Parton, W. J., D. S. Schimel, C. V. Cole, and D. S. Ojima. 1987. Analysis of factors controlling soil organic matter levels in Great Plains grasslands. Soil Science Society America Journal 51:1173–1179.

Pielke, R. A., and R. Avissar. 1990. Influence of landscape structure on local and regional climate. Landscape Ecology 4:133–155.

Post, W. M., W. R. Emanuel, P. J. Zinke, and A. C. Stangenberger. 1982. Soil carbon pools and world life zones. Nature 298:156–159.

Pyne, S. J. 1982. *Fire in America: A Cultural History of Wildland and Rural Fire.* Princeton University Press, Princeton, N.J.

Risser, P. G., and W. J. Parton. 1982. Ecosystem analysis of the tallgrass prairie: nitrogen cycle. Ecology 63:1342–1351.

Risser, P. G., E. C. Birney, H. D. Blocker, S. W. May, W. J. Parton, and J. A. Wiens. 1981. *The True Prairie Ecosystem.* Hutchinson Ross, Stroudsburg, Pa.

Sala, O. E., W. J. Parton, L. A. Joyce, and W. K. Lauenroth. 1988. Primary production of the central grassland region of the United States. Ecology 69:40–45.

Sauer, C. O. 1950. Grassland climax, fire and man. Journal Range Management 3:16–21.

Schlesinger, W. H. 1977. Carbon balance in terrestrial detritus. Annual Review Ecology and Systematics. 8:51–81.

Schlesinger, W. H. 1991. *Biogeochemistry: An Analysis of Global Change.* Academic Press, San Diego, Calif.

Seastedt, T. R. 1988. Mass, nitrogen and phosphorus dynamics of foliage and root detritus of tallgrass prairie. Ecology 69:59–65.

Seastedt, T. R., J. B. Briggs, and D. J. Gibson. 1991. Controls of nitrogen limitation in tallgrass prairie. Oecologia 87:72–79.

Seastedt, T. R., and A. K. Knapp. 1993. Consequences of non-equilibrium resource availability across multiple time scales: the transient maxima hypothesis. American Naturalist, 141:621–633.

Shantz, H. L. 1923. The natural vegetation of the Great Plains region. Annals Association America Geographers 8:81–107.

Stanton, N. L. 1988. The underground in grasslands. Annual Review Ecology and Systematics 19: 573–589.

Tisdall, J. M., and J. M. Oades. 1982. Organic matter and water-stable aggregates in soils. Journal Soil Science 33:141–163.

Turner, C. L., T. R. Seastedt, and M. I. Dyer. 1993. Maximization of aboveground production in grasslands: the role of defoliation frequency, intensity and history. Ecological Applications 3: 175–186.

Vitousek, P. M., and R. W. Howarth. 1991. Nitrogen limitation on land and in the sea: how can it occur? Biogeochemistry 13:87–115.

Weaver, J. E. 1950. Effects of different intensities of grazing on depth and quantity of roots of grasses. Journal Range Management 3:100–113.

Weaver, J. E. 1958. Summary and interpretation of underground development in natural grassland communities. Ecological Monographs 28: 55–78.

Wedin, D. A., and D. Tilman. 1990. Species effects on nitrogen cycling: a test with perennial grasses. Oecologia 84:433–441.

Whittaker, R. H. 1975. *Communities and Ecosystems,* 2nd ed. Macmillan, New York.

Wildung, R. E., T. R. Garland, and R. L. Buchbom. 1975. The interdependent effects of soil temperature and water content on soil respiration rate and plant root decomposition in arid grassland soils. Soil Biology and Biochemistry 7:373–378.

III

CONSERVATION AND RESTORATION

9

Grassland Management: Ecosystem Maintenance and Grazing

LINDA L. WALLACE
M. I. DYER

In the Great Plains the vistas look like music,
like Kyries of grass . . .

G. Ehrlich, *The Solace of Open Spaces*

Grasslands in North America evolved in the presence of a diverse consumer community. The Pleistocene megafauna of the central North American grasslands was a diverse group of herbivores with their associated predators. Following the post-Pleistocene extinctions (the cause of which is still hotly debated), the herbivorous fauna of this region was tremendously reduced in diversity (Gould 1991). However, using scouting reports and accounts from early settlers, estimates have been made of the most numerous herbivorous species, *Bison bison,* prior to the development of European culture throughout North America. These estimates range from 20 to 60 million animals (Larson 1940). In addition to bison, there were large numbers of elk, deer, and antelope as well as small herbivores, ranging from gophers to geese to grasshoppers (Axelrod 1985). Thus even in the absence of the large Pleistocene fauna, the grasslands of central North America still supported large numbers of a diverse herbivore assemblage. The key word here is "diverse."

What is occurring on today's grasslands in North America? Most of the native large-bodied herbivores are gone, and current management philosophies seek to minimize the presence of herbivores other than domestic livestock. Managers who are in charge of this new landscape now wish to maximize the production of only one herbivore species or, at most, two or three. In addition, because of economic pressures, management practices aim to maximize this production on a finite parcel of land, rather than dealing with large spaces previously utilized by native herbivores prior to European cultural expansion. Are such shifts in land-use patterns significant? We believe so and argue that managers need to understand ecosystem dynamics in grasslands in order to design rational land-use plans themselves. With this information, they can better understand how the herbivores they are producing influence grassland function and their own potential economic productivity on a long-term basis.

Although most of the grasslands in question evolved in the presence of the diverse fauna just described, the face of the grasslands changed following its replacement by domestic livestock. Plant species diversity was altered, plant cover was reduced, and other classic signs of overgrazing, such as erosion, were prevalent. As these signs increased in intensity and extent, managers became more aware of a need to develop sustainable grazing strategies. Some looked to the system that preceded domestic livestock; that is, how the native ungulate species used the grassland system (Savory 1979). The logic behind this decision relied on the premise that equally large populations of these native herbivores were readily supported by the grassland over the ages. If management of cattle and other domestic species could mimic the grazing behavior of free-roaming native ungulates, not only could domestic production be maximized over time, but perhaps some of the negative impacts of prior grazing activities could be reversed.

This chapter addresses the decision processes facing grassland managers. Although this subject has been covered far more extensively in other chapters, we will briefly review basic grassland ecosystem processes. Then we will examine different management options that domestic-livestock producers use in light of their influence on ecosystem processes. We also discuss how understanding plant response to herbivory can be an important management tool. Finally, we will examine how grasslands are managed for wild herbivore habitat.

GRASSLAND ECOSYSTEM FUNCTION

Due to the relatively dry climate in grasslands, nutrient availability is typically low. During periods of wet, relatively warm weather, nutrients become available through the process of mineralization. In the grasslands of middle North America, summer drought is common. This mid-growing-season drought can severely limit plant growth, not only because of water limitations, but also because of the reduction of nutrient availability. Climate patterns of the middle North American grasslands are shown in Figure 9.1 (Mock 1991). Note that actual rainfall during the growing season can be substantial. Drought occurs because air temperature is typically high during this time, causing potential evapotranspiration to be higher than the rainfall received. This then results in drought conditions. Also, analyses of drought occurrences (Mock 1991) have shown that droughts occur in the central North American grasslands most frequently in the spring and summer.

Nutrient mineralization occurs when soil bacteria transform organic material into chemical forms that can be utilized by plants. This mineralization depends on adequate water supplies as well as favorable temperatures. When we examine Figure 9.1 (Mock 1991), we can see that nutrient availability should be greatest in the spring following soil warming. This is when plant growth is greatest and plant tissues have the highest nutrient content. It is also when many plant species are the most tolerant of grazing and when many managers initiate grazing activities on native prairies.

Another critical facet of nutrient cycling is the source of the organic material on which soil microbes act. Figure 9.2 (McNaughton et al. 1988) shows that material enters the decomposer subsystem from the plant subsystem directly via litterfall (dead plant material) and periodically via trampling and fire. Using the terminology devel-

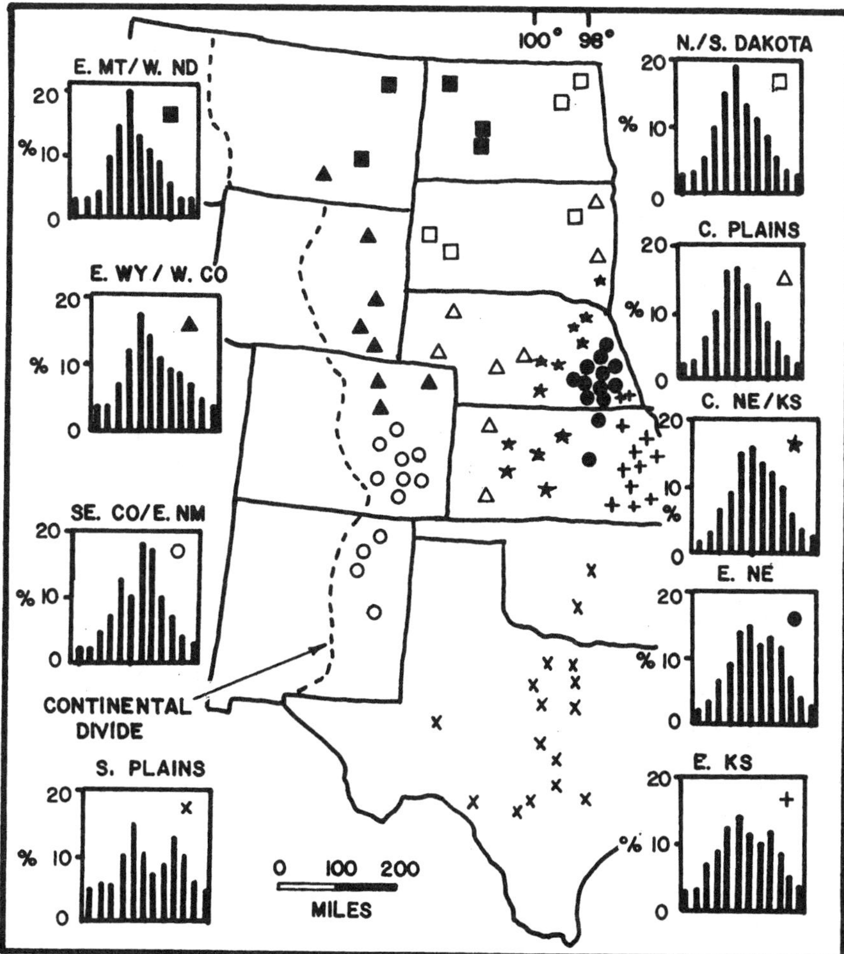

Figure 9.1. Climatic regions of the Great Plains of North America. Climographs give percent annual precipitation for each month from January (left) to December (right). In the more northerly stations, rainfall tends to be condensed into a shorter time period in early summer, whereas the more southerly stations experience rain in late spring and early fall. (From Mock [1991], with permission of Center for Great Plains Studies)

oped by McNaughton et al. (1988), these routes are termed slow and pulsed (periodic) nutrient cycling, respectively. However, when grazing animals deposit waste products, then nutrient cycling can occur much more rapidly. Thus herbivores can create positive feedbacks that indirectly affect plant growth by speeding up the process of plant nutrient uptake.

Herbivores can also reduce the rate of plant-nutrient acquisition. One mechanism whereby this takes place is through soil trampling, which has been shown to reduce plant growth. Soil porosity of any soil containing either clay or high amounts of organic matter is lessened and the rate of diffusion of water, oxygen, and nutrients to

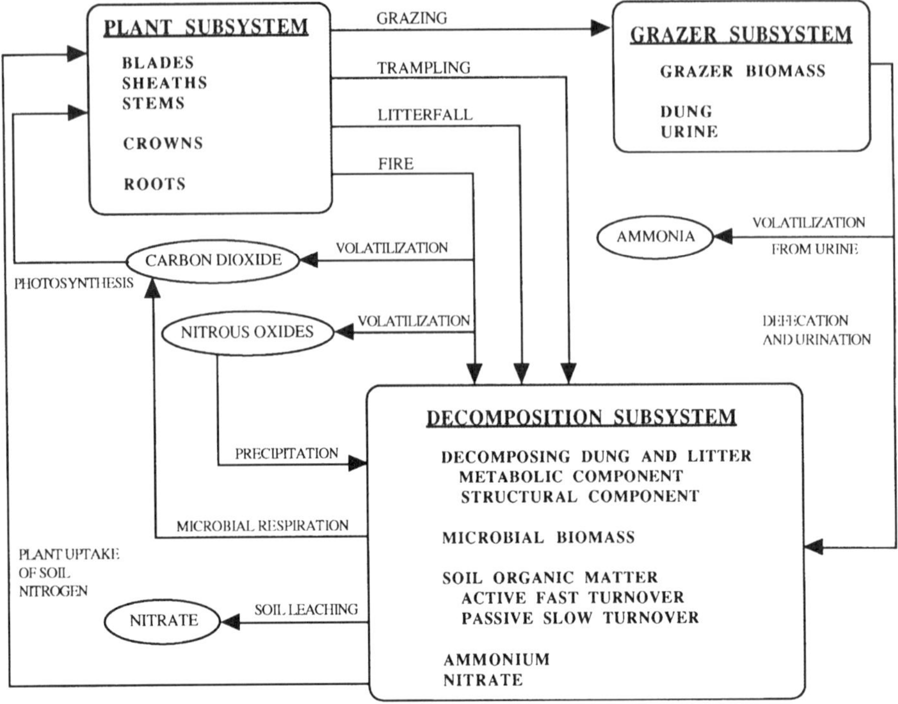

Figure 9.2. Grassland nutrient pathways in ecosystems supporting large-bodied ungulate herbivores. Pathways that are mediated by herbivores tend to be more rapid than ones that are not. Note that in this ecosystem, all the components are highly interconnected, but any effects that grazing animals have on plant nutrient status is modified by organisms in the decomposer subsystem. (From McNaughton et al. [1988], with permission of the American Institute of Biological Sciences)

root surfaces is slowed (Taylor 1989) (Figure 9.3). Root growth itself is slowed in higher density soils produced by trampling, thus reducing the likelihood that roots will grow into areas with higher nutrient concentrations. This is illustrated in Figure 9.4, where individuals of little bluestem were grown in plots that were subjected to clipping (mimicking grazing) and soil compaction (mimicking trampling). The number of simultaneously clipped leaves on plants in compacted soil was significantly fewer than that of other treatments, showing that trampling can have significant negative impacts on plant growth (Wallace 1987). Another mechanism reducing nutrient uptake is a direct reduction of root growth resulting from the loss of aboveground carbon. If grazed plants are unable to transport sufficient food to roots, root growth and nutrient uptake will either cease or slow. This is especially important in plant species that are relatively intolerant of herbivory.

Ungulate grazing often regulates carbon movement throughout a system, effectively controlling ecosystem productivity. Plant species intolerant of grazing often demonstrate lower photosynthetic rates following herbivory. Tolerant species may show no reduction in photosynthetic activity and, indeed, may increase photosynthesis rates (McNaughton 1985; Dyer et al. 1991). Thus the stage is set whereby ungulates

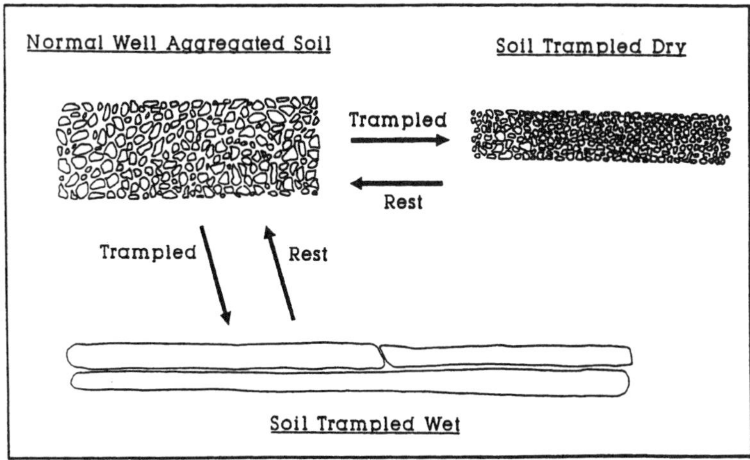

Figure 9.3. Changes in soil structure caused by trampling of either wet or dry soil. Note that the effects of trampling on wet soil are much more severe because soil loses its pore structure as well as volume. (From Taylor [1989], with permission of the Soil and Water Conservation Society)

not only exert control over the average amount of primary production (production by plants), but also control the variance around that mean. We cite several examples of this in Table 9.1. These data demonstrate the degree of control of variance in a large number of grazed systems, ranging from tropical to continental temperate regions, with a variety of grazers, including large ungulates, birds, primates, and insects. Note that we use the verb ''control'' rather than ''change'' since the direction of variance change is uniform in these studies.

What is the importance of this finding? A stable ecosystem will show some mean level of primary productivity that is nonzero. If primary productivity goes to zero, then food supplies for herbivores and decomposers will be either gone or severely limited. Other ecosystem components, such as carnivores and bacterial-feeding nematodes, will eventually be destroyed as well. This ''worse-case'' scenario represents one of the most unstable situations an ecosystem could face. In reality, an ecosystem with a widely fluctuating level of primary production would also be unstable, since herbivore populations would also be subject to wide fluctuations. Thus the more predictable the mean is (the lower the variance around it), the more stable a system will be. However, a small variance can come about only from tightly controlled system-level energy and nutrient flows. Table 9.1 shows that different measures of the variance around mean productivity (as determined in a number of ways from several systems) are lower in grazed conditions than they are in ungrazed.

This means that grazers contribute a considerable degree of control over both the levels of productivity and its variance in total system function during the growing season. It is presently unknown to what degree grazing controls this variability from year to year. Common knowledge suggests that climate is the major controller of ecosystem primary production, but the data in Table 9.1 suggest that the picture is more complex. It is obvious that major climate gradients and weather patterns, com-

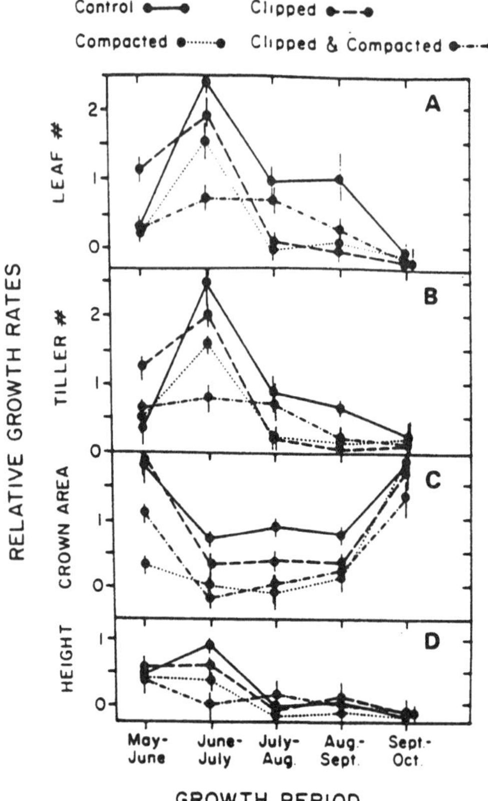

Figure 9.4. Seasonal patterns of the relative rate of leaf growth (leaves/leaf/month), tiller growth (tiller/tiller/month), crown area (cm^2 basal area/basal area/month), and height (cm height/height/month) of little bluestem plants subjected to the treatment combinations indicated. Control (untreated) plants seem to do best, whereas plants that experienced both clipping and soil compaction did poorest. (From Wallace [1987], with permission of *Oecologia*)

bined with various soil types, set the backdrop for grasslands. Such controls operate over long periods of time (years to decades) over large spatial expanses; however, we pose the hypothesis that once this backdrop is set, grazers tend to control the mean and variance of productivity in any given region. Our hypothesis needs to be tested further, but with information available to us at the moment, this seems to be a plausible argument that is important for both developing grassland management strategies and understanding issues in evolutionary and even global-change dynamics.

Variation in primary productivity is also obvious when we look at the spatial patterning of grazing within a pasture. Typically, most herbivores—particularly cattle—tend to regraze the same location, thus creating grazed and ungrazed patches (Ring et al. 1985; Collins 1987). Managers view this patchy type of grazing as a problem, because plant production in the grazed patches will gradually drop from "overgrazing" and the production in the ungrazed patches goes unutilized. However, in an ecosystem context, these ungrazed patches may be important refugia for

Table 9.1. Measures of variance in plant production in comparable grazed and ungrazed systems

System	Units	Type[a]	Grazed	Ungrazed	% diff[b]	Source
Sea plantain/geese	mg/rosette	SD	1.37	2.77	−50.5	Prins et al. 1980
Hay/wild geese	g/m²	SD	15.8	17.0	−7.1	Bedard et al. 1986
Deschampsia/cattle	kg/ha	range	1820	2268	−19.8	Bulow-Olsen 1980
Chaparral/small mammal	g/m²	SE	0.0121[c]	1.54	−99.2	Swank and Oechel 1991
Tragopogon/small mammal	g/plant	SE	0.48	1.0	−52.0	Reichman and Smith 1991
Rhus/deer, beetles	width/ramet	range	2.6	3.0	−13.3	Strauss 1991
Mixed/mountain gorilla	g/m²	SD	363	615	−41.0	Watts 1987
Agropyron/clip	mg/plant	SE	87.5	155	−43.6	Painter and Detling 1981
Bouteloua/clip	mg/plant	SE	24	28	−14.3	Detling et al. 1979
Panicum/grasshopper	g/plant	CV	15.1	21.6	−29.9	Dyer et al. 1991
Panicum/grasshopper	¹¹C model[d]	CV	42.2	53.9	−21.8	Dyer et al. unpublished

[a]The measure of variance where SD = standard deviation, SE = standard error, and CV = coefficient of variation.

[b]Comparison of the grazed–ungrazed variance.

[c]Grazed means went to zero.

[d]Amount of ^{11}C that moved into grazed tissue.

species that are less tolerant of grazing and trampling. These species could be forage plants or even species of soil organisms that are important in decomposition processes.

Finally, what are the effects of different levels of plant growth and production on the production of herbivores? This is the critical question to grazing management. A simple view would state that if there is greater primary productivity, then the productivity of herbivores would also be greater. Although a general pattern supporting this has been found across a wide range of ecosystem types (McNaughton et al. 1991), this may not always be the case. This pattern assumes that the herbivores are limited by the amount of food that they have available to them (Heitschmidt and Taylor 1991).

Herbivores could be limited by competition with other herbivores for food directly or for water or shelter. Water could be limiting, as could be trace elements or shelter. In some cases, predation on herbivores could be the limiting factor to their production or if herbivores have a slow birth rate, their population sizes may not be able to track annual changes in food resources. Grassland managers will attempt to minimize most limitations to herbivore production, but they will not minimize competition from all other herbivores, particularly other members of the herbivore species being raised. If they were to minimize this limitation, they would not maximize total herbivore production on the area of land in use, but would maximize individual herbivore production. Thus individual performance will not be maximized, but the goal of maximum possible herbivore production may be attained. This is where we can see some interesting impacts on ecosystem function caused by different management protocols.

DOMESTIC-LIVESTOCK MANAGEMENT OPTIONS

Nomenclature

First, we must understand the nomenclature used in domestic grazing management. Managers wish to maximize the production of the herbivore(s) in question. To do so, they manipulate not only the numbers of animals placed into their systems, but also the types of plant communities present with fertilization and seeding. The availability of water, trace elements, and shelter can also be enhanced. Grazing management programs deal primarily with manipulation of numbers of animals present as well as the time intervals that these animals are present on given areas of land. The abundance of animals is termed *stocking density* (numbers of animals per unit of land), *grazing pressure* (numbers of animals per unit plant production), or *stocking rate* (numbers of animals per unit land per unit time) (Coughenour 1991).

Maximization of Herbivore Production

The goal of maximization of herbivore production is not easily attained. The performance of the individual animals will be less in most management systems because of intraspecific competition. However, maximizing an individual's performance reduces the total production on the land area in question. Therefore, managers must find a point where the individual's performance is not reduced so much that the total production suffers. Figure 9.5a shows how individual performance declines with a greater density, but total production per area can be maximized (Figure 9.5b). These performance curves will be different for different herbivore species and in different environments and years. Usually, most managers attempt to determine these points by trial and error. However, these curves do not take the grassland itself into consideration. Most managers realize that maximization of production cannot come at the expense of the grassland's ability to regenerate after grazing; thus the concept of sustainable grazing management developed. Here the question is what amount of grazer biomass can be produced in a grassland over the long term.

As we stated at the beginning of the chapter, grasslands of North America evolved in the presence of large numbers of diverse herbivores. These animals were supported by these grasslands year after year. By definition, then, the grazing protocol utilized by these wild herbivores was a sustainable one even though their population levels would have varied. If managers can at least partially mimic these historical patterns of herbivory, perhaps a long-term sustainable (but not necessarily an equilibrium) grazing system can be developed even though we know that an exact replica of historical herbivory patterns cannot be developed. Because managers constrain the activities of livestock to small land areas, the animals cannot move over the entire North American prairies as large herds once did. This spatial constraint forces us to find alternative grazing patterns that may partially mimic successful historical situations. Although no one knows exactly what grazing patterns were employed by free-roaming ungulates, we think that they may have grazed intensely at a site and then left it for some time before returning to graze intensely once again. This concept is the basis of the various grazing systems.

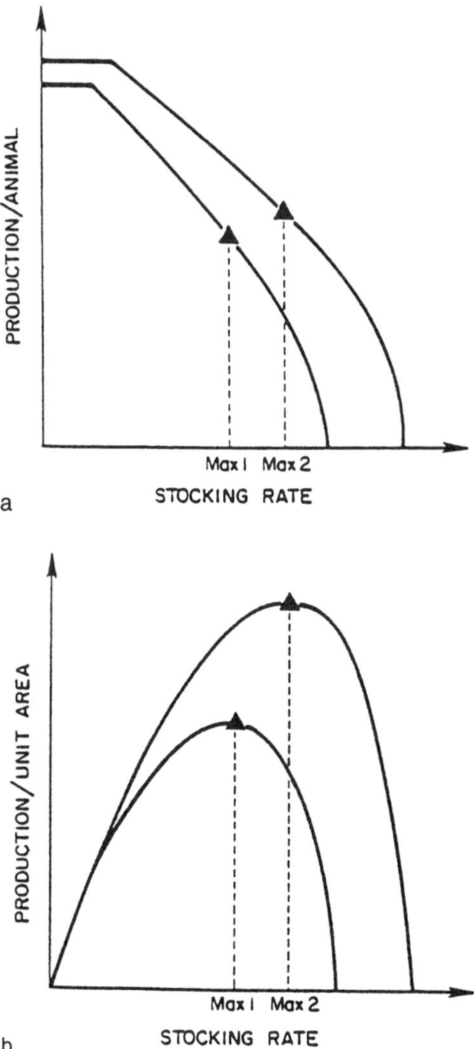

Figure 9.5. (a) Model of the relationship of stocking rate and production of individual animals. The points designated Max 1 and Max 2 correspond with those points. (b) Model of the functional relationship between stocking rate and the production of animals per unit land area. Note that Max 1 and Max 2 do not maximize the production of individuals, but the production of the land area as a whole. Max 1 and Max 2 are the critical production levels that managers hope to achieve. (From Heitschmidt and Taylor [1991], with permission of Timber Press)

Grazing Systems

There are four basic types of grazing systems (grazing system defined as "a specialization of grazing management which define recurring periods of grazing and deferment for two or more pastures or management units" [Heitschmidt and Taylor 1991]). These systems developed from two different schools of thought that eventually blended.

This first school of thought was that animals in a pasture should utilize the pasture

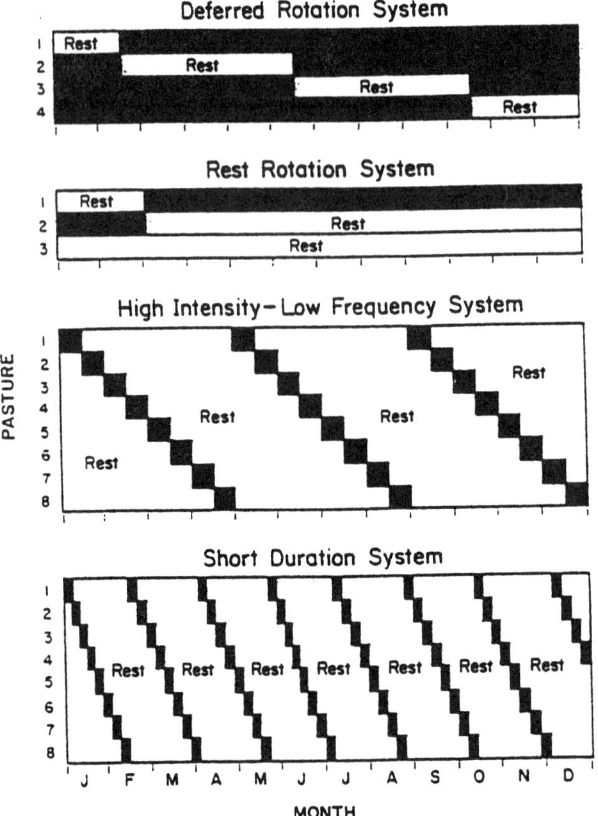

Figure 9.6. Models of the schedule of grazing and resting periods in each of the four grazing systems described in the text. Note the numbers of different pastures that are required for the different methods. (From Heitschmidt and Taylor [1991], with permission of Timber Press)

completely, rather than having heavily grazed spots interspersed with lightly grazed or ungrazed areas. Results showed that animals that grazed large pastures gained less weight than those in small, more completely utilized pastures (Heitschmidt and Taylor 1991). To attempt to ensure more complete usage patterns, the manager would have dispersed salt licks or the location of water supplies.

The second school of thought had to do with the seasonal timing of grazing within a pasture. Some studies had shown that cattle have greater weight gains from a pasture if they are grazed intensively early in the season (the period of peak growth and plant nutritive value), rather than grazed lightly in a pasture over the whole season (Heitschmidt and Taylor 1991). After analyzing these results, managers developed four grazing systems (Figure 9.6). These four systems are analyzed in detail here.

Deferred rotation systems use what is called the *high-performance grazing* tactic. Here plants that are preferred by the herbivores are grazed lightly and species that are not preferred are typically grazed either very little or not at all. In this system, several pastures are grazed lightly, with brief periods of rest allowed during the year.

In the *rest rotation* system, grazing intensity is typically higher than in the de-

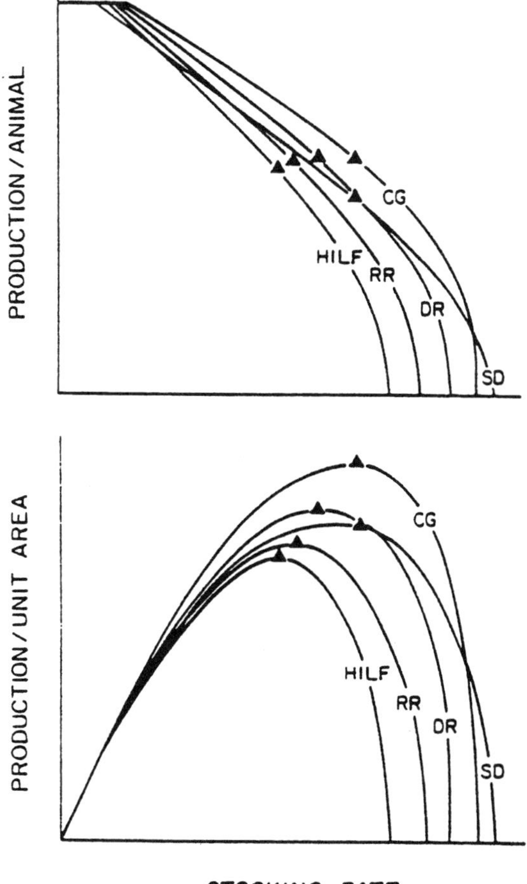

Figure 9.7. Theoretical models of animal production on both an individual and a land-area basis for different rotational grazing systems. Note that the production per animal does not seem to show any differences between systems, but that production per unit land area seems to be higher in continually grazed rather than one of the other grazing systems. G, continuous grazing; DR, deferred rotation; RR, rest rotation; HILF, high-intensity–low-frequency; SD, short duration. (From Heitschmidt and Taylor [1991], with permission of Timber Press)

ferred rotation system, and the periods of grazing or resting are longer. Either the high-performance grazing tactic or the *high-utilization grazing* tactic (both preferred and nonpreferred species grazed moderately) can be used.

High-intensity–low-frequency grazing utilizes several pastures with brief periods of moderate grazing using a high stocking rate. The last grazing system, *short-duration grazing,* has a shorter grazing period than the high-intensity–low-frequency system, and the grazing tactic used is the high-performance rather than the high-utilization grazing.

Theoretically, all these grazing systems can yield similar amounts of herbivore production (Figure 9.7). However, both practical experience and an ecosystems analysis tell us that these curves are going to be dependent not only on which site is used for grazing, but also on the climatic regime of the particular year in question and the availability of herbivore stock.

Ecosystem Functioning in Different Grazing Systems

Few studies have addressed the ecosystem response of grasslands to the grazing systems just described. However, there is a rich literature about various components of ecosystem function. We have summarized a few of these studies in Table 9.2. Few generalizations can be drawn from this compilation, except that no one study appears to have addressed all aspects of ecosystem function. Community structure appears repeatedly to be altered by grazing in those studies that monitored this critical parameter. Basal area appeared to decrease in all studies that monitored it, and grass cover also appeared to decrease. Although some grass species that were important forage species showed an increase, cover in the community as a whole decreased. This phenomenon was observed even in cases where the dominant shrub species was mechanically removed from the grassland, thus removing a major source of competition (Eckert and Spencer 1987).

In systems where these responses have occurred for prolonged periods in the past, grassland community structure has been irrevocably altered (Archer and Smeins 1991). Grassland managers are well aware of this possibility, and try to adjust grazing intensity so that this does not occur. It is possible that at intermediate levels of grazing intensity, higher levels of plant diversity and productivity could occur (Connell 1978; McNaughton 1985). This is shown graphically in Figure 9.8 (Milchunas et al. 1988), where we see higher levels of plant diversity at intermediate levels of grazing. McNaughton (1985) found a similar productivity response in East African grasslands. Thus if the proper grazing intensity is found for each system, then it is *theoretically* possible that grassland productivity and plant diversity could be increased. What form these increases in diversity take (e.g., more palatable species or more weeds) will vary considerably from ecosystem to ecosystem.

The levels and responses of plant productivity monitored in the summarized studies were quite variable (Table 9.2). Four studies noted a change in production, with three studies noting a decrease in production due to either continuous or short-duration grazing. The reasons behind this decrease are not immediately clear. Some measures of nutrient cycling showed a decrease; there was less nitrogen entering the plant component in one study, perhaps slowing growth. However, in three other studies, the rate of nitrogen uptake increased and the level of in vitro digestibility increased. In vitro digestibility, a laboratory measure of how readily a ruminant can digest a particular forage, is higher in plants with high crude protein levels.

In two of these studies, the rates of plant production dropped, indicating that something other than nitrogen uptake was limiting the growth of these plants. One study indicated that soil compaction is increased in short-duration grazing systems, thus limiting root growth and nutrient movement (Schlepers and Lantinga 1985). However, the management goal of higher animal production was met in all but one study. Therefore, on the surface it appears that animal production may be enhanced in several of these systems, but at the expense of long-term primary production.

Secondary production is sometimes maximized, but no significant differences were found in plant production (Hart et al. 1988; White et al. 1991). These systems may have an evolutionary history of grazing, but more likely, the managers found a stocking rate that would maximize production over the short term. In general, the benefit of these different types of grazing systems is being questioned, with few types

Table 9.2. Grazing system impacts on various ecosystem-level parameters

System	Stocking rate	Soil compaction	Nutrient cycling	Production		Location	Community structure
				Plant	Animal		
Hart et al. 1988							
Continuous	0.33 and			1173 kg/ha	.8 kg/day[a]	Wyoming, U.S.	No species change, but increase
Rotational (25%)	0.44		Litter production decreased	1173 kg/ha			in bare soil
Short duration (12.5%)							
White et al. 1991							
Continuous	Moderate			1200 kg/ha		New Mexico, U.S.	Decrease in basal cover
Short duration (11.1%)	Moderate			1200 kg/ha			Increase in basal cover
Schlepers and Lantinga 1985							
Rotational	3.0–3.4	''More serious''		663 kg/ha	148 kg/ha[a]	Flevo Ploders, Netherlands	
Continuous		''Less serious''		157 kg/ha	126 kg/ha		
Parton and Risser 1979							
Continuous	6.5 ha/AU		+1.6%[b]	−5%[c]		Oklahoma, U.S.	Decrease in grass biomass;
	4.0 ha/AU			−28%			increase in forb biomass
	3.2 ha/AU				−32%[c]		

Table 9.2. (*continued*)

System	Stocking rate	Soil compaction	Nutrient cycling	Production		Location	Community structure
				Plant	Animal		
Heitschmidt et al. 1982a, 1982b, 1982c							
Ungrazed				215 g/m^2		Texas, U.S.	
Short duration			−37.4%[e]	125 g/m^2	30–69 kg/ha		
Walton et al. 1981							
Continuous	1.97				0.91 kg/day[a]	Alberta, Canada	Planted pasture
Rotational (25%)	2.42		Increase in Ca, Mg, Cu, P, and protein[f]		1.09 kg/day		
Eckert and Spencer 1986							
Rotational	152–1193 AUM					Nevada, U.S.	Shrub cover increase, forb increase, grass decrease
Eckert and Spencer 1987							
Short duration (33%)	525–742 AUM					Nevada, U.S.	Basal area decrease, despite sagebrush removal

Note: Rotational systems are followed by the percentage figure giving the percent time spent by cattle in each pasture. All systems were grazed by cattle. Stocking rates are given in numbers of animals/hectare unless otherwise noted. AU = animal unit (any combination of animals that utilize 12 kilograms of dry matter/day [Heitschmidt and Taylor 1991]); AUM = animal unit month.

[a]Increase in live weight.

[b]Percent change in crude protein after grazing.

[c]Compared with ungrazed.

[d]Weight gain/animal compared with light stocking.

[e]In vitro digestibility.

[f]Relative to continuous grazing.

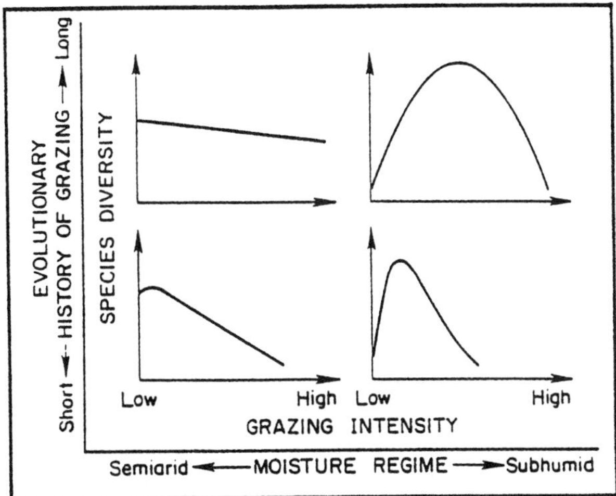

Figure 9.8. Theoretical relationships of plant species diversity to grazing intensity in different grassland types. In ecosystems with long histories of grazing, species diversity either is relatively unaffected by grazing (dry environment) or is actually maximized by middle levels of herbivory (wet system). In ecosystems with a short or no history of grazing, species diversity is strongly reduced by herbivory. (From Milchunas et al. [1988], with permission of the University of Chicago Press)

showing long-term gains from these management protocols (Coughenour 1991). The one exception in the studies listed here (Walton et al. 1981) is a planted pasture rather than natural grassland. It may be that we need to consider these grazing systems (as defined by Heitschmidt and Taylor [1991]) only in the context of manipulated plant systems.

Ungulate Effects on Nutrient Distribution in Grazing Systems

To offset some of these problems, managers have also introduced fertilizer and supplemental feeding into different grazing systems. Fertilizer may act to reduce plant stress, and may actually decrease changes in plant community structure. In theory, it was hoped that animal nutrient inputs via urine and feces could offset any nutrient limitations in the system. Given the large amount of waste products produced by large-bodied ungulate herbivores, their potential impact on nutrient budgets of grazed areas is not trivial. However, grazing animals do not deposit urine and feces in an even distribution pattern across the landscape. Cattle tend to defecate when they resume a standing position after ruminating, and rumination is preferentially done in the shade. This behavior (and others) results in large quantities of fecal material being deposited in some areas and none in others. Furthermore, cattle tend not to graze in areas adjacent to fecal pats.

Grazing will occur in one area (removing nutrients) and deposition of feces and urine may occur more frequently in other areas. Large-bodied ungulates then move nutrients across a system, resulting in nutrient-spiralling in space and time. Short, intense grazing bouts are designed to reduce this patchy aspect of grazing, but if shelter and water are localized in a grazing system, a patchy distribution of nutrients could still occur. This may be a key reason for the patchy grazing patterns seen even in rotational grazing systems designed to eliminate it (Kirby et al. 1986).

Ecosystem Functioning in Continuously Grazed Pastures

Most managers who use either continuous grazing or season-long grazing actually carry out some intense management protocols to maximize herbivore production and minimize negative impacts on the grasslands they use. Managers watch for symptoms of overgrazing, and then manipulate stocking rates. The symptoms of overgrazing that are typically used are the appearance of large bare areas with concomitant soil erosion, the appearance of "increaser species" (plant species that are tolerant of grazing or are preferentially avoided by herbivores), the disappearance of forage species, and the reduction of plant species diversity in general. Some of these symptoms take time to develop, and once developed, it may be too late to rectify the problems of that particular pasture (Archer and Smeins 1991).

Some short-term symptoms for which managers can monitor include reductions in the standing stock (weight/area) of available forage. If reductions of standing stock occur too quickly, then the system may be overgrazed. If this seems to be the case, then managers can implement the *put-and-take* method of grazing management, where, after noting pasture condition, the manager may either add (put) or remove (take) animals from the pasture. In essence, what is being managed for is an ecosystem equilibrium as well as maximum herbivore production. When done correctly, this tactic may even mimic the historical migratory patterns of the large herds more successfully than the grazing systems described earlier, because even though we do not know precise population sizes of historical herds, we can rest assured that their numbers in any one location varied substantially both annually and seasonally.

Plant Response to Herbivory

As mentioned earlier, McNaughton (1985) and others noted that grazing is not necessarily a negative phenomenon for all forage species. Many plants experience an increase in photosynthesis, transpiration, nutrient uptake, and growth following herbivory (Ruess 1984; McNaughton 1985; Detling 1988). Thus following herbivory, these plants rapidly grow highly nutritious forage. This attracts herbivores to utilize the same plants again, and a grazed patch is born. In the interim, ungrazed plants may have grown as well through the production of less nutritious and less palatable stems and flowering culms rather than the production of more leaves. These plants would be subsequently avoided by herbivores, resulting in a patch structure that would decrease system utilization by herbivores, and perhaps decrease their production. The grazed patches could be overgrazed, with useful forage in ungrazed patches going unused. Dry-season fires can reset this picture and may be used as an important management tool in both domestic and wild grazing systems (Collins and Wallace 1990; Coughenour 1991).

There are several possible patch development scenarios based on plant response. Species may show an increased growth rate following one defoliation event, but not repeated events. This would speed the deterioration of grazed patches. Species may not show increased growth at all following herbivory. Patch formation may be less likely to occur in a system populated uniformly with such grazing-intolerant plants. Thus an understanding of plant performance will give managers a greater opportunity to predict the outcome of various grazing treatments on their systems.

MANAGEMENT FOR WILD HERBIVORES

Usually the herbivores in question in the central North-American grasslands are the large-bodied wild ungulates, such as bison, elk, deer, and pronghorn. Management for the latter two groups generally involves management of favored browse species and the development of hunting-season plans that will allow maximum sustained harvest. Both of these groups are very mobile, being able to either leap or crawl under fences easily. They may compete with domestic herbivores for food. For example, a high dietary overlap was found between pronghorn and sheep (Schwartz and Ellis 1981), and fall cattle grazing has been found to displace deer (Willms et al. 1981). Because of their mobility, direct management is usually not done. Management of the other two groups may be more direct.

Both elk and bison can be confined by certain types of fencing; thus fencing may restrict these populations if placed across travel corridors, and so forth. In some cases, such as on wildlife refuges, fencing is necessary to keep animals within the protected confines of the refuge. Both species are known to compete with domestic ungulates, since both are grazers. Bison utilize graminoids (grasses and grasslike plants) predominantly (Schwartz and Ellis 1981; Van Vuren 1984), whereas elk may use both graminoids and forbs (Stephenson et al. 1985). Both species are capable of moving large distances and, because of the large amount of forage needed to sustain them, usually need large areas of land.

Consider two scenarios: one where the manager wishes to manage for a wild ungulate and domestic livestock, and the other where the manager wishes to manage only wild ungulates. In the former, one of the factors limiting domestic-livestock production (competition with other herbivores) will no longer be removed by human intervention. Thus the manager needs to understand the amount of dietary overlap between the two species and take steps to minimize competition through the manipulation of either plant communities or herd movement and size.

Dietary overlap can be quite large between bison and cattle. In one shrub–steppe community, diets overlap by approximately 95% (Van Vuren 1984); high dietary overlap was also noted in shortgrass prairie (Schwartz and Ellis 1981). Although diets are similar between these two species, bison digest their forage more thoroughly than do cattle (Richmond et al. 1977). Dietary similarities of 71% were found between elk and wild horses (Stephenson et al. 1985), and presumably domestic, pasture-fed horses may experience similar overlaps. Diets may also be quite similar between elk and cattle (Wydeven and Dahlgren 1983).

The greater the amount of overlap between species, the more likely that range deterioration could occur. Management would need to be intensive in this situation, and the steps listed by de Vos (1969) may be particularly important: (1) deciding on an appropriate level of grazing intensity, (2) taking steps to improve forage production, (3) increasing the usability of the range by examining water and shelter locations and altering them if need be, and (4) managing the numbers of animals on the range either by hunting or by other removal techniques.

Because these steps are really not substantially different from what one would do while ranching domestic livestock, the key issues (aside from dietary overlaps) include behavioral similarities and differences. Domestic ungulates have largely been bred to remain somewhat sedentary, whereas wild ungulates move a great deal more. If fencing is confining, then pasture size needs to be sufficiently large to accommodate these movement patterns. Pasture size also needs to be sufficiently large to reduce agonistic behavior between species. This is particularly true if bison are involved. Bison tend to be aggressive, particularly to horses and cattle (Wallace, personal observation).

The structure of the grassland vegetation needs to facilitate maximum foraging rates by large-bodied ungulates. Foraging rate in bison, elk, and deer decreases when forage biomass declines below approximately 800 kilograms per hectare (Wickstrom et al. 1984; Hudson and Frank 1987). Thus there need to be large areas of highly concentrated forage unless managers wish to feed supplements to animals.

Another important aspect of grassland structure is the presence of cover (hiding areas, shade, etc.). Grover and Thompson (1986) have shown that feeding-site selection by elk is negatively correlated with cattle feeding and positively correlated with the proximity of cover, favoring a habitat interspersed with wooded areas that also is away from human disturbances such as roads. Bison favor large contiguous habitats that are highly interconnected (Van Vuren 1987). These two criteria are not necessarily mutually exclusive. Bison could forage in the center of large grassy areas, with elk in edge habitats. Such foraging behavior has been noted and has also been found to be bioenergetically feasible for bison (Christopherson et al. 1979), since these animals can tolerate extremes of temperature that would be found in the middle of large open spaces. Cattle utilize areas that are more open than those used by elk, but will favor cover more than bison (Wallace, personal observation).

What if one wishes to manage a grassland strictly for wild ungulates? If a species mixture is to be used, then the same concerns outlined earlier will apply. If only one species is to be managed, then the manager needs to know whether the available plant communities can support the dietary needs of the target species. In addition, the role of predators needs to be understood. If the area under consideration is large, will natural predation be allowed, and if so, what effects will the presence of predators have on neighboring land users? Neighboring land users may also be concerned with the potential of disease transmission by wildlife (Meagher 1973), since both elk and bison carry *Brucellosis* and have been implicated in its transfer to domestic livestock.

As has been outlined, bison and elk may segregate themselves on the landscape based on the presence of cover. In areas where there is a mixture of C_3 and C_4 plant species, bison tend to utilize sites dominated by the cool-season C_3 plants (Wydeven and Dahlgren 1985). Elk and deer utilize sites dominated by warm-season forage, whereas pronghorn and bison utilize disturbed sites, such as prairie dog towns (Coppock et al. 1983; Wydeven and Dahlgren 1985).

In locations where ungulates are to be kept all year, attention must be paid to the seasonal use of the landscape as well. Bison and elk are migratory in many ecosystems, with seasonal movements between winter and summer ranges (Shaw and

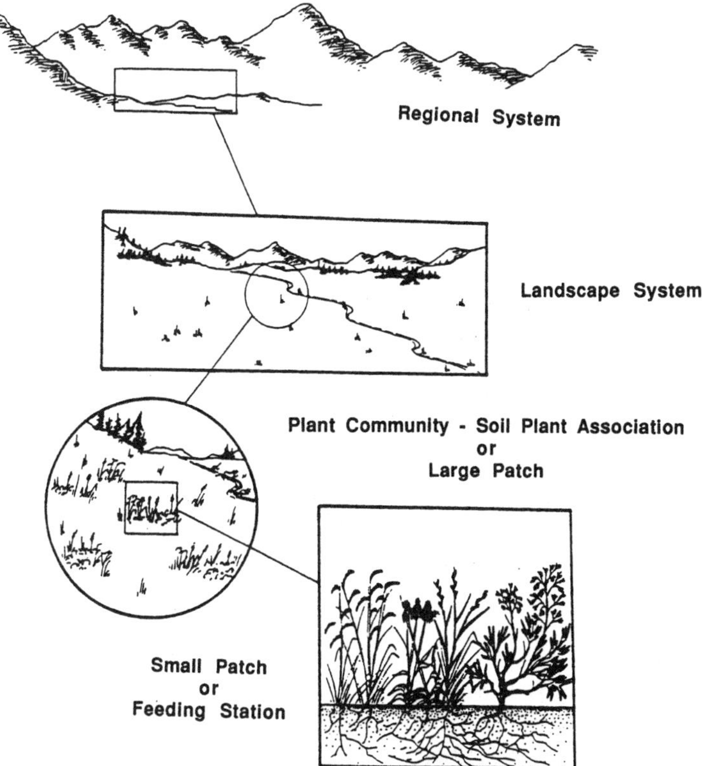

Figure 9.9. Large-bodied herbivores must make foraging choices at a number of ecological scales, ranging from extremely large regional and landscape scales to that of the individual feeding site. Grassland managers will need to be particularly aware of the effects that patterns at the landscape, community, and small-patch scale have on animal behavior and foraging choices. (From Senft et al. [1987], with permission of the American Institute of Biological Sciences)

Carter 1990; Boyce 1991). Once a migratory route becomes established, animals tend to show a great deal of fidelity to it, regardless of weather or forage conditions (Meagher 1989). The existence of such routes needs to be recognized because the possibility of overgrazing in these predictably reused sites may be high.

Thus the overall structure of the landscape is critical in managing for wildlife species. Domestic species may be more dependent on our artificial structures, but wildlife have been shown to be very responsive to slight changes in landscape. Indeed, Senft et al. (1987) have elegantly shown that wild herbivores respond to the landscape at a number of ecological scales (Figure 9.9). A complete knowledge of these interactions is required in order to manage these animals effectively. For example, knowledge of which individual species are preferred and their responses to grazing is important as well as knowledge of their distribution across the landscape and how other landscape factors will influence ungulate habitat and grazing choices.

CONCLUSIONS

We have not offered any ultimate solution to the difficulties of managing grasslands for the production of either domestic livestock or wildlife. The issues are complex and need solutions at the site-specific level, not some regional prescription. However, some general conclusions can be drawn.

Managers must understand the fundamental ecology and basic functioning of their particular system. They need to understand grazing-tolerance patterns of the plant species in the grasslands they are managing in order to predict the outcome of different grazing systems. They need to understand how the structure of their system will influence herbivore behaviors, including forage choice, patterns of nutrient deposition, and resting locations. If mixtures of herbivore species are involved, then their dietary and behavioral interactions must be known. So we see that grassland managers have an extremely complex task involving elements of most of the subdisciplines of ecology. Given that most grassland managers are not trained in all these areas, it is important that range ecologists and range management schools as well as ecologists in the various subdisciplines simultaneously develop comprehensive management plans for individual areas, and then compare them. Until such integrated efforts are undertaken, grassland managers may not have the tools necessary to complete such a complex task successfully.

Ecosystem analysis needs to become a fundamental tool of rangeland managers seeking to develop sustainable management protocols. As we could see from our abbreviated analysis of the grazing system literature (Table 9.2), continuance of the various grazing systems employed would not be possible over the long-term because of deterioration of any of various ecosystem components. This also points out our misconceptions of how grasslands really function.

Acknowledgments

The preparation of this chapter was aided by an award from the National Science Foundation, BSR 9108715, to L. L. Wallace

LITERATURE CITED

Archer, S., and F. E. Smeins. 1991. Ecosystem-level processes. Pp. 109–139 in R. K. Heitschmidt and J. W. Stuth (eds.). *Grazing Management: An Ecological Perspective.* Timber Press, Portland, Ore.

Axelrod, D. I. 1985. Rise of the Grassland Biome. Botanical Review 51:163–201.

Bedard, J., A. Nadeau, and G. Gauthier. 1986. Effects of spring grazing by Greater Snow Geese on hay production. Journal Applied Ecology 23:65–75.

Boyce, M. S. 1991. Migratory behavior and management of elk (*Cervus elaphus*). Applied Animal Behaviour Science 29:239–250.

Bulow-Olsen, A. 1980. Net primary production and net secondary production from grazing an area dominated by *Deschampsia flexuosa* Trin. by nursing cows. Agro-Ecosystems 6:51–66.

Christopherson, R. J., R. J. Hudson, and M. K. Christophersen. 1979. Seasonal energy expenditures and thermoregulatory responses of bison and cattle. Canadian Journal Animal Science 59:611–617.

Collins, S. L. 1987. Interaction of disturbances in tallgrass prairie: a field experiment. Ecology 68:1243–1250.

Collins, S. L., and L. L. Wallace (eds.). 1990. *Fire in North American Tallgrass Prairies.* University of Oklahoma Press, Norman.

Connell, J. H. 1978. Diversity in tropical rain forests and coral reefs. Science 199:1302–1310.

Coppock, D. L., J. E. Ellis, J. K. Detling, and M. I. Dyer. 1983. Plant–herbivore interactions in a North American mixed-grass prairie. II. Response of bison to modification of vegetation by prairie dogs. Oecologia 56:10–15.

Coughenour, M. B. 1991. Spatial components of plant–herbivore interactions in pastoral, ranching and native ungulate ecosystems. Journal Range Management 44:530–542.

de Vos, A. 1969. Ecological conditions affecting the production of wild herbivorous mammals on grasslands. Advances Ecological Research 6: 127–183.

Detling, J. K., M. I. Dyer, and D. T. Winn. 1979. Net photosynthesis, root respiration, and regrowth of *Bouteloua gracilis* following simulated grazing. Oecologia 41:127–134.

Detling, J. K. 1988. Grasslands and savannas: regulation of energy flow and nutrient cycling by herbivores. Pp. 131–148 in L. R. Pomeroy and J. J. Alberts (eds.). *Concepts of Ecosystem Ecology: A Comparative View.* Springer-Verlag, New York.

Dyer, M. I., M. A. Acra, G. M. Wang, D. C. Coleman, D. W. Freckman, S. J. McNaughton, and B. R. Strain. 1991. Source–sink carbon relations in two *Panicum coloratum* ecotypes in response to herbivory. Ecology 72:1472–1483.

Dyer, M. I., D. C. Coleman, D. W. Freckman, and S. J. McNaughton. In preparation. Influence of heterotrophs on above- to belowground source–sink carbon dynamics in *Panicum coloratum* L. Ecological Applications.

Eckert, R. E., Jr., and J. S. Spencer. 1986. Vegetation response on allotments grazed under rest-rotation management. Journal Range Management 39: 166–174.

Eckert, R. E., Jr., and J. S. Spencer. 1987. Growth and reproduction of grasses heavily grazed under rest-rotation management. Journal Range Management 40:156–159.

Gould, S. J. 1991. Abolish the recent. Natural History. May: 16–21.

Grover, K. E., and M. J. Thompson. 1986. Factors influencing spring feeding site selection by elk in the Elkhorn Mountains, Montana. Journal Wildlife Management 50:466–470.

Hart, R. H., M. J. Samuel, P. S. Test, and M. A. Smith. 1988. Cattle, vegetation, and economic responses to grazing systems and grazing pressure. Journal Range Management 41:282–286.

Heitschmidt, R. K., J. R. Frasure, D. L. Price, and L. R. Rittenhouse. 1982a. Short duration grazing at the Texas Experimental Ranch: weight gains of growing heifers. Journal Range Management 35:375–379.

Heitschmidt, R. K., D. L. Price, R. A. Gordon, and J. R. Frasure. 1982b. Short duration grazing at the Texas Experimental Ranch: effects on aboveground net primary production and seasonal growth dynamics. Journal Range Management 35:367–372.

Heitschmidt, R. K., R. A. Gordon, and J. S. Bluntzer. 1982c. Short duration grazing at the Texas Experimental Ranch: effects on forage quality. Journal Range Management 35:372–374.

Heitschmidt, R. K., and C. A. Taylor, Jr. 1991. Livestock production. Pp. 161–177 in R. K.

Heitschmidt and J. W. Stuth (eds.). *Grazing Management: An Ecological Perspective.* Timber Press, Portland, Ore.

Hudson, R. J., and S. Frank. 1987. Foraging ecology of bison in aspen boreal habitats. Journal Range Management 40:71–75.

Kirby, D. R., M. F. Pessin, and G. K. Clambey. 1986. Disappearance of forage under short duration and season-long grazing. Journal Range Management 39:496–500.

Larson, F. 1940. The role of the bison in maintaining the short grass plains. Ecology 21:113–121.

McNaughton, S. J. 1985. Ecology of a grazing ecosystem: the Serengeti. Ecological Monographs 55:259–294.

McNaughton, S. J., R. W. Ruess, and S. W. Seagle. 1988. Large mammals and process dynamics in African ecosystems. BioScience 38:794–800.

McNaughton, S. J., M. Oesterheld, D. A. Frank, and K. J. Williams. 1991. Primary and secondary production in terrestrial ecosystems. Pp. 120–139 in J. Cole, G. Lovett, and S. Findlay (eds.). *Comparative Analyses of Ecosystems.* Springer-Verlag, Berlin.

Meagher, M. 1973. The Department of Agriculture, the Department of Interior, *Brucellosis,* and the farmer. BioScience 32:311–312.

Meagher, M. 1989. Range expansion by bison of Yellowstone National Park. Journal Mammalogy 70:670–675.

Milchunas, D. G., O. E. Sala, and W. K. Lauenroth. 1988. A generalized model of the effects of grazing by large herbivores on grassland community structure. American Naturalist 132:87–106.

Mock, C. J. 1991. Drought and precipitation fluctuations in the Great Plains during the late nineteenth century. Great Plains Research 1:26–57.

Painter, E. L., and J. K. Detling. 1981. Effects of defoliation on net photosynthesis and regrowth of western wheatgrass. Journal Range Management 34:68–71.

Parton, W. J., and P. G. Risser. 1979. Simulated impact of management practices upon the tallgrass prairie. Pp. 135–155 in N. R. French (ed.). *Perspectives in Grassland Ecology.* Springer-Verlag, Berlin.

Prins, H.H.T., R. C. Ydenberg, and R. H. Drent. 1980. The interaction of Brent geese, *Branta bernicla,* and sea plantain, *Plantago maritima,* during spring staging; field observations and experiments. Acta Botanica Neerlandica 29:585–596.

Reichman, O. J., and S. C. Smith. 1991. Responses to simulated leaf and root herbivory by a biennial, *Tragopogon dubius.* Ecology 72:116–124.

Richmond, R. J., R. J. Hudson, and R. J. Christopherson. 1977. Comparison of forage intake and digestibility by American bison, yak and cattle. Acta Theriologica 22:225–230.

Ring, C. H., R. A. Nicholson, and J. L. Launchbaugh. 1985. Vegetational traits of patch-grazed rangeland in west-central Kansas. Journal Range Management 38:51–55.

Ruess, R. W. 1984. Nutrient movement and grazing: experimental effects of clipping and nitrogen source on nutrient uptake in *Kyllinga nervosa.* Oikos 43:183–188.

Savory, A. 1979. A holistic approach to ranch management using short duration grazing. Proceedings of the First International Grassland Congress 1:555–557.

Schlepers, H., and E. A. Lantinga. 1985. Comparison of net pasture yield with continuous and rotational grazing at a high level of nitrogen fertilization. Netherlands Journal Agricultural Science 33:429–432.

Schwartz, C. C., and J. E. Ellis. 1981. Feeding ecology and niche separation in some native and domestic ungulates on the shortgrass prairie. Journal Ecology 18:343–353.

Senft, R. L., M. B. Coughenour, D. W. Bailey, L. R. Rittenhouse, O. E. Sala, and D. M. Swift. 1987. Large herbivore foraging and ecological hierarchies. BioScience 37:789–799.

Shaw, J. H., and T. S. Carter. 1990. Bison movements in relation to fire and seasonality. Wildlife Society Bulletin 18:426–430.

Stephenson, T. E., J. L. Holechek, and C. B. Kuykendall. 1985. Diets of four wild ungulates on winter range in north-central New Mexico. Southwestern Naturalist 30:437–441.

Strauss, S. Y. 1991. Direct, indirect and cumulative effects of three native herbivores on a shared host plant. Ecology 72:543–558.

Swank, S. E., and W. C. Oechel. 1991. Interactions among the effects of herbivory, competition and resource limitation on chaparral herbs. Ecology 72:104–115.

Taylor, C. A., Jr. 1989. Short-duration grazing: experiences from the Edwards Plateau region in Texas. Journal Soil and Water Conservation 44:297–302.

Turner, C. L. 1990. The influence of mowing and grazing on plant productivity and canopy spectral reflectance characteristics of tallgrass prairie. Ph.D. Dissertation, Division of Biology, Kansas State University, Manhattan.

Turner, C. L., T. R. Seastedt, and M. I. Dyer. 1993. Maximization of aboveground production in grasslands: the role of defoliation frequency and defoliation history. Ecological Applications 3:175–186.

Van Vuren, D. 1984. Summer diets of bison and cattle in southern Utah. Journal Range Management 37:260–261.

Van Vuren, D. 1987. Bison west of the Rocky Mountains: an alternative explanation. Northwest Science 61:65–69.

Wallace, L. L. 1987. Effects of clipping and soil compaction on growth, morphology and mycorrhizal colonization of *Schizachyrium scoparium,* a C$_4$ bunchgrass. Oecologia 72:423–428.

Walton, P. D., R. Martinez, and A. W. Bailey. 1981. A comparison of continuous and rotational grazing. Journal Range Management 34:19–21.

Watts, D. P. 1987. Effects of mountain gorilla foraging activities on the productivity of their food plant species. African Journal Ecology 25:155–163.

White, M. R., R. D. Pieper, G. B. Donart, and L. W. Trifaro. 1991. Vegetational response to short-duration and continuous grazing in south-central New Mexico. Journal Range Management 44:399–403.

Wickstrom, M. L., C. T. Robbins, T. A. Hanley, D. E. Spalinger, and S. M. Parish. 1984. Food intake and foraging energetics of elk and mule deer. Journal Wildlife Management 48:1285–1301.

Willms, W., A. W. Bailey, A. McLean, and R. Tucker. 1981. The effects of fall defoliation on the utilization of bluebunch wheatgrass and its influence on the distribution of deer in spring. Journal Range Management 34:16–18.

Wydeven, A. P., and R. B. Dahlgren. 1983. Food habits of elk in the northern Great Plains. Journal Wildlife Management 47:918–923.

Wydeven, A. P., and R. B. Dahlgren. 1985. Ungulate habitat relationships in Wind Cave National Park. Journal Wildlife Management 49:805–813.

10

The Challenges of Grassland Conservation

JANE H. BOCK

CARL E. BOCK

> The grand simplicity of the prairie is its peculiar beauty, and its occurring events are peculiar and of their own kind.
>
> J. C. Frémont, *Memoirs of My Life*

On June 3, 1805, the expedition led by Meriwether Lewis and William Clark reached the junction of Maria's River and the Missouri, on the High Plains of what is now central Montana. The two men climbed out of the river valley onto adjacent highlands. In their own words,

> We ascended together the high grounds in the fork of these two rivers, whence we had a very extensive prospect of the surrounding country. On every side it was spread into one vast plain, covered with verdure, in which innumerable herds of buffalo were roaming, attended by their enemies the wolves; some flocks of elk also were seen, and the solitary antelopes were scattered with their young over the face of the plain. (Lewis and Clark 1893, p. 344)

These were the North American High Plains as they existed less than two centuries ago: vast open grasslands, well populated by large numbers of herbivorous animals and their predators. Preceding chapters in this book review what is known about the ecological glue that holds (or once held) Plains and prairie ecosystems together. Careful field research has revealed much about factors such as the role of fire in maintaining certain grasslands, the nature of prairie soils, and the importance of predators in controlling populations of grassland herbivores. Yet many aspects of Great Plains ecology remain elusive, because the Plains cannot be studied as they once existed. Bison and prairie dogs doubtless were major ecological forces, but they and their most important predators are largely gone. Fires that swept the Plains no longer burn with anything like their prehistoric frequency or extent. Fertile soils, once plowed, have blown away.

The North American Great Plains evolved as one of the largest uninterrupted grasslands on earth. Despite all the subsequent human intrusion, much remains of the native Great Plains flora and fauna. The challenge is to craft a realistic agenda for conservation of what remains, and for restoration of what can be recovered. In this chapter, we first review two aspects of Great Plains ecology that make their conservation a special challenge. We then examine in more detail two issues that have proved especially vexing: livestock grazing and the introduction of exotic vegetation. We conclude with three suggestions for ways to sustain more of the Plains' natural biodiversity.

WHY PLAINS GRASSLANDS ARE DIFFICULT TO CONSERVE AND RESTORE

Grasslands Are Vulnerable to Environmental Change

Undisturbed grasslands may be as durable as any natural ecosystem. However, both fossil and current field evidence suggests that they can persist only in a rather narrow environmental window, and that they may be especially vulnerable to human perturbation. The grasslands of central North America are of rather recent origin, and they have advanced and retreated several times in response to changing climate. Uplift of the Rocky Mountains blocked flows of moist Pacific air, thereby creating the arid climate in which the Great Plains evolved. While this process likely began in the Oligocene epoch, over 30 million years ago (Daubenmire 1978), fossil evidence strongly suggests that the grasslands did not achieve their maximum treeless extent until just the past few thousand years (Axelrod 1985). Grasslands certainly were present during the Miocene and Pliocene epochs (25–2 million years ago), but they were parts of open woodlands and savannas, much as is the case along the eastern margins of the Plains today. Global cooling that began in the Oligocene culminated in the Pleistocene Ice Age, beginning about 2 million years ago. Ice covered the northernmost Plains to varying degrees during the different glacial advances. Of greater importance, the climate south of the ice was cooler and wetter than either before or after glaciation. As a result, forest replaced grassland with each glacial advance (e.g., Wells and Stewart 1987). The last glacial front retreated 10,000 to 12,000 years ago, but fossil evidence suggests that extensive woodlands persisted across much of the Plains as recently as 5000 years ago. All of this makes clear the "youthfulness" (Axelrod 1985, p. 164) of the Great Plains, and the frequency and apparent ease with which they have been replaced by other sorts of ecosystems over geologic time.

There is evidence that North American grasslands remain vulnerable to virtually permanent invasions by woody plants (Figure 10.1), although this is more the case in peripheral than in central parts of the Great Plains (Milchunas et al. 1988). One of the best studied examples involves invasion and establishment of trees and shrubs in grasslands of western and southern Texas (Archer 1989). Field observations and experiments indicate that mesquite trees initiate these invasions, as individual trees appear in the grassland and become foci for clusters of other shrubs that develop in their vicinity. Livestock apparently play a dual role in mesquite invasion. First, they eat mesquite beans and disperse the seeds in their dung. Second, they graze away grass cover that otherwise can (1) outcompete mesquite seedlings and (2) carry fires that kill the young woody plants. Archer and his co-workers hypothesize that the mes-

Figure 10.1. A formerly open semidesert grassland in southeastern Arizona that has been permanently invaded by mesquite trees. These invasions were triggered by the direct and indirect effects of livestock grazing.

quites, once established, attract birds that use them as perches. The birds, in turn, deposit seeds of other woody plants with their feces. Eventually, the resulting clusters of woody plants grow together to form a nearly continuous woodland. Adult mesquite are not killed by fire, so these woodlands, once established, are essentially permanent.

Invasions by woody plants bring into question whether there actually is such a thing as a grassland climate—that is, some combination of temperature and precipitation that allows grassland to replace desert, but that is insufficient for woodland to replace grassland (Sauer 1950). Prevailing opinion is that such a climate may not exist, or at least that the present Great Plains extend beyond its boundaries into areas where factors other than climate are necessary to maintain the grasslands. By far the most important of these factors is fire. Even today, small woodlands of juniper and other species occur on Great Plains escarpments that are naturally protected from fire by their topography (Wells 1965). The persistence of these woodlands, plus historical and paleoecological evidence, suggests that fire, both lightning- and human-caused, was largely responsible for creating the vast treeless plains as we believe they existed just prior to European colonization (e.g., Axelrod 1985).

There is much current concern about the possible effects of global climate change on structure and function of the world's ecosystems (Smith and Tirpak 1988). We are not yet in a position to make confident predictions about the nature and extent of these effects. However, it is reasonable to assume that ecosystems already in a climatically precarious position, such as the Great Plains, will be especially vulnerable.

Some models do indicate that the Plains will be affected dramatically by predicted increases in temperature and reductions in effective precipitation (Burke et al. 1991). Any prudent agenda for Plains conservation must include provision for such climatic shifts. Because undisturbed grasslands probably are more resistant to the consequences of climate change (e.g., invasions by woody and/or alien species), wise land-use policy dictates a curtailment of practices such as tilling the sod or overgrazing by livestock (Bock et al. 1991).

Pristine Great Plains Grasslands No Longer Exist

Even if we agree in theory that undisturbed grasslands are more likely to survive in a changing climate, we cannot be certain just what "undisturbed" means. Most North American grasslands were occupied and modified by Europeans so quickly and so thoroughly that we have only a faded image of their prehistoric condition. The challenge to grassland ecologists is to use scant historic information, coupled with careful studies of existing grasslands, in order to estimate how these ecosystems used to function and what they looked like.

Important changes to the Plains following European colonization included (1) conversion of grassland to cropland, (2) introduction and spread of exotic plants, (3) reduction in extent and frequency of fire, (4) complete or virtual extermination of certain large mammals (the megafauna), and (5) introduction of domestic grazers. There is no doubt that plowed grassland is disturbed grassland, and the lesson of the Dust Bowl made clear the consequences of indiscriminate sodbusting (Worster 1979). Similarly, a grassland dominated by plants and animals native to Eurasia or Africa cannot possibly qualify as pristine or undisturbed. But how important were fire and grazing by native animals in determining the nature of prehistoric "undisturbed" Plains ecosystems?

There is abundant historical evidence that fires were common events on the Great Plains in the early 1840s (e.g., Frémont 1886). Field experimental evidence makes clear that fires are essential for the persistence of tallgrass prairies (e.g., Gibson and Hulbert 1987), without which they are quickly invaded and replaced by forest trees. Fire likely played a similar role in setting boundaries between arid grasslands and desert scrublands in the Southwest (Humphrey 1987; Bahre 1991), and in preventing invasions of woody plants such as ponderosa pine (Figure 10.2) and big sagebrush (Figure 10.3) into grasslands of the northern Plains and along the Rocky Mountain front (Bock and Bock 1984, 1987; Steinauer and Bragg 1987; Veblen and Lorenz 1991). However, the plow and the cow have reduced grass cover so extensively that fires no longer occur on a regular basis throughout much of the central Great Plains. We cannot be certain about the role of fire in many of these areas, despite its likely historical importance (Wright and Bailey 1982; Axelrod 1985).

Imagine the reaction of members of the Lewis and Clark expedition, if they somehow were transported forward onto the late-twentieth-century Great Plains. Certainly they would be unprepared for the consequences of a petroleum-based economy. However, in the context of their own experiences, these early Plains travelers probably would be most startled by the loss of the megafauna; those large Plains herbivores and their predators that once were so characteristic of the region.

During the Pleistocene, North America supported a megafauna rivaling that of

Figure 10.2. Boundaries between pine and grassland, such as this one in Wind Cave National Park in western South Dakota, are determined largely by the frequency of fire.

East Africa (Martin and Guilday 1967). Most of these large mammals disappeared at the end of the last glacial episode, under rather mysterious circumstances. The Pleistocene overkill hypothesis holds that they were exterminated by the first humans to colonize the New World (Martin and Klein 1984). Of those large mammals that survived the Pleistocene extinctions, by far the most numerous and ecologically important was the bison (Figure 10.4). We can never know just how many bison once occupied the Great Plains, but an estimate of 30 million is reasonable and gives some impression of their prehistoric dominance (McHugh 1972). Their near extermination in the nineteenth century leaves us forever in the dark about their true ecological role in driving the dynamics of North American grasslands. Even the largest national parks offer but a fragment of the landscape where bison and other megafauna once roamed at will.

Bison are grazing animals, dependent almost entirely on grasses and sedges (Peden 1976; Popp 1981; Krueger 1986). These and other diet studies suggest that bison prefer certain grass and sedge species when they are abundant and of high quality. Overall, however, bison are both opportunistic and generalized grass-feeders, a fact consistent with their ubiquity on the Plains prior to the slaughters of the nineteenth century.

Bison persist in certain protected areas of the Plains, and one of the best studied of these is Wind Cave National Park, in the southern Black Hills of South Dakota. In this midgrass prairie, bison are involved in strong and generally positive interactions with two other dominant Plains forces: fire (Figure 10.5). and black-tailed prairie dogs (Figure 10.6). Prescribed burning is conducted in Wind Cave National Park in an

Figure 10.3. (a) A sagebrush grassland at Little Bighorn National Battlefield in southeastern Montana. (b) The same general area two years after a fire eliminated all the sagebrush.

Figure 10.4. A mature bison bull at Wind Cave National Park in western South Dakota.

Figure 10.5. Wildfire and grazing by native ungulates were two of the most important ecological forces in the prehistoric Great Plains. This prescribed burn in Wind Cave National Park, South Dakota, attracted the bison in the foreground.

Figure 10.6. Prairie dogs reduce grass cover and encourage the growth of broad-leaved herbs.

effort to restore or maintain historical boundaries between woodland and plains (Bock and Bock 1984). Bison are attracted to burned areas for one or two postfire years, probably because of increased nutritive quality of grasses (Bock and Bock 1989). Prairie dogs are conspicuous colonial rodents that also feed preferentially on grasses and that drastically modify the structure of Plains ecosystems by reducing or eliminating grasses, and increasing forbs and shrubs (Whicker and Detling 1988). Despite potentially competing for food, bison and prairie dogs have a mutually beneficial relationship. This results from the fact that grasses can respond to moderate grazing by producing new tissue that is rich in both energy and nitrogen (Coppock et al. 1983; Krueger 1986). At least in summer, bison in Wind Cave National Park are attracted to the edges of prairie dog towns, where they feed on the nutritious grasses and use bare areas of prairie dog towns as wallows. Bison, in turn, facilitate the expansion of prairie dog colonies into new habitat, by grazing down taller grasses adjacent to existing towns. Prairie dogs are dependent on vision and an elaborate social warning system to escape predators such as coyotes and raptorial birds (Hoogland 1981). They apparently are unable to colonize dense grasslands, where, among other things, predator detection is more difficult.

Studies such as those at Wind Cave provide important insights into the dynamics of Plains grasslands and their native herbivores. However, they leave many questions unanswered. If bison facilitated the spread of prairie dog colonies, what prevented both species from simply overwhelming the entire Plains? One possibility is predation, but important predators such as the wolf and the black-footed ferret are gone, even from protected places such as Wind Cave.

Probably the major organizing force on the prehistoric Great Plains was a complex interaction among drought, fire, and grazing (Anderson 1982). Bison may have moved great distances when they were attracted to recently burned areas, when they were required to abandon other areas after the rains failed, or when prairie dogs locally depleted the grasses. The result would have been a large-scale spatial and temporal mosaic of grasslands in various stages of ecological succession. These patterns can no longer manifest themselves on the Great Plains in their present state of agricultural modification. It is difficult, if not impossible, to manage remnants of Great Plains ecosystems as they once existed. The challenge to conservation biologists is to use available areas to preserve or re-create something like the historical mosaic. Given the near ubiquity of domestic grazing animals over much of the Plains, what is rare today are grasslands long protected from the powerful influence of grazing by large mammals. That is the subject of the following section.

EFFECTS OF LIVESTOCK GRAZING

Of all the issues facing students and stewards of grasslands, none is more contentious than grazing by domestic livestock. Part of the problem derives from the economic and social issues involved, but another part is related to some genuine ambiguities about the actual ecological consequences of grazing. If bison once roamed the plains in the millions, have not livestock simply replaced a once-natural force that the surviving native flora and fauna tolerate or perhaps even require? We believe the answer to this question is a qualified no, for the following reasons.

Not All Grasses and Grasslands Respond Similarly to Grazing

Different grasses and grasslands have been affected differently by grazing. Mack and Thompson (1982) showed that perennial bunchgrasses of the Intermountain West were impacted severely by livestock, in part because the last 10,000 years of their evolution occurred in the virtual absence of bison. The same is true of certain arid grasslands in the American Southwest. By contrast, grasslands of the Great Plains evolved with millions of bison present, and these ecosystems have proved relatively tolerant of livestock grazing.

Milchunas et al. (1988) included precipitation as well as historic association with native ungulates to account for differences in responses of various grasses and grasslands to grazing. They found that relatively arid grasslands of the western Great Plains may not change substantially following livestock exclusion, because the same grass species that can survive grazing also are the species best able to survive droughts.

Perennial grass species vary widely in their ability to withstand grazing because of a variety of morphological, biochemical, and physiological traits. The best studied of these is growth form. Bunchgrasses are deep-rooted species whose lateral growth is restricted to formation of upright stems, called tillers, near the root crown. By contrast, sodgrasses spread laterally by the formation of horizontal stems that grow in the soil (rhizomes) or on the soil surface (stolons). As these species grow, they can cover the ground and enmesh the soil with a mat of vegetation called a sod. Sodgrasses, such as buffalo grass, and short-stature bunchgrasses, such as blue grama, generally are more tolerant of grazing than are taller bunchgrasses (Mack and Thompson 1982; Detling 1988), probably because being clipped off near ground level does them less anatomical damage. In many parts of the Great Plains with sufficient precipitation, livestock grazing determines whether the grassland will be dominated by short and/or sod-forming species, or whether taller bunchgrasses such as bluestems, wheatgrasses, and needle-grasses become dominant.

Livestock Have Permanently Degraded Some Grasslands

Certain grasslands, especially those in the southern Plains and Southwest, may have been so altered by historic overgrazing that they will not recover in a time span meaningful to humans. The Jornada Experimental Range in southern New Mexico is one likely example. In 1858, all but the eastern third of this site was open High Plains grassland (Buffington and Herbel 1965) dominated by black grama. Three woody species common on the eastern part of the Jornada were mesquite, creosotebush, and tarbush. By 1915, these woody plants were spreading into the grassland, and by 1963 these invasions were so extensive that none of the Jornada was classified as open grassland. Over 100 years, grass cover had declined from 90% to 25% of the range (Figure 10.7).

What caused these dramatic changes at the Jornada Experimental Range? Subtle climate changes may have played a role (Neilson 1986); however, Schlesinger et al. (1990) concluded that livestock grazing was the major factor causing this desertification. Heavy cover of black grama prevents soil erosion, so that water and nutrients are deposited uniformly near the soil surface, where they are available to the shallow-rooted grasses. However, once the grass cover was broken by grazing, horizontal erosion moved water and nutrients from most areas of the soil surface, while concen-

Figure 10.7. This former desert grassland in southern Arizona was converted by historical overgrazing into an essentially permanent desert shrubland, dominated by creosotebush.

trating them in others. Deep-rooted shrubs became established in these "islands of fertility," while intervening areas became virtually barren, being too depleted of water and nutrients to support a grassland, even if livestock subsequently were removed from the ecosystem.

Some Grasslands Increase in Cover and Stature After Livestock Removal

In many areas, especially those with more precipitation than the Jornada (annual mean = 21 centimeters), livestock removal causes significant changes in grassland cover and species composition. In such places, permanent livestock exclosure may be all that is necessary to re-create formerly abundant grasslands that have declined or disappeared because of the ubiquity of domestic grazers. The following are four of many possible examples.

In Alberta, rough fescue increased under livestock exclusion between 1949 and 1967, whereas it declined over the same period in the presence of livestock (Johnston et al. 1971). In a study in western Kansas, grazed lands were dominated by blue grama and buffalo grass, while taller bunchgrasses such as sideoats grama and little bluestem became abundant in livestock exclosures (Tomanek and Albertson 1957). On the Edwards Plateau of Texas, grazing favored short grasses such as hairy grama and curley mesquite grass, whereas taller bunchgrasses such as sideoats grama and Texas wintergrass were more common on land protected from livestock (Thurow et al. 1988). The Appleton–Whittell Research Sanctuary of the National Audubon Society is a grassland preserve in southeastern Arizona, from which livestock have been excluded since 1968. By the mid-1980s, perennial grass cover had increased on the sanctuary to the point where it was substantially higher than on most adjacent grazed pastures (Bock et al. 1984; Brady et al. 1989). While grazed sites were dominated by a variety of shorter grama grass species, other taxa such as plains lovegrass and wolftail grass were more common on the sanctuary (Figure 10.8).

Grazing Systems Do Not Eliminate Grazing Effects

Careful livestock management certainly can moderate the impacts of grazing, and it is not our purpose here to compare and criticize these management practices. We agree that herbivory is a powerful selective force, and that grasses surviving and reproducing for thousands or even millions of years in the presence of grazing mammals frequently have evolved means of withstanding their impacts. We also agree that grazing frequently stimulates grassland productivity, and that some grasses (especially shorter sodgrasses) can grow and reproduce better in the presence of livestock than in their absence (although perhaps in many cases only because livestock remove their taller competitors). We do not agree that grasslands somehow will deteriorate in the absence of livestock, and we are wary of assertions that the problems associated with grazing are purely a consequence of improper livestock management.

Short-duration grazing is a controversial livestock management scheme, purported to both increase livestock production and improve the quality of a grassland ecosystem over what it would be even in the absence of grazing (Savory 1988). It involves subjecting arid grasslands to short periods of very intense, high-density grazing, followed by much longer periods of rest. Benefits to the grassland are supposed

Figure 10.8. Fenceline along one border of the Appleton–Whittell Research Sanctuary of the National Audubon Society in southeastern Arizona. Grasslands on the right have been ungrazed for 20 years, while those on the left continue to be grazed by cattle.

to derive from (1) more even (nonselective) grazing of all species of edible plants, (2) removal of litter and standing dead vegetation that otherwise will choke grass plants, block sunlight, and reduce productivity, and (3) increased water infiltration and litter incorporation into the soil through the physical impact of livestock hooves. These are clear scientific hypotheses, and the range management community has been busy testing them. The results of studies published to date generally do not support Savory's assertions. Here are some examples.

In fescue grasslands of Alberta, Dormaar et al. (1989) found that areas subjected to short-duration grazing lost grass cover and had lower soil moisture content than ungrazed sites. In a study without livestock exclosures, Hart et al. (1988) found no vegetation or livestock production differences between continuously grazed and short-duration-grazed lands near Cheyenne, Wyoming. Weltz and Wood (1986) compared soil water infiltration rates on New Mexico blue grama grasslands subjected to continuous grazing, short-duration grazing, and livestock exclusion. They found that water-infiltration rates were highest in the absence of livestock, and that "the low infiltration rates after the short-duration grazing period are alarming" (p. 368). Studies at the Texas Experimental Ranch showed that short-duration grazing at high livestock densities decreased soil water infiltration, increased soil erosion, and benefited neither grass production nor livestock foraging efficiency (Heitschmidt et al. 1987; Walker et al. 1989).

Evidence such as this casts extreme doubt on the supposed benefits of short-duration grazing. We agree with Holechek (1991) that overall stocking density, and not the grazing system, is the major determinant of the impact of livestock on grasslands. Nevertheless, there is a large body of mostly unpublished testimony about marked improvements on various ranches following adoption of the Savory method. Resolution of this seeming paradox may lie in understanding that short-duration grazing is only a part of what Savory (1988) describes as "holistic resource management." When one cuts to the heart of holistic resource management, it is essentially that livestock growers must pay *very* close attention to the condition of the lands they are grazing and must handle their herds accordingly. We suspect that the benefits of the holistic method are less in the grazing system than in a heightened environmental awareness on the part of some ranchers. We have no problem with this sort of "holism," as long as it does not translate to a view that grazing is universally good for grasslands.

Native Grassland Animals Respond Variably to Grazing

It is a fallacy that livestock grazing always benefits wildlife. Such claims usually center on the observation that grazing increases grassland productivity by clearing away accumulated dead organic matter. However, many native grasslands animals are highly dependent on such litter and on taller ungrazed grasses for cover. Far from creating habitat for these species, grazing excludes them from the ecosystem. We conclude this section with a few examples of the response of wildlife to livestock grazing.

Mule deer in southern Arizona preferred ungrazed to grazed pastures in semiarid grassland/shrubland, although invasions of woody plants into these grasslands may have benefited deer historically (Ragotzkie and Bailey 1991). Short-duration grazing in southern Texas caused white-tailed deer to move more often to avoid cattle than they did in areas with lower-density continuous livestock grazing; short-duration grazing also reduced some forage species important to the deer (Cohen et al. 1989).

Sometimes whole species distributions can be altered by livestock activity. For example, the bunchgrass lizard is a grassland specialist found north of Mexico largely in meadows high in isolated southwestern mountain ranges. However, in 1989 we found that it was the most abundant reptile in ungrazed semiarid grasslands on the Sonoita Plain in southeastern Arizona (Bock et al. 1990). This discovery led us to conclude that the apparent virtual restriction of bunchgrass lizards to montane meadows was an historical artifact of livestock grazing at lower elevations.

Certain birds prefer the heavy cover of ungrazed grasslands, while other species are characteristic of more open habitats created by livestock or other disturbances. For example, Kantrud (1981) found that heavily grazed North Dakota grasslands supported the highest densities of killdeer, horned larks, and chestnut-collared longspurs, whereas bobolink and savannah sparrows were most abundant in ungrazed areas. Ungrazed grasslands in southeastern Arizona were dominated by Montezuma quail (Figure 10.9), Cassin's sparrows, Botteri's sparrows, and grasshopper sparrows, while nearby grazed areas supported mostly scaled quail, horned larks, and lark sparrows (Bock and Bock 1988). Habitat structure, as affected by such variables as grazing by large mammals, also plays a critical role in determining distribution patterns of Great Plains rodents (e.g., Moulton et al. 1981; Grant et al. 1982; Kaufman and Kaufman 1990).

Probably every type of North American grassland includes a fauna of grazing-

Figure 10.9. A female Montezuma quail fitted with a radio transmitter and antenna. This south-western grassland and savanna bird requires heavy cover as a refuge from predators. It does not thrive on heavily grazed land.

intolerant wildlife, and another fauna dependent on grazing. Given the near ubiquity of domestic grazers across most of North America, the grazing-intolerant species have comparatively few places left to live. The challenge is to preserve such places, and to create more of them.

EXOTIC VEGETATION

In the 1820s at the start of western exploration, the High Plains were covered with a tight sod of grasses with interlocking roots. Interspersed in this sod were colorful wildflowers and occasional small shrubs. Only in the riparian habitats did this vege-tation change to a predominantly woody flora. Successful colonization of a continuous grassland sod is difficult for new species because there is little open space for the establishment of new seedlings.

Sources of Sod-Breaking

Large-scale colonization by exotic plants has occurred in just over 150 years through purposeful and accidental human activities. Crops are one important source of exotics, but before crops could be planted, the sod had to be broken. The most important sod-breaking came with the introduction of the metal plow. Before the plow, farming by

Native Americans and colonists usually was restricted to the more easily tilled bottomlands along the waterways.

Not all sod disturbances are caused by people. Native fauna also cause openings in the sod. For example, prairie dogs build mounds and eat away vegetation cover (Whicker and Detling 1988), and remnant buffalo wallows are reminders of past bison dust baths (Uno 1989). Grassland sod destruction today by herbivores comes primarily from intensive grazing by fenced livestock. This disrupts the grassland flora over great expanses of the High Plains (Ham and Higham 1987). Fire, on the contrary, seldom destroys prairie sod because wildfires rarely kill the plants. Rather, they burn back the vegetation to the soil surface, leaving the ground-level shoot meristems and the underground roots intact (Bock et al. 1986). Fire also may encourage germination of short-lived postfire succession species that live on the burned area until the grass canopy is closed once more.

Exotic Plants Have Been Introduced for Crops, Fodder, and Range Restoration

Today, between 30 and 60% of the plant species on the Great Plains are not native to the region (Santanachote 1992). Disturbed sod invites colonization by non-native plants. Such plants can be divided into two categories: (1) those that were purposefully introduced, and (2) those, usually called weeds, that came unbidden into the disturbed lands. The best-known exotics are the cultivated crop plants of the High Plains, especially wheat and corn. Their cultivation of necessity destroys the prairie sod.

A second source of purposeful introduction of non-native species is sometimes referred to as range restoration. The goals here are to stop soil erosion, reestablish fodder, and restore vegetation cover in areas that have been plowed or overgrazed and no longer support profitable livestock husbandry. For example, the exotic annual cheatgrass (*Bromus tectorum*) has proved to be an exceptionally successful colonizer and competitor in much of the Great Basin and Great Plains (Mack 1981). The ecological genetics of this species have been well studied (Rice and Mack 1991). Two perennial grass genera, *Agropyron* and *Eragrostis,* also include examples of the problems created by exotic grasses in grassland restoration.

Several species of wheatgrass have been introduced by public-land managers and private individuals to restore overgrazed or plowed habitat, provide forage for grazing stock, and stabilize roadsides and building sites. The introduced species—for example, crested wheatgrass, tall wheatgrass, *Agropyron intermedium,* and quackgrass—all are Eurasian in origin (McGregor and Barkley 1986). Crested wheatgrass is especially widespread throughout the northern half of the High Plains. It is highly valued as a cattle fodder and soil binder. The genus *Agropyron* produces interspecific and intergeneric hybrids with both native and exotic species. Western wheatgrass (*Agropyron smithii*), one of the most important native wheatgrasses, has been shown to be of hybrid origin (Dewey 1975). Both native and exotic *Agropyron,* once planted, are prone to colonize nearby disturbed places including railroad rights of way, roadsides, and waste ground. *Agropyron* species provide a relatively high-grade forage for domestic livestock, but they rarely are superseded by mixed native vegetation once they are introduced, especially if disturbance is ongoing.

In the southern High Plains, various South African lovegrasses (*Eragrostis* spp.)

have been introduced, primarily to restore overgrazed grassland. Two of these exotics are Lehmann's lovegrass and Boer's lovegrass. They were brought into the southern High Plains several decades ago (Cox et al. 1990), and they now occupy vast expanses once covered with native grasses, including the native plains lovegrass (*Eragrostis intermedia*). Below elevations of approximately 1300 meters, the African species spread readily into any openings in adjacent lands. Above this elevation, plants grow readily, but tend not to spread to native grasslands. African lovegrasses, once estab lished, grow in nearly pure stands and support a significantly lower variety and abundance of native plants and animals than do adjacent sites covered with native plants (Bock et al. 1986) (Figure 10.10). The exotic lovegrasses persist through natural wildfires (Bock and Bock 1992). Although the South African lovegrasses have excellent soil-holding capacity, they are no longer favored either by land managers interested in wildlife and plant diversity or by ranchers seeking cattle feed, because these lovegrasses provide very low-quality forage for native or domestic grazers. The exotic lovegrasses are likely to persist into the future, even if there are no more purposeful introductions. In the future, another source of purposeful introduction of exotic plants may come from genetically engineered organisms. These organisms present the same set of liabilities as other exotic species (Keeler 1988).

"Typical" and Atypical Weeds Thrive on the High Plains

Exotic plants also can enter the High Plains unbidden. Two examples will illustrate: Russian thistle and Canada thistle. Their success in foreign habitats is related to their reproductive patterns.

Russian thistle is considered a nuisance throughout most of the western Great Plains. It is a tumbleweed; that is, this annual plant dies after setting seed, breaks off at ground level, and scatters its seeds as it is blown across the countryside. At the next growing season, if the seeds land in a suitable disturbed site, they produce new plants that complete their life cycle in one growing season, die, and tumble off to sow the seeds of the next generation.

Russian thistle was introduced into the United States by accident when its seeds were present in flax seed imported from Russia and sown in South Dakota in 1873 (Dewey 1895). It has spread throughout the western two-thirds of the United States and has invaded Mexico and Canada as well. It grows abundantly throughout the Great Plains in places where vegetation cover is sparse and soil is infertile. It does well on roadsides and overgrazed rangeland. Once disturbance ceases in an area infested with Russian thistle, the plants can be replaced by more long-lived native plants through plant succession. However, this replacement process can be slow. It took eight years after cattle were removed from land in southeastern Arizona before Russian thistle disappeared from a mesa top (J. Bock, personal observation). Entrepeneurs from one southwestern state seriously suggested treating Russian thistle as a crop plant for an inexpensive source of fuel. However, not figured into the costs was the expense of ongoing abuse of the land necessary to maintain the crop.

A second unbidden weed is Canada thistle, native to Eurasia, not Canada. It is distributed in moister habitats throughout the northern half of the United States and adjacent Canada.

Baker (1974) has identified the traits usually associated with weedy species. Can-

Figure 10.10. (a) Monocultures of exotic African lovegrasses support a greatly reduced variety and abundance of native plants and animals, compared with (b) the diverse grasslands dominated by native species.

ada thistle breaks these rules in two ways: It is perennial, and it has separate male and female plants. Most weeds are annual and hermaphroditic. Canada thistle has unusually efficient means of both sexual (seed) and vegetative reproduction. Each year, the female plants produce thousands of seeds with varying genotypes, the result of enforced outbreeding that separate sexes require. It is highly resistant to most herbicides; in part, this is due to the production of many genetic combinations in each seed crop (LeBaron and Gressel 1982). Among these are likely to be some genetic combinations that are resistant to any given pesticide. Once the plant has become established in an appropriate habitat, it reproduces vegetatively by frequent budding on brittle underground roots. Each bud can give rise to a new plant. Nadeau and Vanden Born (1989) found, "On average, an 18-wk-old plant had the potential of producing 930 shoots if its root system was cut into 10-cm long pieces" (p. 1199). This species readily invades the smallest openings in the grassland cover. Even moderate grazing by livestock invites the entry of Canada thistle.

Before the introduction of exotic weeds, native short-lived species filled the role of colonizers, and plant succession consisted of one group of native plants replacing another. The exotic species, because of their differing evolutionary histories, often do not fit into the succession patterns in the same way as their native predecessors. Enormous sums of money continue to be spent on weed control and eradication in the Great Plains. However, the weeds have met these challenges with their genetic flexibility and their morphological and physiological endowments. Many of these exotic species appear to be here to stay in spite of weed-control efforts.

The established exotic species have changed the face of the Great Plains forever, whether these introductions were purposeful or accidental. For those who value conservation of the native plants and animals of the Great Plains, vigilance to prevent further degradation of the native prairie is essential.

THREE PROPOSALS FOR CREATING MORE UNDISTURBED GRASSLANDS

There is no chance that the North American Great Plains can be returned as a whole to anything resembling their prehistoric extent or condition. Certain of the megafauna (e.g., bison and wolves) and some ecological processes (e.g., large-scale fires) can never exist and function as they once did. However, because Great Plains ecosystems were always dynamic, both spatially and temporally, endemic flora and fauna evolved to varying degrees to deal with a heterogeneous environment. Therefore, a mosaic of grasslands subject to various levels of perturbation by such factors as grazing, burning, and even plowing should provide opportunities for a significant proportion of native plants and animals to survive and even thrive. What is missing or rare in the present landscape, among the things that can realistically be created, are habitat blocks of sufficient size and number that can remain or be restored to a relatively undisturbed state—permanently free of livestock, cultivation, and exotic vegetation. Without such a system, those surviving plants and animals most vulnerable to the effects of habitat disturbance will remain rare, and vulnerable to extinction.

We have three specific proposals for ways of restoring and maintaining more relatively undisturbed grassland on the Great Plains.

1. *Continue a modified Conservation Reserve Program.* In 1985, Congress authorized the Conservation Reserve Program (CRP), which pays farmers an annual rental fee per acre to plant erosion-prone cropland into grass and leave it undisturbed for 10 years (Joyce et al. 1991). The primary objective of the program, which through 1990 enrolled more than 13 million hectares (34 million acres) across the United States, was to reduce soil erosion. A secondary objective was to improve habitat for wildlife (Hays et al. 1989). For example, data from the northern Great Plains indicate that a variety of grassland birds generally declining in the region were much more abundant on CRP grasslands than on adjacent tilled cropland (Johnson et al. 1991). Species responding positively to CRP lands included bobolink, Baird's sparrow, grasshopper sparrow, clay-colored sparrow, dickcissel, and lark bunting.

The CRP program has powerful implications for grassland conservation on the Great Plains, because about two-thirds of present CRP lands are in this region. The program should be extended in some form, but it also should probably be changed. One of the problems with the CRP is that retired croplands have been seeded with exotic grasses about 2.5 times more often than they have with native grasses (Joyce et al. 1991). While seeding with exotic grasses conserves soil and some wildlife, it clearly does little for the indigenous Plains flora. Use of non-native grasses has been dictated by cost and availability of seed, but any plans for continuation or future expansion of the CRP should include incentives for using native seed. There also is a need for more research on ecological succession on CRP lands, including the status of native plant species (Goetz 1989).

A major problem with the CRP is vulnerability of the land to recultivation for row crops (Heimlich and Kula 1989; Joyce et al. 1991). These decisions will be in the hands of the landowners, as they should be. However, it would be unfortunate if CRP lands were tilled (setting the successional clock back to zero), only to be returned to grassland when crop prices or future government programs made it economically attractive to the farmers. A better strategy might be to find ways to encourage landowners to graze their relatively durable CRP grasslands, thereby retaining at least part of their conservation value, as well as reducing the need for government subsidy. We would find such a plan particularly attractive if it could be coupled with establishment of more ungrazed public grassland.

2. *Dedicate a portion of each federal grazing lease as a livestock exclosure.* There are at least 86 million hectares (212 million acres) of public federal land being grazed by domestic livestock in 17 western states (Sabadell 1982). Most are managed by the U.S. Bureau of Land Management and the U.S. Forest Service. We urge establishment of a system of federal livestock exclosures, whereby 20% of each parcel of land presently leased to a livestock grower would be set aside as a permanently ungrazed reserve. Such a system would serve two purposes.

1. Each would function as an ecological benchmark, against which the actual consequences of livestock grazing on that particular lease or allotment could be measured objectively and permanently.
2. The exclosure system would create more than 15 million hectares of previously unavailable habitat for those plants and animals that are intolerant of the activities of large, hooved mammals.

Opposition to the exclosures will take at least four specific forms. First, some ranchers and range managers point to evidence that livestock either do not affect or are beneficial to rangelands and their wildlife. We considered these issues in this chapter, and simply repeat here our main point: Some plants and animals respond positively to grazing, whereas others clearly do not. A livestock-exclosure system imposes on the landscape a mosaic of habitats meeting the needs of both groups of species.

The second difficulty with the establishment of the exclosure system is the assertion that livestock grazing is important economically. Estimates of the proportion of U.S. beef produced on western public lands range from 3 to 10% (e.g., Martin 1981; Field 1990). A 20% loss in that amount (less than 2%) will not cause any American to be deprived of red meat, nor will it require us to import more of the product from foreign markets. Conversion of marginal croplands to grazing lands in more productive parts of the Great Plains would more than compensate for this loss (see our first proposal). Nevertheless, it is true that some western livestock growers, and their local economies, would be affected by a 20% loss in income. We have two suggestions for ways to reduce these costs:

1. For some relatively productive rangelands, it should be possible to increase stocking levels on the remaining 80% of the allotment without causing unacceptable ecological damage. It is those plants and animals intolerant of anything but economically unfeasible levels of grazing that are in the greatest danger. We believe a slight increase in grazing elsewhere is an acceptable trade-off for creation of habitat for these species.
2. In arid southwestern regions, many grazing allotments may be hanging so close to the ecological brink that any increase in grazing would permanently degrade the grassland. In these cases, we suggest substantial reductions in grazing fees, in lieu of increased stocking levels, as a means of making exclosure more palatable. This may become a particularly attractive approach when and if federal grazing fees increase.

The third objection is that other, more costly, range-restoration methods work better than livestock exclusion. This may be true in some cases. However, most range-improvement efforts, like grazing itself, have been experiments without permanently ungrazed controls. Fertilizing, bulldozing, root plowing, chaining, mowing, shredding, prescribed burning, contour furrowing, applying pesticides, reseeding with native or exotic grasses, and implementing various grazing systems may well cause grasslands to change. However, these activities should never take place on the exclosures, one of whose precise purposes would be to monitor grassland condition in the absence of such intrusive manipulations.

Finally, some conservation groups are calling for removal of all livestock from public rangelands. This is understandable, given the obvious destruction of many grasslands by livestock historically, especially in the Great Basin and the Southwest. Nevertheless, it is impractical, insensitive, and probably unnecessary to so completely disregard the human element, especially on relatively grazing-tolerant grasslands of the Great Plains. As we have argued throughout this essay, the problem is not so much with the existence of livestock as it is with their virtual ubiquity.

3. *Permanently remove all livestock from the U.S. national grasslands.* We strongly advocate adoption of the livestock-exclosure system just outlined, as a means of creating new and much-needed ungrazed grassland in the American West. However, it is a fact that relatively little of the Great Plains themselves are federal land. One exception is the national grasslands, presently managed by the U.S. Forest Service largely for purposes of livestock production, within the general framework of multiple use typical of the national forests (Lewis 1989; West 1990).

The national grasslands include 19 administrative units, 17 of which are on the western Great Plains, from Montana to Texas (Peek and Risser 1979; West 1990). Together, they comprise more than 1.5 million hectares (3.8 million acres), and many include grassland types little represented on any other public lands (Lewis 1989). Some individual grassland blocks are large enough to sustain populations of even the most wide-ranging native Plains animals, such as bison and pronghorn, while most are dispersed among private range and croplands as a series of smaller landscape units.

We are naive neither to the difficulties involved in designating Forest Service land as biological preserves nor to the strength of the opposition to such a change. However, we also are aware of the declining agricultural value of these lands, and of their likely increase in value as natural landscape (e.g., Popper and Popper 1991). Many ecosystems represented among the national grasslands are comparatively resistant to the impacts of grazing, and probably could be managed successfully for certain components of the native flora and fauna under conditions of moderate livestock use. However, grazed grasslands are anything but scarce across the central United States, and they will be even more abundant if CRP lands are converted to livestock production. We therefore call for absolute protection of the national grasslands from livestock, as well as from exotic vegetation, as an essential part of a long-term strategy to re-create the extraordinary habitat mosaic that once composed the North American Great Plains.

Acknowledgments

We thank Yan Linhart, Kathleen Keeler, and Anthony Joern for their constructive comments on an earlier version of this essay. The following agencies and organizations have supported our work in North American grasslands over the past 20 years: the U.S. Forest Service, Bureau of Land Management, National Park Service, National Biological Service, and National Science Foundation; the National Geographic Society; the Charles Lindbergh Foundation; the National Audubon Society; and the University of Colorado.

LITERATURE CITED

Anderson, R. C. 1982. An evolutionary model summarizing the roles of fire, climate, and grazing animals in the origin and maintenance of grasslands: an end paper. Pp. 297–308 in J. R. Estes, R. J. Tyrl, and J. N. Brunken (eds.). *Grasses and Grasslands: Systematics and Ecology.* University of Oklahoma Press, Norman.

Archer, S. 1989. Have southern Texas savannas been converted to woodlands in recent history? American Naturalist 134:545–561.

Axelrod, D. I. 1985. Rise of the grassland biome, central North America. Botanical Review 51:163–201.

Bahre, C. J. 1991. *A Legacy of Change.* University of Arizona Press, Tucson.

Baker, H. G. 1974. The evolution of weeds. Annual Review Ecology and Systematics 5:1–24.

Bock, C. E., and J. H. Bock. 1987. Avian habitat occupancy following fire in a Montana shrub-steppe. Prairie Naturalist 19:153–158.

Bock, C. E., and J. H. Bock. 1988. Grassland birds in southeastern Arizona: impacts of fire, grazing, and alien vegetation. Pp. 43–58 in P. Goriup (ed.). *Ecology and Conservation of Grassland Birds*. Technical Publication No. 7. International Council for Bird Preservation, Cambridge, England.

Bock, C. E., J. H. Bock, K. L. Jepson, and J. C. Ortega. 1986. The ecological effects of planting African lovegrasses in Arizona. National Geographic Research 2:456–463.

Bock, C. E., J. H. Bock, W. R. Kenney, and V. M. Hawthorne. 1984. Responses of birds, rodents, and vegetation to livestock exclosure in a semidesert grassland site. Journal Range Management 37:239–242.

Bock, C. E., H. M. Smith, and J. H. Bock. 1990. The effect of cattle grazing upon abundance of the lizard *Sceloporus scalaris* in southeastern Arizona. Journal Herpetology 24:445–446.

Bock, J. H., and C. E. Bock. 1984. Effects of fires on woody vegetation in the pine–grassland ecotone of the southern Black Hills. American Midland Naturalist 112:35–42.

Bock, J. H., and C. E. Bock. 1989. Ecology and evolution in the Great Plains. Pp. 551–577 in J. H. Bock and Y. B. Linhart (eds.). *The Evolutionary Ecology of Plants*. Westview Press, Boulder, Colo.

Bock, J. H., and C. E. Bock. 1992. Vegetation responses to wildfire in native vs. exotic Arizona grassland. Journal Vegetation Science 3:439–446.

Bock, J. H., W. D. Bowman, and C. E. Bock. 1991. Global change in the high plains of North America. Great Plains Research 1:283–301.

Brady, W. W., M. R. Stromberg, E. F. Aldon, C. D. Bonham, and S. H. Henry. 1989. Response of a semidesert grassland to 16 years of rest from grazing. Journal Range Management 42:284–288.

Buffington, L. C., and C. H. Herbel. 1965. Vegetation changes on semidesert grassland range from 1858 to 1963. Ecological Monographs 35:139–164.

Burke, I. C., T.G.F. Kittel, W. K. Lauenroth, P. Snook, C. M. Yonker, and W. J. Parton. 1991. Regional analysis of the central Great Plains. Bioscience 41:685–692.

Cohen, W. E., D. L. Drawe, F. C. Bryant, and L. C. Bradley. 1989. Observations on white-tailed deer and habitat response to livestock grazing in south Texas. Journal Range Management 42:361–365.

Coppock, D. L., J. E. Ellis, J. K. Detling, and M. I. Dyer. 1983. Plant–herbivore interactions in a North American mixed-grass prairie. Oecologia 56:10–15.

Cox, J. R., G. B. Ruyle, and R. B. Roundy. 1990. Lehmann lovegrass in southeastern Arizona: biomass production and disappearance. Journal Range Management 43:367–372.

Daubenmire, R. 1978. *Plant Geography*. Academic Press, New York.

Detling, J. K. 1988. Grasslands and savannas: regulation of energy flow and nutrient cycling by herbivores. Pp. 131–148 in L. R. Pomeroy and

J. J. Alberts (eds.). *Concepts of Ecosystem Ecology: A Comparative View*. Springer-Verlag, New York.

Dewey, D. R. 1975. The origin of *Agropyron smithii*. American Journal Botany 62:524–530.

Dewey, L. H. 1895. *The Russian Thistle*. Circular No. 3, rev. ed. U.S. Department of Agriculture, Division of Botany. Government Printing Office, Washington, D.C.

Dormaar, J. F., S. Smoliak, and W. D. Willms. 1989. Vegetation and soil responses to short-duration grazing on fescue grasslands. Journal Range Management 42:252–256.

Field, T. G. 1990. The future of grazing on public lands. Rangelands 12:217–219.

Frémont, J. C. 1886. *Memoirs of My Life*, vol. 1. Belford Clarke, Chicago.

Gibson, D. J., and L. C. Hulbert. 1987. Effects of fire, topography and year-to-year climate variation on species composition in tallgrass prairie. Vegetatio 72:175–185.

Goetz, H. 1989. The Conservation Reserve Program—where are we heading? Rangelands 11:251–252.

Grant, W. E., E. C. Birney, N. R. French, and D. M. Swift. 1982. Structure and productivity of grassland small mammal communities related to grazing-induced changes in vegetative cover. Journal Mammalogy 63:248–260.

Ham, G. E., and R. Higham (eds.). 1987. *The Rise of the Wheat State: A History of Kansas Agriculture*. Sunflower University Press, Manhattan, Kans.

Hart, R. H., M. J. Samuel, P. S. Test, and M. A. Smith. 1988. Cattle, vegetation, and economic response to grazing systems and grazing pressure. Journal Range Management 41:282–286.

Hays, R. L., R. P. Webb, and A. H. Farmer. 1989. Effects of the Conservation Reserve Program on wildlife habitat: results of 1988 monitoring. Transactions North American Wildlife and Natural Resources Conference 54:365–376.

Heimlich, R. E., and O. E. Kula. 1989. Grazing lands: how much CRP land will remain in grass? Rangelands 11:253–257.

Heitschmidt, R. K., S. L. Dowhower, and J. W. Walker. 1987. 14- vs. 42-paddock rotational grazing: aboveground biomass dynamics, forage production, and harvest efficiency. Journal Range Management 40:216–223.

Holechek, J. L. 1991. Chihuahuan desert rangeland, livestock grazing, and sustainability. Rangelands 13:115–121.

Hoogland, J. L. 1981. The evolution of coloniality in white-tailed and black-tailed prairie dogs (Sciuridae: *Cynomys leucurus* and *C. ludovicianus*). Ecology 62:252–272.

Humphrey, R. R. 1987. *90 Years and 535 Miles: Vegetation Changes Along the Mexican Border*. University of New Mexico Press, Albuquerque.

Johnson, D. H., M. D. Schwartz, and K. L. Richardson. 1991. Bird use of CRP lands in the prairie pothole region. Unpublished manuscript.

U.S. Fish and Wildlife Service, Jamestown, N.D.

Johnston, A., J. F. Dormaar, and S. Smoliak. 1971. Long-term grazing effects on fescue grassland soils. Journal Range Management 24:185–188.

Joyce, L. A., J. E. Mitchell, and M. D. Skold (eds.). 1991. *Proceedings, The Conservation Reserve—Yesterday, Today, and Tomorrow.* General Technical Report RM-203. U.S. Department of Agriculture, Forest Service.

Kantrud, H. A. 1981. Grazing intensity effects on the breeding avifauna of North Dakota native grasslands. Canadian Field Naturalist 95:404–417.

Kaufman, D. W., and G. A. Kaufman. 1990. Influence of plant litter on patch use by foraging *Peromyscus maniculatus* and *Reithrodontomys megalotis*. American Midland Naturalist 124:195–198.

Keeler, K. H. 1988. Can we guarantee the safety of genetically engineered organisms in the environment? Critical Reviews Biotechnology 8:85–97.

Krueger, K. 1986. Feeding relationships among bison, pronghorn, and prairie dogs: an experimental analysis. Ecology 67:760–770.

LeBaron, H. M., and J. Gressel (eds.). 1982. *Herbicide Resistance in Plants.* Wiley, New York.

Lewis, M. E. 1989. National Grasslands in the Dust Bowl. Geographical Review 79:161–171.

Lewis, M., and W. Clark. 1893. *History of the Expedition Under the Command of Lewis and Clark.* Ed. E. Coues. Harper, New York.

Mack, R. N. 1981. The invasion of *Bromus tectorum* L. into western North America: an ecological chronicle. Agro-Ecosystems 7:145–165.

Mack, R. N., and J. N. Thompson. 1982. Evolution in steppe with few large, hooved mammals. American Naturalist 119:757–773.

Martin, P. S., and J. E. Guilday. 1967. A bestiary for Pleistocene biologists. Pp. 1–62 in P. S. Martin and H. E. Wright, Jr. (eds.). *Pleistocene Extinctions: The Search for a Cause.* Yale University Press, New Haven, Conn.

Martin, P. S., and R. G. Klein (eds.). 1984. *Quaternary Extinctions: A Prehistoric Revolution.* University of Arizona Press, Tucson.

Martin, W. E. 1981. The distribution of benefits and costs associated with public rangelands. Pp. 229–256 in G. M. Johnston and P. M. Emerson (eds.). *Public Lands and the U.S. Economy.* Westview Press, Boulder, Colo.

McGregor, R. L., and T. M. Barkley (eds.). 1986. *Flora of the Great Plains.* University Press of Kansas, Lawrence.

McHugh, T. 1972. *The Time of the Buffalo.* Knopf, New York.

Milchunas, D. G., O. E. Sala, and W. K. Lauenroth. 1988. A generalized model of the effects of grazing by large herbivores on grassland community structure. American Naturalist 132:87–106.

Moulton, M. P., J. R. Choate, S. J. Bissell, and R. A. Nicholson. 1981. Associations of small mammals on the central plains of eastern Colorado. Southwestern Naturalist 26:53–57.

Nadeau, L. B., and W. H. Vanden Born. 1989. The root system of Canada thistle. Canadian Journal Plant Science 69:1199–1206.

Neilson, R. P. 1986. High-resolution climatic analysis and Southwest biogeography. Science 232:27–34.

Peden, D. G. 1976. Botanical composition of bison diets on shortgrass plains. American Midland Naturalist 96:225–229.

Peek, J. M., and P. G. Risser. 1979. Wildlife and range management on the National Grasslands. Unpublished report. National Audubon Society, New York.

Popp, J. K. 1981. Range ecology of bison on mixed grass prairie at Wind Cave National Park. Ph.D. Dissertation. Iowa State University, Ames.

Popper, F. J., and D. E. Popper. 1991. The reinvention of the American frontier. Amicus Journal 13:4–8.

Ragotzkie, K. E., and J. A. Bailey. 1991. Desert mule deer use of grazed and ungrazed habitats. Journal Range Management 44:487–490.

Rice, K. J., and R. N. Mack. 1991. Ecological genetics of *Bromus tectorum.* Oecologia 88:77–83.

Sabadell, J. E. 1982. *Desertification in the United States.* U.S. Department of the Interior, Bureau of Land Management, Washington, D.C.

Santanachote, K. 1992. Vegetation cover, seed bank, and seed rain of the relictual Tallgrass Prairie of Boulder County, Colorado. Ph.D. Dissertation, Department of Environmental, Population, and Organismic Biology, University of Colorado, Boulder.

Sauer, C. O. 1950. Grassland, climax, fire, and man. Journal Range Management 3:16–22.

Savory, A. 1988. *Holistic Resource Management.* Island Press, Washington, D.C.

Schlesinger, W. H., J. F. Reynolds, G. L. Cunningham, L. F. Huenneke, W. M. Jarrell, R. A. Virginia, and W. G. Whitford. 1990. Biological feedbacks in global desertification. Science 247:1043–1048.

Smith, J. B., and D. A. Tirpak (eds.). 1988. *The Potential Effects of Global Climate Change on the United States.* Vol. 1: *Regional Studies.* Government Printing Office, Washington, D.C.

Steinauer, E. M., and T. B. Bragg. 1987. Ponderosa pine (*Pinus ponderosa*) invasion of Nebraska Sandhills Prairie. American Midland Naturalist 118:358–365.

Thurow, T. L., W. H. Blackburn, and C. A. Taylor, Jr. 1988. Some vegetation responses to selected livestock grazing strategies, Edwards Plateau, Texas. Journal Range Management 41:108–114.

Tomanek, G. W., and F. W. Albertson. 1957. Variations in cover, production, and roots of vegetation on two prairies in western Kansas. Ecological Monographs 27:267–281.

Uno, G. E. 1989. Dynamics of plants in buffalo wallows. Pp. 431–444 in J. H. Bock and Y. B. Linhart (eds.). *The Evolutionary Ecology of Plants.* Westview Press, Boulder, Colo.

Veblen, T. T., and D. C. Lorenz. 1991. *The Colo-

rado Front Range: A Century of Ecological Change. University of Utah Press, Salt Lake City.

Walker, J. W., R. K. Heitschmidt, and S. L. Dowhower. 1989. Some effects of a rotational grazing treatment on cattle preference for plant communities. Journal Range Management 42:143–148.

Wells, P. V. 1965. Scarp woodlands, transported soils, and concept of grassland climate in the Great Plains region. Science 148:246–249.

Wells, P. V., and J. D. Stewart. 1987. Cordilleran-boreal tiaga and fauna on the central Great Plains of North America, 14,000–18,000 years ago. American Midland Naturalist 118:94–106.

Weltz, M., and M. K. Wood. 1986. Short duration grazing in central New Mexico: effects on infiltration rates. Journal Range Management 39:365–368.

West, T. 1990. USDA Forest Service management of the National Grasslands. Agricultural History 64:86–98.

Whicker, A. D., and J. K. Detling. 1988. Ecological consequences of prairie dog disturbances. Bioscience 38:778–785.

Worster, D. 1979. *The Dust Bowl: The Southern Plains in the 1930s*. Oxford University Press, Oxford.

Wright, H. A., and A. W. Bailey. 1982. *Fire Ecology: United States and Southern Canada*. Wiley, New York.

Scientific Names

Scientific names of plant and animal species mentioned in this book, alphabetized by common name, are listed. Families are also given for plants in accordance with convention. # indicates that the plant was introduced from outside the region.

Common Name	Latin Name	Family
Plants		
Ash	*Fraxinus*	Oleaceae
Aster	*Aster* spp.	Asteraceae
Aster, white	*Aster ericoides*	Asteraceae
Barley, little	*Hordeum pusillum*	Poaceae #
Bergamot, wild	*Monarda fistulosa*	Labiatae
Birch, paper	*Betula papyrifera*	Betulaceae
Bluegrass, Kentucky	*Poa praetensis*	Poaceae #
Bluestem, big	*Andropogon gerardii*	Poaceae
Bluestem, little	*Schizachyrum scoparium*	Poaceae #
Bluestem, sand	*Andropogon hallii*	Poaceae
Briar, cat's claw sensitive	*Schrankia nuttalii*	Fabaceae
Brome, downy	*Bromus tectorum*	Poaceae #
Brome, smooth	*Bromus inermis*	Poaceae #
Bromegrass, annual	*Bromus* sp.	Poaceae #
Buffalo grass	*Buchloë dactyloides*	Poaceae
Clover, sweet	*Melilotus* spp.	Fabaceae #
Cocklebur	*Xanthium strumarium*	Asteraceae
Compass plant	*Sylphium perfoliatum*	Asteraceae
Coneflower, prairie	*Ratibida columnifera*	Asteraceae
Coneflower, purple	*Echinacea purpurea*	Asteraceae
Cordgrass, prairie	*Spartina pectinata*	Poaceae
Corn. *See* Maize		
Cottonwood	*Populus deltoides*	Salicaceae
Creosotebush	*Larrea tridentata*	Zygophyllaceae
Dogwood, rough-leaved	*Cornus drummondii*	Cornaceae
Dropseed, sand	*Sporobolus cryptandrus*	Poaceae
Elm	*Ulmus* spp.	Ulmaceae
Evening primrose, Missouri	*Oenothera missouriensis*	Onagraceae
Fescue, rough	*Festuca scabrella*	Poaceae
Flax	*Linum* sp.	Linaceae
Four o'clock	*Mirabilis hirsuta*	
Gayfeather, tall	*Liatris aspera*	Asteraceae
Goldenrod, Canada	*Solidago canadensis*	Asteraceae
Grama, side-oats	*Bouteloua curtipendula*	Poaceae

Common Name	Latin Name	Family
Plants		
Grama grass, annual	*Bouteloua breviseta*	Poaceae
Grama grass, black	*Bouteloua eriopoda*	Poaceae
Grama grass, blue	*Bouteloua gracilis*	Poaceae
Grama grass, hairy	*Bouteloua hirsuta*	Poaceae
Grama grasses	*Bouteloua* spp.	Poaceae
Grapes	*Vitus* spp.	Vitaceae
Hackberry	*Celtis* spp.	Ulmaceae
Indian grass	*Sorghastrum nutans*	Poaceae
Ironweed, western	*Vernonia baldwinii*	Asteraceae
Lamb's quarters	*Chenopodium album*	Chenopodiaceae #
Leadplant	*Amorpha canescens*	Fabaceae
Lovegrass, plains	*Eragrostis intermedia*	Poaceae
Lovegrass, sand	*Eragrostis trichodes*	Poaceae
Maize	*Zea mays*	Poaceae
Maple	*Acer* spp.	Aceraceae
Maple, sugar	*Acer saccharum*	Aceraceae
Marjoram, wild		Lamiaceae
Mesquite	*Prosopis glandulosa*	Fabaceae
Mesquite, curley	*Hilaria belangeri*	Fabaceae
Morning-glory, bush	*Ipomoea leptophylla*	Convolvulaceae
Muhly, sand	*Muhlenbergia pungens*	Poaceae
Mulberry	*Morus* sp.	Moraceae
Needle-and-thread grass	*Stipa comata*	Poaceae
New Jersey tea	*Ceanothus* spp.	Rhamnacecae
Oak	*Quercus* spp.	Fagaceae
Oats, wild	*Avena* spp.	Poaceae #
Pennyroyal	*Hedeoma*	Lamiaceae
Pigweed	*Amaranthus retroflexus*	Amaranthaceae
Pine, jack	*Pinus banksiana*	Pinaceae
Pine, ponderosa	*Pinus ponderosa*	Pinaceae
Pine, slash	*Pinus palustris* (Florida)	Pinaceae
Plantain, English	*Plantago lanceolata*	Plantaginaceae
Plantain, sea	*Plantago* ?	Plantaginaceae
Porcupine grass	*Stipa spartea*	Poaceae
Prickl(e)y pear	*Opuntia* spp.	Cactaceae
Puccoon, Carolina	*Lithospermum caroliniense*	Boraginaceae
Quackgrass	*Agropyron repens*	Poaceae
Ragweed, giant	*Ambrosia trifida*	Asteraceae
Ragweed, little	*Ambrosia artemisifolia*	Asteraceae
Ragweed, western	*Ambrosia psilostachya*	Asteraceae
Red cedar, eastern	*Juniperus virginiana*	Cupressaceae
Richweed	*Pilea* sp.	Urticaceae
Rye grass	*Lolium* sp.	Poaceae
Sagebrush, big	*Artemisia tridentata*	Asteraceae
Sandreed, prairie	*Calamovilfa longifolia*	Poaceae
Sedge	*Carex* spp.	Cyperaceae
Sloughgrass	*Beckmannia syzigachne*	Poaceae
Spruce	*Picea* spp.	Pinaceae
Spurge, mat	*Euphorbia glyptosperma*	Euphorbiaceae
Stickleaf	*Mentzelia nuda*	Loasaceae
Stickseed	*Lappula reddowskii*	Boraginaceae
Sumac	*Rhus glabra*	Anacardiaceae

Common Name	Latin Name	Family
Plants		
Sunflower, annual	*Helianthus annuus*	Asteraceae
Sunflower, common	*Helianthus annuus*	Asteraceae
Sweetgrass	*Hierocholoe odorata*	Poaceae
Switchgrass	*Panicum virgatum*	Poaceae
Sycamore	*Platanus* spp.	Platanaceae
Tarbush	*Flourensia cernua*	Asteraceae
Thistle, Canada	*Cirsium arvense*	Asteraceae #
Thistle, Platte	*Cirsium plattensis*	Asteraceae
Thistle, Russian	*Salsola iberica*	Chenopodiaceae #
Three-awn grass	*Aristida oligantha*	Poaceae
Timothy	*Phleum pratense*	Poaceae #
Walnut	*Juglans niger*	Juglandaceae
Wheat	*Triticum aestivum*	Poaceae #
Wheatgrass, bluebunch	*Agropyron spicatum*	Poaceae
Wheatgrass, crested	*Agropyron cristatum*	Poaceae
Wheatgrass, western	*Agropyron smithii*	Poaceae
Wintergrass, Texas	*Stipa leucotricha*	Poaceae
Wolftail grass	*Lycurus phleoides*	Poaceae
Yarrow, western	*Achillea millifolium*	Asteraceae
Reptiles		
Bunchgrass lizard	*Scelporus scalaris*	
Northern prairie skink	*Eumeces septenrionalis*	
Ringnecked snake	*Diadophis punctatus*	
Western box turtle	*Terrapene ornata*	
Western rattlesnake	*Crotalus viridis*	
Birds		
Baird's sparrow	*Ammodramus bairdii*	
Bobolink	*Dolichonyx oryzivorus*	
Bobwhite quail	*Colinus virginianus*	
Botteri's sparrow	*Aimophila botterii*	
Cardinal	*Richmondena cardinalis*	
Cassin's sparrow	*Aimophila cassinii*	
Chestnut-collared longspur	*Calcarius ornatus*	
Dickcissel	*Spiza americana*	
Grasshopper sparrow	*Ammodramus savannarum*	
Horned lark	*Eremophila alpestris*	
Killdeer	*Charadrius vociferans*	
Lark bunting	*Calamospiza melanochorys*	
Lark sparrow	*Chondestes grammacus*	
LeConte's sparrow	*Ammospiza leconteii*	
Lesser prairie chicken	*Tympanuchus pallidiccinctus*	
Montezuma quail	*Cyrtonyx montezumae*	
Savanna sparrow	*Passerculus sandwichensis*	
Scaled quail	*Callipepla squamata*	
Sharp-tailed grouse	*Tympnuchus phasianellus*	
Upland sandpiper	*Bartramia longicauda*	
Western meadowlark	*Sturnella neglecta*	
Mammals		
Antelope. *See* pronghorn		
Badger	*Taxidea taxus*	
Bison	*Bison bison*	
Blackfooted ferret	*Mustela nigripes*	

Common Name	Latin Name	Family
Mammals		
Blacktailed prairie dog	*Cynomys ludovicianus*	
Buffalo. *See* bison		
Cattle	*Bos domesticus*	
Coyote	*Canis latrans*	
Deer mouse	*Peromyscus maniculatus*	
Elk	*Cervus canadensis*	
Grizzly bear	*Ursus americanus horridus*	
Horse	*Equus equus*	
Kangaroo rat	*Dipodomys* spp.	
Meadow jumping mouse	*Zapus hudsonius*	
Mule deer	*Odocoileus hemionus*	
Opossum	*Didelphis virginiana*	
Ord's kangaroo rat	*Dipodomys ordii*	
Packrat, also called woodrats	*Neotoma* spp.	
Pocket gopher	*Geomys* spp.	
Prairie vole	*Microtus ochrogaster*	
Pronghorn antelope	*Antilocapra americana*	
Short-tailed shrew	*Blarina brevicauda*	
Thirteen-lined ground squirrel	*Spermophilus tridecemlineatus*	
Western harvest mouse	*Reithrodontomys megalotis*	
White-tailed deer	*Odocoileus virginianus*	
Wolf	*Canus lupus*	

Glossary

abiotic Nonliving (never living) features, especially physical (climate) or chemical factors that influence ecological responses. Compare **biotic.**

allelopathy Negative effect of metabolic products of plants on the growth and development of other nearby plants.

ammonium (NH_4^+) Ionic form of ammonia (NH_3), such as ammonia dissolved in water, that can combine with negatively charged ions. It is important in the N-cycle.

angiosperms Flowering plants. These are the dominant plants of the world, characterized by having both a flower with anthers, stigma, and stamen for transporting the pollen containing the sperm to the ovary, and an ovary that develops into a hard-coated seed. Grasses, sedges, elms, maples, roses, tomatoes, cattails, and cottonwoods are all angiosperms. Pine trees, red cedar, and ferns are not.

autotrophic organism Organism that synthesizes large molecules from small inorganic molecules—that is, makes its own food. The vast majority of living autotrophs depend on photosynthesis to produce glucose from CO_2 and water as an autotrophic process, but other pathways exist, such as those seen in many chemosynthetic bacteria. Compare **heterotrophic organism.**

bacterium (plural, **bacteria**) Unicellular organism whose cells lack nuclei or protein-wrapped chromosomes; member of the kingdom Protista. Most bacteria are very small and called microorganisms.

biodiversity Biological diversity, a measure of the number of species in an area or region.

biomass Total weight (e.g., grams) of all living things in a given system.

biome Group of ecosystems characterized by similar vegetation and climate. The two are directly related.

biosphere That part of the earth and its atmosphere that can support life.

biota All the living things (plants, animals, microorganisms) of a region.

biotic Of living things. Compare **abiotic.**

blowout   In sandhills prairie, a break in the plant cover, resulting in an area of active, blowing sand, often appearing as a depression.

C_3 plant   Any plant that produces a 3-carbon compound as its first step in photosynthesis; this pathway is universal. Ecologically, C_3 plants exhibit minimal biochemical or physiological modifications for drought tolerance.

C_4 plant   Any plant that produces as the first step in photosynthesis a 4-carbon compound. C_4 plants have compartmentalized photosynthesis, with the effect of becoming more drought and heat tolerant because they lose less water through transpiration.

calcification   Soil formation in which soluble nutrients and salts are not removed from the upper layers of the soil. Characteristic of low-rainfall areas. Under extreme conditions a thick layer of salts ends up at the soil surface.

chronosequence   Time series.

class   In biological classification, a group within a phylum; made up of several to many orders.

climax community   Stable, final community in a region, when a steady state is reached and the species composition stays the same for long periods of time.

community   Group of species that coexist and interact in an area.

competition   Use or defense of an essential, limiting resource by one individual that reduces the availability of that limiting resource to other individuals.

continental drift   Movement of large land masses (tectonic plates) across the globe, over geologic time.

discrete generations   Populations (species) in which all the individuals alive at the same time are the same age (no carryover between generations).

ecosystem   Group of interacting organisms occurring together in association with exchanges with the physical environment. An ecosystem is often defined to be nearly self-contained so that the matter that flows into and out of it is small compared with the quantity that is internally recycled. Energy flows into and out of an ecosystem, with all energy originating from the sun in almost all cases.

ecotone   Boundary or transition between two or more different communities.

endemic   Organism restricted to a particular region.

endotherm   Organism capable of maintaining body temperature within a predetermined range as a result of the internal release of energy through metabolism.

environment   Aggregate of external biotic and abiotic conditions that influence the life of an individual or a population.

Eocene   Geologic time division, approximately 54 million years ago to about 27 million years ago. It is the second epoch in the Tertiary period, the Age of Mammals.

evapotranspiration   Evaporation plus transpiration, the two processes by which water is returned to the atmosphere and which results in water becoming limiting to plants.

exponential growth   In populations, the increasingly accelerated increase in the number of organisms present as a result of the increasing number of individuals added to the reproductive base. Populations increase exponentially much as money in a bank savings account increases through compound interest.

family   In biological classification, a group within an order; made up of several to many genera.

fauna   Animals; generally used to refer to the list of all animals found in an area.

flora   Plants; generally used to refer to the list of all plants found in an area.

forb   Herbaceous (nonwoody) plant that is not a grass. (Sometimes grasslike plants such as sedges are excluded, so forbs are herbaceous nongraminoids.)

fossorial   Adapted for burrowing or digging (e.g., fossorial small mammal).

fungus (plural, **fungi**)   Mostly nonmobile organism that is nonphotosynthetic (i.e., heterotrophic). Fungi live by breaking down large molecules to extract the energy. A member of the kingdom Fungi. Mushrooms, molds, and mycorrhizae are fungi.

gallery forest   Strip of forest along a river; also called riparian forest.

genus (plural, **genera**)   In biological classification, a group within a family; made up of several to many species.

glacial   Interglacial cycle.

grass   Member of the grass family (Poaceae or Gramineae); characterized by long linear leaves that are initiated underground.

grassland   Biological community containing few trees or shrubs; characterized by mixed herbaceous vegetation and dominated by plants of the grass family.

greenhouse gas   Gaseous form of an element or a compound (e.g., methane, carbon dioxide) contributing to trapping heat within the earth's atmosphere.

habitat   Environment, especially the immediate area, in which an organism lives.

heterotrophic organism   Organism that obtains food by consuming other organisms. They cannot manufacture food directly from air and water. Compare **autotrophic organism.**

High Plains   Area of central North America that generally refers to shortgrass prairie areas from western Texas to southeastern Saskatchewan, eastern Wyoming, and Montana. In general, this area is of fairly high elevation because of the uplift of the Rocky Mountains.

humus   Complex mixture of organic and inorganic compounds in soils, sufficiently broken up so that the source of the material is not recognizable. Humus serves as a major source of plant nutrients.

hybrid   Offspring of two parents who differ in one or more heritable characteristics.

hypha (plural, **hyphae**)   Filament of a fungus's vegetative body (thallus).

hypsothermal   Warmest part of a geologic time period.

inorganic compound   Compound that is not made up of a carbon–hydrogen skeleton. Calling a compound inorganic serves to indicate that it was synthesized by nonliving processes.

instar   Developmental stage in insect or other invertebrates between successive molts.

irruption   "Outbreak" of a species. Sudden, often unpredictable occurrence of very high densities (numbers) of a species.

landscape   Expanse of natural scenery seen by the eye at one view; in a more technical sense, an area with all its physical variation that contributes to ecological processes in an area. This notion emphasizes the spatially explicit heterogeneous processes responsible for actually occurring processes.

latent heat   Additional heat required to change a solid to a liquid or a liquid to a gas; also, heat "stored" within an object that can be released at a later time.

lateritic soils   Soils produced under conditions of leaching and high temperatures (called laterite). While not all tropical soils are lateritic, all lateritic soils are tropical or subtropical.

leaching   Downward movement of soluble nutrients and solids through a soil by water seeping through the soil.

litter   Animal and plant matter found on the soil surface, partly decayed but still having some recognizable structure; also, the name of that layer of the plant community and containing surface organisms such as very low flowering plants, mosses, and lichens.

loam   Soil consisting of a mixture of about 40% sand, 40% silt, and 20% clay-size particles.

loess   Soil produced by the deposition of windblown particles onto a standing grassland. Particles are very fine-textured, and good organic matter content is produced.

megafauna   Big animals, especially larger than 5 kilograms (size of a collie dog or badger); also, the listing of all such animals found in an area.

micelles   Tiny bubblelike spheres of one substance suspended in another.

model   Abstraction or simplification of a natural phenomenon developed to explain or predict a new phenomenon or to provide insights into existing ones.

mollisols   Prairie soils formed by calcification; characterized by the accumulation of calcium carbonate in lower horizons and high organic matter in upper horizons.

monocarpic   Reproducing only once in a lifetime, even if it lives many years; used only for plants. Compare **polycarpic.**

mutualism   Relationship between two individuals (usually members of different species) that benefits both.

mycorrhiza (plural, **mycorrhizae**)   Association of fungi with the roots of plants that improves the plants' uptake of nutrients from the soil.

N   Chemical symbol for nitrogen; often used to refer generally to nitrogen without specifying the form the element takes.

Neogene   Geologic time division; the later half of the Tertiary period, including the Miocene and Pliocene epochs.

nitrate   Compound form of nitrogen (NO_3).

order   In biological classification, a group within a class; made up of a number of families.

organic compound   Compound made of a carbon and hydrogen backbone. It can also contain other elements. On earth, these are all produced only by the action of living (''organic'') things.

overlapping generations   Population (or species) in which individuals of different ages are present at the same time, and in which individuals from successive generations are breeding at the same time.

Paleocene   Geologic time division; approximately 65 million years ago to 54 million years ago. It is the first epoch in the Tertiary period.

paleoenvironment   Environment of the past, often no longer in existence.

Paleogene   Geologic time division; the earlier half of the Tertiary period, including the Paleocene, Eocene, and Oligocene epochs.

paleosol   Buried soil or fossil soil; a soil horizon that was buried in the geologic past.

parasite   Organism that obtains its food by absorbing (eating, digesting) tissues of a living organism (the host), without killing the host organism. The relationship between parasite and host often lasts a long time.

parent material   Inorganic material, whether rock or river sand, that is the source of the mineral portion of a soil.

patch   Small isolated piece of ground distinguished from that about it by appearance or by the vegetation on it; also used for any other isolated area an organism recognizes or uses as different.

photosynthesis   Autotrophic transformation of sunlight into chemical energy usable by life by forming glucose from carbon dioxide and water with energy stored in the chemical bonds.

plate tectonics   Geologic mechanisms that are responsible for the movements of continents, known as continental drift.

Pleistocene   Unit of geologic time, 1.5 million years ago to 10,000 years ago. It is the first epoch in the Quaternary period and was characterized by recurrent intervals of glaciation (at least four).

podsolic soils   Soils formed under conditions in which precipitation exceeds evapotranspiration (leaching) and cool conditions so that a sandy, acidic, nutrient-poor soil is produced.

podsolization   Soil formation in podsolic soils.

polycarpic   Reproducing repeatedly in a lifetime; applied to plants not animals. Compare **monocarpic.**

polymorphism   Occurrence together in the same population of two or more morphologically or genetically distinct forms. Often used in population genetics to denote a group in which more than one version of a trait (allele) is present.

population   Basic unit of individuals within a species that interact or influence one another. These individuals have a high potential to breed with one another.

primary producers   Organisms such as plants that synthesize their own organic substances from inorganic substances.

primary succession   Ecological succession that begins in an area that has not been previously occupied by a community or organisms (i.e., from an area lacking soil).

Quaternary   Unit of geologic time that contains the Pleistocene and Recent epochs. This period began about 1.5 million years ago and is still going.

race   a subspecies; a morphologically distinct set of populations within the range of a species.

Recent   Unit of geologic time beginning about 10,000 years ago (i.e., since the last glacier retreated). Also known as the Holocene, it is the second epoch in the Quaternary period.

refugia (singular, **refugium**)   Refuges; areas to which individuals or populations of plants or animals retreat to escape problems, whether glaciers or predators.

respiration   Metabolic process in all living things by which large molecules are broken down into small molecules, generally water and carbon dioxide, releasing useful energy.

rhizome   Underground shoot (stem) of a plant from which leaves and roots develop.

riparian   Of rivers or streams; as in riparian forest, a forest along a river.

Rocky Mountain uplift   Rising of the land in the west-central United States to form the Rocky Mountain range.

sand   Mineral particle 0.2–2.0 millimeters in diameter.

saphrophage   Organism that eats dead and decaying matter.

savannah   Area of mixed forest and grassland, with sizable distances between the trees; technically, having less than 10% tree cover.

secondary succession Ecological succession that begins on an area, such as an abandoned field, that has been previously occupied by organisms (i.e., with a well-developed soil but no organisms).

shrub Woody plant, generally less than 2–3 meters (6–9 feet) tall, often with many trunks.

silt Soil particles between the sizes of sand particles and clay particles, 0.002–0.2 millimeter in diameter.

species (both singular and plural; abbreviated **sp.** for the singular, **spp.** for the plural) Group of populations of organisms forming a basic, unique evolutionary lineage that often interbreed freely with one another but not with any other.

species richness Number of species within a community or sample.

stolon Shoot (stem) of plant that travels aboveground to initiate a new plant (leaves, roots).

subspecies Group within a species that can be distinguished from others of the species (found in a particular area, somewhat different in size or color), but is able to breed freely with the others of the same species.

succession Predictable, sometimes orderly, sequence of changes in plant (and animal) communities that eventually leads to a community that changes only slightly except when disturbed.

symbiosis Intimate, often obligatory association of two species; usually involving complex adaptations of each organism to the others. Often symbioses are mutualistic, although they can be parasitic.

territory In animal behavior, any area that an animal defends.

Tertiary Unit of geologic time, 65 million years ago to 1.5 million years ago, that contains the Paleocene, Eocene, Oligocene, Miocene, and Pliocene epochs.

tiller Upright stem of grass.

transpiration Loss of water vapor from the leaves of plants.

trophic level Feeding level; level at which energy (food) is transferred from one organism to another.

true prairie Tallgrass prairie.

variety Subspecies.

volatilization Conversion of a material into a vapor.

Wisconsin ice Glaciers of the most recent glacial advance (Wisconsin), about 10,000 years ago.

woodland Area dominated by sparse trees; sometimes used synonymously with forest; sometimes reflecting very thin forest.

Subject Index

Aesthetics, 25, 42–45
Agriculture, 94, 125
Allelopathy, 136, 227
Ammonium, 136, 159, 164, 168
Annual plants, 83–84, 89–91, 116, 121, 136, 137, 213–214, 216
Anthropogenic effects, 189
Appleton–Whittell Research Sanctuary, 209–210
Aquatic habitats, 119, 121–123, 152

Baker, H. G., 214, 216
Bessey, C., 12
Bidwell, T. G., 61
Bierstadt, Albert, 36
Biodiversity, 150, 152, 153, 178, 199, 227
Biotic interactions, 100–124
Biomass, 121, 139, 140, 170, 184, 227
Black Elk, 28, 29, 30
Black Hills, 203
Bodmer, Karl, 34, 35, 36
Buffalo wallow, 205, 213
Bunchgrass prairie, 50, 52, 64, 136

C_3 plants. See Cool-season plants
C_4 plants. See Warm-season plants
Calcium, 68, 159, 164
Callicott, Baird, 29
Carbon dioxide, 104, 149–151, 163, 164
Castañeda (Spanish explorer), 29–31
Cather, Willa, 3, 38–39, 100
Cheyenne, 27
Chronosequence, 135, 137, 228
Clark, George Rogers, 31–34, 59, 199, 202
Clements, Frederick, 12, 21, 103, 135
Climate, 52
    change in, 150, 201–202
    grassland and, 49, 178–179, 181, 201
    patterns of, 54–56, 178–179
Climax, 135, 228
C:N ratio, 116, 150
Colonizing ability, 90, 137, 212
Colonizing species, 13, 90–99, 212, 213, 216
Colorado prairie, 144

Comanche, 27
Community, 128–151, 228
    bottom–up organization of, 121, 122
    composition of, 145
    core–satellite hypothesis of, 111, 148–149
    dynamic equilibria of, 146
    dynamics of, 129, 135, 138
    immigration–extinction model of, 146–147
    regulation of, 121, 122
    structure of, 123, 129, 141, 188–189, 190, 194
    top–down organization of, 121
Competition, 84–85, 91, 100, 103–113, 116, 119, 122, 137, 199, 228
    apparent, 103, 115–116
    belowground, 86–88
    Centrifugal Community Organization model of, 111, 112
    community-level consequences of, 111–112, 120, 122, 123
    competition–stress tolerator–ruderal model of, 109–111
    constraints on, 104
    diffuse, 116
    disturbance and, 90, 112–113
    effect of growth rates on, 109
    hierarchy model of, 111
    resource limitation and, 104, 107–109, 137
    resource-ratio model of, 107–109, 137
    trophic cascades and, 121–122
Competitive ability, 85, 109, 110, 111, 112, 137
Conservation, 3, 6, 13, 76–78, 152–153, 200, 202, 207, 216–219
    and grassland vulnerability, 200–202, 216
Conservation Reserve Program, 217, 219
Cool-season plants, 54, 56, 62, 65, 66–67, 76, 91, 150, 194
Core–satellite hypothesis, 111, 148–149
Coronado, Francisco de, 29–31
Corridors, 152, 193
Cover, 57, 63, 69, 188, 194, 202, 207–208, 211, 213
Crops, 212–214, 217, 219
Cultural perception, 25–26

Davis, M., 151
Decomposer/decomposition, 73, 74, 75, 129, 138,
    165–167, 171, 173, 178
Defoliation, 116
Degradation, 123, 195, 218
Demography, 92, 103
Density-dependence, 101, 103
Desert shrubland, 208, 211, 218
Desertification, 208
Digestibility, in vitro, 188, 190
Disease, 194
Dispersal, 83, 88
Disturbance, 89, 90, 107, 110, 112, 121, 123,
    128–137, 138, 142–145, 147, 152, 194,
    202, 213, 216. See also Community
Dormaar, J. F., 210
Drought, 20, 56, 57, 66, 86–88, 92, 94, 141, 158–
    159, 162, 178
Dust, 57
Dust Bowl, 202

Ecosystem, functioning of, 120, 164–173, 178–
    183, 188–191, 192, 196, 228
Ecotone, 132, 228
Edwards Plateau, 209
Endemic species, 20, 83, 134
Energy flow, 120, 121
Engle, D. M., 61
Eocene epoch, 131, 228
Equilibrium, 107, 108, 109, 145, 173, 192
Erosion, 50, 178, 192, 208, 210, 213, 217
European colonization, 57–58, 65, 66, 83, 202,
    213
European exploration, 29, 30, 39, 57, 212
Evaporation, 57, 75, 162, 163
Evapotranspiration, 54–55, 57, 229
    potential, 55–57, 159
Exotic species (vegetation), 17, 65, 116, 121, 153,
    200, 202, 212–216, 217, 219
Extinction, 11, 94, 120, 145–147, 148–149, 172,
    177, 202–203, 216
Extirpation, 71

Fencing, of large herbivores, 193–194
Fertilization, 184, 191, 218
Fire, 57–76, 91–93, 128, 141–142, 151–152, 192,
    206, 213, 216
    ash runoff and, 75
    backfire, 61, 77–78
    "behavior" of, 61
    drought and, 65, 128, 158
    effect of, 49, 56, 61–76, 88–89, 112–146, 167,
        199
        on animals, 68–74, 141, 202
        on fungi, 68, 74
        on grasslands, 61–68, 92
    on myccorhizae, 74
    on populations, 68–74, 91–92
    on production, 61, 62, 64, 65, 66, 67, 75, 77,
        138, 139, 171, 173
    on woody vegetation, 62, 63–66, 78, 138–
        139, 142, 158, 201–203
    erosion and, 76–77
    exotic species and, 213–214
    frequency of, 58, 65–68, 71, 77, 142, 158, 168,
        202–203
    grassland–desert transition and, 201
    grassland–forest transition and, 62, 201, 203,
        206
    grazing and, 62, 67–69, 142, 202–203
    headfire, 61, 77–78
    intensity of, 58, 61, 65, 75
    invertebrates and, 71–72, 92
    lightning-caused, 58–60, 62, 65, 75, 158, 201
    litter and, 65, 67, 75–76, 138–139, 167
    as management tool, 65, 68, 71, 77–78, 144,
        203, 218
    in mixed-grass prairie, 65–66, 203
    Native Americans and, 58, 65, 77, 158, 201
    plant disease and, 74
    in sandhills prairie, 66, 76, 145–146
    scale of, 138–142
    seasonality of, 58, 60, 62, 65, 66, 67, 70–71,
        77, 138, 145
    seedlings and, 65, 213
    in shortgrass prairie, 67, 144, 147
    short-term effects of, 141
    smoke emissions and, 76–77
    soil chemistry and, 68, 74–75, 138–139, 144,
        167, 168–169, 171
    soil moisture and, 73, 74, 75–76, 144
    soil organic matter and, 68, 74, 75, 138, 165,
        167, 168–171
    in southern mixed-grass prairie, 67
    species composition and, 62, 65, 66, 71–72,
        139, 140, 146, 170
    spread rate of, 61, 71
    suppression of, 66, 139, 202, 216
    in tallgrass prairie, 56, 62–65, 138–139, 142,
        145–146, 202
    temperature of, 58, 61, 72–73, 74, 75
    thunderstorms and, 59
Florida forest, 71
Food webs, 119–123
Forage, 68, 210, 212–214
Forbs, 62, 67, 84–92, 102, 105, 116, 117,
    137, 142, 206, 229. See also individual
    species
Forest (includes woodland), 52. 56–57, 74, 75,
    132, 133, 134, 152, 157, 158–159, 162–
    165, 194, 200, 201
Fragmentation, 13, 199

Frequency-dependence, 101, 103
Fugitive species, 101

Geology, processes of, 50, 131
Gill, S. D., 27
Gilmore, M. R., 26
Glacial periods, 32, 131–135, 158, 203, 229
Glaciation, 50, 58, 132, 133–134, 158–159, 200
Glacers, 131, 158, 164
Global climate, 52, 149–152, 159, 163, 181, 201
Gradients, 50, 54, 55, 102, 108, 112, 161–162
Grass cover, 188, 202, 207–208, 211, 213
Grassland
    age of, 131–135, 158–159, 200–202
    in Alberta, 64, 209–210
    annual, 16–17, 64
    in Arizona, 209, 210, 211, 214
    belowground, 73–74, 113–114
    boundaries of, 14–17, 50, 57, 199, 201
    in California, 16–18, 52, 53, 64, 116, 119
    in Colorado, 89
    community descriptions of, 15–16, 102
    definition of, 14–15, 50, 229
    degradation of, 178, 208–216
    desert, 16, 52, 53, 74, 118, 121, 159, 202, 207–
        208, 211
    eastern, 62, 161, 165, 172
    fescue, 209, 210
    herbivores in, 113, 206
    in Illinois, 62, 64, 71–72
    in Iowa, 62, 64
    in Kansas, 19, 29, 30, 31, 49, 56–57, 58, 62, 64, 71,
        84, 91, 105, 139, 140, 142, 145–147
    in Minnesota, 71
    in Missouri, 62, 89
    in Montana, 50, 204, 219
    in Nebraska, 49, 50, 56–57, 60, 62, 64, 67, 71,
        74
    in North Dakota, 66, 101, 102, 211
    northern, 64–65, 70, 75
    in Ohio, 74
    in Oklahoma, 56–57, 64, 67, 70, 143, 145, 148
    prehistoric, 57, 66, 206, 216
    preservation of, 3, 15, 152, 216–219
    sage, 202, 204
    sand plain, 137
    in Saskatchewan, 65, 75
    soils in, 50–52
    in South Dakota, 64, 65, 70
    southern, 57, 64, 67, 68
    species composition of, 19–21, 56, 57, 62, 65,
        114, 117, 120, 134–135, 145–148, 151
    in Texas, 56–57, 58, 92
        transition of, to desert, 18, 121, 201–202
        transition of, to forest, 13, 50, 62, 132, 201,
            203, 206

    in Washington, 74
    western, 50, 67, 165, 167, 212, 218
    in Wisconsin, 72
    in Wyoming, 210
Grazing
    belowground, 113, 114, 116–117, 165, 171
    continuous, 210–211
    deferred rotation, 186, 188, 190
    as disturbance, 89–91, 142–143
    by domestic livestock, 66, 184–187, 193–196,
        201–202, 207–209, 210–211, 212, 216–219
    drought and, 57, 207
    effects of
        on forbs, 83, 92–93, 193, 206
        on grasses, 92–93, 171, 188, 193, 200, 206,
            208, 216
        on nutrient cycling, 178–183, 189–190, 191,
            206
        on productivity, 116, 172, 181, 192, 207
        on soil, 88, 137, 158–159, 160, 171–172,
            179–180, 189, 210
    exclusion of, 209–210, 217–219
    high-performance, 186
    high-utilization, 187
    interaction of herbivores on, 206
    leases for, 217–218
    management of, 209, 211, 218–219
    overgrazing, 202, 207–208, 212–214
    plant response to, 116–117, 171, 178, 188, 206
    put-and-take, 192
    rest rotation, 186, 189, 190
    Savory system of, 178, 211
    short-duration, 187, 188, 189–190, 209–211
    species diversity and, 142, 178, 191
    species intolerant of, 212
    succession and, 137
    systems of, 184, 187, 209–211
    by ungulates, 113, 123, 191, 193–194, 207
    wildlife and, 68, 211, 216, 218
Grazing pressure, 184, 206, 210
Greenhouse gas, 159, 229
Grime, J. P., 109–112
Group living, 96, 107
Growth rates, of plants, 71, 75, 109, 110, 112,
    138, 182, 192

Habitat Structure, 68, 121, 141, 211, 229
Hart, R. H., 210
Herbicide, 216
Herbivory, 92–93, 100, 107, 113–118, 123, 177,
    191
    animal responses to, 114, 206–207, 211
    belowground, 113–114, 116–118
    and C:N ratio, 116, 150
    domestic herbivores and, 182–187, 193–194,
        202, 207–212, 216–219

Herbivory (*continued*)
    effect of, on primary productivity, 114, 180–183
    and herbivore production, 184–185, 209, 218
    insects and, 116, 122, 139
    invertebrates and, 114, 116, 171
    management of, 183–187, 188–193, 209
    nutrient cycling rates and, 178–184, 189–191
    plant population dynamics and, 88, 92–93, 114–115, 188
    plant response to, 116–117, 178, 192, 207, 209
    succession and, 117–118
Herpetofauna, 70–71
Higher-order interactions, 119–120
Holechek, J. L., 211
Holistic resource management, 211
Holocene epoch, 132
Hopkins, H. H., 89
Humus, 74, 164–165, 166, 168–172, 229
Hunting, 194
Hybrids, 213, 230
Hypsithermal period, 132–133, 158

Immigration, 71, 83, 145–147, 148–149, 152, 153
Increaser species, 189–190, 192
Indirect interactions, 88, 119, 121
Interaction webs, 88, 101, 120–123, 161
Introduced species, 116, 172, 202, 213–216
Invasion, 17, 62, 65, 78, 151, 153, 200

Jenny, H., 157, 165–166, 167
Jornada Experimental Range, 208–209

Kantrud, H. A., 211
Keystone species, 120–121, 128
Konza Prairie Research Natural Area, 139–140, 141, 142, 146, 147, 148

Lakota, 25, 26, 27, 28, 29, 30, 36, 39
Land use, 137, 139, 144, 158, 177, 194, 202
Landscape, 139, 149–153, 194–195, 216, 219, 230
Language, 3, 25–26
Latent heat, 162, 230
Leaching, 163, 230
Legends, 27–29
Leopold, A., 157
Lévi-Strauss, C., 26
Lewis, Meriwether, 31–34, 59, 199, 202
Life history, 83–84, 118
Light, 84, 104, 107–111, 114, 137, 157, 171, 202
Limiting resources, 104
Lippert, R. D., 89
Litter, 65–67, 68–69, 71, 73, 75, 138, 139, 158, 159, 165, 167, 169, 170, 210, 230
Little Bighorn National Battlefield, 204
Livestock, 207–212
    deferred rotation and, 186, 189–190
    grazing pressure of, 184
    rest rotation and, 186, 189–190
    short-duration grazing by, 187–190, 209–211
    stocking density of, 184, 189–190, 211, 218
    stocking rate of, 184, 189–90, 218
Lorrain, Claude, 31, 33–34

Magnesium, 68, 159
Manganese, 68
Marshall, W. H., 70
Maximilian (prince of Wied), 34
Megafauna, 177, 202–203, 216, 230
Microbes, 68, 74–75, 136, 157, 164, 165, 166–167, 169–171, 173, 178
Migration, 21, 133, 194–195
Milkunas, R. G., 207, 221
Miocene epoch, 131, 200
Mixed prairie (including midgrass), 16, 50–76, 86, 89, 118, 142, 143, 203
    northern, 50, 64, 70, 75
    southern, 50, 64, 67
Moisture index, 56
Monocarpic plant species, 83
Mutualism, 100, 120
Myccorhizae, 87, 231

National Audubon Society, 209–210
National grasslands, 219
Native Americans, 25, 26–29, 36, 58, 65, 72, 212
Natural selection, 114
Nebraska sandhills prairie, 6, 11, 12, 16, 49, 50, 64, 66, 76, 88, 90, 118–119, 145, 160
Nitrate, 168, 231
Nitrogen, 74, 75, 105, 108, 109, 111, 116, 136, 137, 138, 139, 142, 151, 158, 164–171, 172–173, 188, 206
Nitrogen fixation, 74, 75, 169, 170
Nitrous oxide, 169
Nutrient cycles and dynamics, 120, 128, 136, 157–173, 178–181, 189–190, 191–192
Nutrients, 68, 74–75, 84, 87, 104, 107, 109, 110, 116, 136, 157, 159–162, 163, 171, 172, 178–181, 191–192, 206, 208–209

Oklahoma prairie, 145, 148
Oligocene epoch, 131
Omaha–Ponca, 26, 27

Paleontology, 131, 201
Pasture, 184–187, 209
Patch
    dynamics of, 89, 113, 114, 138, 139, 142, 144, 145, 149, 153, 182, 192
    structure of, 182, 192–193
Pawnee, 27, 30

Perceptions, of Plains
  georgic, 25, 39
  epic myth, 37–39
  as ocean, 25, 37–39
  picturesque, 25, 31–39
Permanent plots, 146
Perturbation, 112, 122, 200, 216
Phosphorus, 68, 87, 170
Photosynthesis, 54, 71, 87, 132, 158, 168, 180, 192, 232
Plant mortality, 83–84, 115
Pleistocene epoch, 132, 151, 158, 177, 200, 202, 232
Pliocene epoch, 131, 200
Pollen analysis, 21, 131, 132, 133
Pollination, 84, 87, 92, 100, 116
Polycarpic plant species, 83, 232
Populations, 232
  belowground, 88, 89, 117
  clonal reproduction of, 84–85, 86
  demography of, 82–93
  effects of grazing on, 87, 92–93, 114–115, 117
  fire and, 68–74, 91–93, 139
  of forbs, 84–92
  of grasses, 92
  herbivory and, 88, 114–115, 122
  life histories of, 83
  of perennial plants, 83–84, 89, 216
  rhizomes as link among, 85, 91, 207
  seed bank and, 86, 88, 89–90, 113
  seedling establishment and, 86–87
  sexual reproduction of, 84–88
  small-scale disturbance of, 89–91
  vegetative reproduction of, 84–86, 216
Potassium, 68, 159, 164
Pound, Roscoe, 11, 12, 13
Prairie, remnants of, 13, 62, 71, 151, 206, 207
Prairie dogs, 6, 93, 95, 113, 114, 123, 203, 206, 207, 213
Prairie Peninsula, 6
Precipitation, 17, 52–55, 56, 57, 65, 75, 145, 150, 151, 160–161, 162–163, 178, 201–202, 209
  nutrient dynamics and, 128, 163
  productivity and, 180, 185, 207
  rain shadow and, 53
  soil formation and, 161–162
Predation, 68, 73, 88, 93, 95–96, 100, 118–120, 121, 122, 177, 194, 206, 212
Preservation, 3, 152–153, 207, 212
Productivity, 64, 91, 112, 122, 150–151, 160–161, 163–165, 166–167, 171–173, 177, 181–189, 190–193, 196, 209–210, 211, 232
  primary, 122, 218, 232
  secondary, 183, 185, 188–191
  soil, 159, 165, 167, 171
Propagules, 138

Pulse-phase dominance, 145

Quaternary period, 131, 232

R*, 107, 109
Rabinowitz, D., 89
Ramets, 84–85
Recruitment, 118
Refuge (refugium), 111, 134, 182, 193, 212, 232
Rescue effect, 151
Resilience, 17
Resources
  allocation of, 105–107
  light, as limiting, 104, 107–111
  limiting, 104, 105, 107, 112, 116, 166, 173
  soil, 114
  trade-offs of, 105–107
Restoration, 21, 152, 157, 172–173, 213, 218
Reynolds, Sir Joshua, 34
Rhizome, 85, 90–91
Riparian habitat, 212, 232
Rocky Mountains, 51–53
Rölvaag, Ole, 39
Roots, 87, 104–107, 116, 117, 157, 159, 163, 165, 167–169, 171, 180
Root:shoot ratio, 105–106
Runoff, 57, 75, 159
Rydberg, Per, 12

Sandoz, Marie, 3
Sandsage prairie, 145
Sandford, Dorsey, 38
Scale, 129, 130, 148, 195
Seed bank, 88–89
Seeds, 69, 83–84, 86, 88, 91–92, 138, 139, 158, 184, 200, 201, 214, 216–217
  predation of, 88, 114, 121
  production of, 86, 114, 216
Seedlings, 86–87, 110, 200, 212
Sensible heat, 163
Settlement, of Plains, 37–39
Sexual reproduction, 86–88
Shade, 107, 110, 115, 167, 191, 194
Shelter, 94, 183, 184, 193
Shortgrass prairie, 16, 17, 49–76, 86, 89–90, 94, 105, 117, 144–147, 150, 169, 193
Shrublands, 163, 202, 208
Shrub-steppe, 50, 193
Sioux, 27
Smith, Jared, 11, 12, 13
Sod, 211–213
Sodgrass, 207, 209
Soil
  Alfisols, 51, 52
  Aridisols, 51, 52
  carbon storage in, 158, 161, 163, 164, 165,

172–173, 178–180
chemistry of, 74, 159, 161, 162, 164, 167, 168–
    169, 180
clay in, 67, 158, 159
climate and, 160–163, 165, 178–180
compaction of, 159, 179–180, 181, 188–189,
    190, 210
disturbance of, 83, 89–91, 101
effect of drought on, 158
effects of grazing on, 117
Entisols, 50, 51
fertility of, 138, 139, 150, 157, 158, 160, 163,
    164, 165–170, 171, 209
fire and, 66, 73–76, 138–139, 157, 165, 167–170
formation of, 50, 114, 138, 157, 158, 159, 163–
    167, 170–171, 172
geologic history of, 50, 158–159
infiltration in, 159–160
invertebrates in, 71
leaching in, 163, 166, 168, 169, 230
loess, 50, 158
microbes in, 136, 157, 164, 165, 166–167, 169–
    171, 173, 178
mineralization of, 150, 168–169, 178, 180
moisture in, 54, 55, 73–75, 102, 114, 144, 160,
    162, 164–165, 166, 210
Mollisols, 50, 51, 158, 167
nitrogen in, 75, 107–111, 136, 137, 138–139,
    144, 150, 165, 166, 168–169, 173
nitrogen fixation in, 169, 170
nutrient cycles and, 171, 178, 189–191
nutrients in, 68, 87, 114, 163, 167
orders of, 50–52
organisms in, 164–165, 166, 169, 171, 183
paleosols, 158, 231
pore space of, 159, 179–180
sand in, 66–67, 158, 159, 232
stabilization of, 212–214
temperature of, 68, 73, 74, 75, 138, 164, 166,
    167
texture of, 158–160, 161
types of, 50–51
Vertisols, 50, 51, 52
Solar energy, 162, 163, 172
Spatial mosaics, 85, 113, 123, 207, 215, 216
Species
    diversity of, 139–142, 151, 161, 178, 188, 190,
        192, 214–215
    richness of, 117, 129, 140–141, 143, 147, 151,
        233
Stella, Frank, 44

Stolons, 207, 233
Stomates, 105
Storage effect, 113
Succession, 108–109, 117–118, 135–138, 161,
    207, 214, 216, 233
    old-field, 135, 136
    postfire, 57, 135, 213
    primary, 135
Sulfur, 75

Tallgrass prairie, 5–6, 49–76, 89, 91, 93–94, 116,
    117, 129, 132, 135, 136, 137, 138–139,
    142–143, 145, 150, 152, 171
Temperature, 50, 52, 54, 68, 71, 73, 95, 178, 194,
    201–202
Tester, J. E., 70
Texas Experimental Ranch, 210
Thornwaite, C. W., 56
Tillers, 84–85, 87, 91–93, 138, 181
Tilman, David, 107–109, 112, 137, 170
Trade-offs, 105–108, 110, 137
Trampling, 68, 178, 179–182
Transpiration, 54, 56, 233
Trophic cascade, 121–122
True prairie. *See* Tallgrass prairie
Tundra, 133, 166

U. S. Bureau of Land Management, 217
U. S. Forest Service, 217, 219

Variability, of community, 142–143
    productivity and, 181, 183
Vegetative reproduction, 84–86

Walker, J. R., 28
Warm-season plants, 54, 56, 65–67, 91, 138, 150,
    170, 194
Weaver, John E., 12, 57, 89
Weeds, 88–89, 137, 151, 172, 188, 213–214, 216
Weltz, M., 210
Whorf, Benjamin, 25
Wichita, 27–29, 143
Wilder, L. J., 3
Wildlife, 210–212, 217–218
Wind Cave National Park, 203, 205, 206
Wister, Owen, 38
Wittgenstein, Ludwig, 37
Wood, M. K., 210
Woody plants, 62–63, 66, 78

Zinc, 68

Species Index

Algae, 74, 102
Amaranthaceae, 21
Ambrosia, 21
Amphibian, 70, 71
Ant, 4, 6, 71, 72, 73, 89, 113, 118
Antelope. *See* Pronghorn antelope
Artemisia, 21, 33
Arthropod, 117, 159
Ash (*Fraxinus*), 33, 62, 131
Aster (*Aster* spp.), 62, 102
    white (*Aster ericoides*), 62, 65
Auchenorrhyncha, 117

Bacteria, 68, 74, 117, 136, 164, 178, 181
Badger, 113
Baltic rush, 102
Bass, 119
Beetle (Coleoptera), 71, 72, 73, 84, 88, 91, 183
Bergamot, wild (*Monarda Fistulosa*), 91
Birch, paper (*Betula papyrifera*), 133
Bird, 20, 58, 61, 69, 100–101, 118, 119, 121, 149,
    181, 201, 211, 217
Bison
    Bison antiquus, 132
    Bison bison, vii, 13, 26, 34, 58, 66, 68, 69, 78,
        83, 93–94, 96, 138, 142, 169, 172, 177, 193–
        194, 199, 202–203, 205, 206, 207, 213, 216,
        219
Blackbird
    red-winged (*Agelaius phoeniceus*), 70
    yellow-headed (*Xanthocephalus
        xanthocephalus*), 70
Bladderwort, 102
Blowout penstemon (*Penstemon haydeni*), 83
Bluegill, 119
Bluegrass (*Poa*), 26
    Kentucky (*Poa praetensis*), 65, 66, 67
Blue-green algae (cyanobacteria), 74
Bluestem, 26, 208,
    big (*Andropogon gerardii*), 15, 38, 62, 65, 84,
        102, 107, 111, 134, 137, 162
    little (*Schizachyrum scoparium*), 16, 65, 67, 68,
        111, 136, 137, 145, 146, 180, 182, 209
    sand (*Andropogon hallii*), 66, 67

Bobolink (*Dolichonyx oryzivorus*), 70, 211, 217
Brome
    downy (*Bromus tectorum*), 213
    smooth (*Bromus inermis*), 65
Bromegrass, annual *(Bromus* sp.), 16
Brucellosis, 194
Buffalo. *See* Bison
Buffalo grass (*Buchloë dactyloides*), 16, 19, 30,
    34, 57, 67, 90, 105, 207, 209
Bulrush, 102
Bunchgrass, 85, 92, 207–209
Butterfly, 71, 100
    buckeye, 151

Candle anenome, 102
Cattle (*Bos domesticus*), 18, 68, 77, 142, 165,
    178, 183, 184–187, 191, 192, 193, 202, 207,
    209, 210–211, 217, 218, 219
Centipede, 73
Cheatgrass, 213
Chenopodiaceae, 21
Clover, sweet (*Melilotus* spp.), 65, 84
Cocklebur (*Xanthium strumarium*), 33
Compositae, 21
Coneflower
    prairie (*Ratibida columnifera*), 91
    purple (*Echinacea purpurea*), 65
Cordgrass, prairie (*Spartina pectinata*), 62
Corn. *See* Maize
Cottonwood (*Populus deltoides*), 33
Cowbird, brown-headed (*Molothrus ater*), 70
Coyote (*Canis latrans*), 93
Cranefly (Diptera), 73
Creosote bush (*Larrea tridentata*), 16
Cryptogram, 74

Deer, 177, 183, 193, 194
    mule, 211
    white-tailed (*Odocoileus virginianus*),
        211
Deschampsia, 183
Dickcissel (*Spiza americana*), 217
Digger wasp (Hymenoptera), 117
Dogwood, rough-leaved (*Cornus drummondii*), 62

Dropseed, sand (*Sporobolus cryptandrus*), 66, 84
Duck, 70

Earthworm, 71, 73, 159, 164, 165
Elk (*Cervus canadensis*), vii, 13, 34, 69, 177, 193, 194, 199
Elm (*Ulmus* spp.), 62
Erodium sp., 116
Evening primrose, Missouri (*Oenothera missouriensis*), 89

Ferret, black-footed (*Mustela nigripes*), 83, 295
Fescue, rough (*Festuca scabrella*), 210
Fish, 20, 122
Flax (*Linum* sp.), 30, 214
Flea hopper, 94
Fly, 100
Four o'clock (*Mirabilis hirsuta*), 83
Fungi, 68, 74, 84, 87, 88, 164

Gall-forming insects, 92
Gayfeather, tall (*Liatris aspera*), 65
Golden alexander, 102
Goldenrod
    Canada (*Solidago canadensis*), 90, 105
    soft, 105
Goose, 177, 183
Gopher. *See* Pocket gopher
Gorrilla, 183
Grama, sideoats (*Bouteloua curtipendula*), 57, 67, 209
Grama grass
    annual (*Bouteloua breviseta*), 114
    black (*Bouteloua eriopoda*), 208
    blue (*Bouteloua gracilis*), 16, 19, 30, 34, 57, 66, 90, 93, 105, 107, 117, 134, 207, 209, 210
    hairy (*Bouteloua hirsuta*), 209
Grama grasses (*Bouteloua* spp.), 16, 26, 183
Grape (*Vitus* spp.), 30
Grasshopper (Orthoptera), 18, 71, 72, 73, 84, 88, 100, 118, 139, 141, 149, 177, 183, 194
Ground squirrel, thirteen-lined (*Spermophilus tridecemlineatus*), 69, 141
Grouse, sharp-tailed (*Tympanuchus phasianellus*), 70

Hackberry (*Celtis* spp.), 131
Harvest mouse, western (*Reithrodontomys megalotis*), 69
Horned lark (*Eremophila alpestris*), 70, 211
Horse (*Equus equus*), 193

Indiangrass (*Sorghastrum nutans*), 15, 62, 65
Insect, 20, 58, 68, 71, 114, 117, 141, 181

Invertebrate, 61, 71, 114, 118, 122, 159, 164, 165, 166
Ironweed, western (*Vernonia baldwinii*), 91

June beetle, 91, 117
Junegrass (*Koeleria pyrimidata*), 91
Juniper. *See* Red cedar, eastern
Kangaroo rat, 121
Killdeer (*Charadrius vociferans*), 70, 211
Kingbird, western (*Tyrannus verticalis*), 70

Lamb's quarters (*Chenopodium album*), 33
Lark bunting (*Calamospiza melanocorys*), 217
Leadplant (*Amorpha canescens*), 62, 84
Leafhopper (Homoptera), 71
Lichen, 74
Lizard, 118
Longspur, chestnut-collared (*Calcarius ornatus*), 70, 211
Lovegrass
    African, 213, 215
    Boer's, 214
    Lehmann's, 214
    plains (*Eragrostis intermedia*), 209, 214
    sand (*Eragrostis trichodes*), 66

Maize (*Zea mays*), 27
Malvastrom, 105
Mammal, 20, 69, 141, 149, 183
Maple (*Acer* spp.), 131, 132
    sugar (*Acer saccharum*), 132
Marjoram, wild, 30
Meadowlark, western (*Sturnella neglecta*), 70
Melanoplus keeleri, 72
Melanoplus scudderi, 72
Melanoplus spretus, 72
Mesquite (*Prosopis glandulosa*), 16, 201, 208
    curley (*Hilaria belangeri*), 209
Microbe, 117, 164, 165, 166, 167, 169, 170, 178
Millipede, 73
Mint, 102
Mite, 73
Morning glory, bush (*Ipomoea leptophylla*), 84
Moss, 74
Mouse, 6
    deer (*Peromyscus maniculatus*), 69, 211
    white-footed, 69
Muhly
    Richardson's, 102
    sand (*Muhlenbergia pungens*), 66
Mulberry (*Morus* sp.), 30

Needle-and-thread grass (*Stipa comata*), 65, 66, 102

Needle grasses (*Stipa* spp.), 16, 208
Nematode, 114, 117, 159, 165, 181

Oak (*Quercus* spp.), 62, 131
Opossum (*Didelphis virginiana*), 94
Orchid (*Spiranthes cernua*), 84
Orphullela, 72

Peccary, 131
Pennyroyal (*Hedeoma*), 30
Phoetaliotes nebrascensis, 72
Phyllophaga sp. (Coleoptera), 72
Phytoplankton, 122
Pine
   jack (*Pinus banksiana*), 132, 134
   ponderosa (*Pinus ponderosa*), 202
Plantain
   English (*Plantago lanceolata*), 151
   sea, 183
Pocket gopher (*Geomys* sp.), 6, 101, 113, 114,
   123, 138, 177
Porcupine grass (*Stipa spartea*), 26, 62, 65, 102
Prairie chicken, lesser (*Tympanuchus
   pallidicinctus*), 70
Prairie dog, black-tailed (*Cynomys ludovicianus*,
   6, 38, 93, 95, 113, 114, 123, 203, 206, 207,
   213
Prairie vole (*Microtus ochrogaster*), 69
Prickl(e)y pear (*Opuntia* spp.), 33
Pronghorn antelope (*Antilocapra americana*), 13,
   34, 38, 69, 193, 199, 219
Puccoon, Carolina (*Lithospermum caroliniense*),
   84

Quackgrass (*Agropyron repens*), 102
Quail
   bobwhite (*Colinus virginianus*), 70
   montezuma (*Cyrtonyx montezumae*), 211, 212
   scaled (*Callipepla squamata*), 211

Rabbit, 69
Ragweed, 21
   giant (*Ambrosia trifida*), 89
   little (*Ambrosia artemisifolia*), 89
   western (*Ambrosia psilostachya*), 67, 105
Rattlesnake, western (*Crotalus viridis*), 94
Red cedar, eastern (*Juniperus virginiana*), 62, 201
Rhinocerus, 131
Richweed (*Pilea* sp.),
Ringneck snake (*Diadophis punctatus*), 71
Robber fly (Diptera), 118, 119
Ryegrass (*Lolium* sp.), 30

Sage, 33, 102
Sagebrush, big (*Artemisia tridentata*), 16, 38, 204

Sandpiper, upland (*Bartramia longicauda*), 70
Sandreed, prairie (*Calamovilfa longifolia*), 16
Sarcophagid (Diptera), 118
Scelionid (Hymenoptera), 118
Scurfy pea, 105
Sedge (*Carex* spp.), 66
Sheep, 117, 193
Shrew, short-tailed (*Blarina brevicauda*), 69
Skink, 71
Sloughgrass (*Beckmannia syzigachne*), 26
Snake, 71
Southern bog lemming, 69
Sow thistle, 102
Sparrow
   Baird's (*Ammodramus bairdi*), 217
   Botteri's (*Aimophila botterii*), 211
   Cassin's (*Aimophila cassinii*), 211
   clay-colored (*Spizella pallida*), 70, 217
   grasshopper (*Ammodramus savannarum*), 70,
      211, 217
   lark (*Chondestes grammacus*), 211
   Le Conte's (*Ammospiza lecontei*), 70
   savannah (*Passerculus sandwichensis*), 70,
      211
   vesper (*Pooecetes gramineus*), 70
Spider, 71, 73, 118
Sprangletop marshgrass, 102
Springtail (Collembola), 73
Spruce (*Picea* spp.), 132, 134
Spurge, mat (*Euphorbia glyptosperma*),
   89
Stickleaf (*Mentzelia nuda*), 90
Stickseed (*Lappula reddowskii*), 89
Sumac (*Rhus glabra*), 62, 183
Sunflower, 62
   annual (*Helianthus annuus*), 136
Sweet-grass (*Hierocholoe odorata*), 26, 28
Switchgrass (*Panicum virgatum*), 15, 62, 67,
   183
Sycamore (*Platanus* spp.), 131

Tarbush (*Flourensia cernua*), 208
Thistle
   Canada (*Cirsium arvense*), 214, 216
   musk, 65
   Platte (*Cirsium plattensis*), 88
   Russian (*Salsola iberica*), 214
Three-awn grass (*Aristida*), 73
   Fendler, 105
Ticklegrass, 111
Timothy (*Phleum pratense*), 33
Tortoise, 131
Tragopogon, 65
Tumbleweed. *See* Thistle, Russian
Turtle, 118

Ungulate, 123, 178, 180, 191, 193, 194

Walnut (*Juglans niger*), 33
Wheatgrass (*Agropyron*), 16, 150, 183, 208, 213
  bluebunch (*Agropyron spicatum*, 16
  crested (*Agropyron cristatum*), 92, 213
  western (*Agropyron smithii*), 51, 65, 93, 213
White grub. *See* June beetle
Willow, 33

Wintergrass, Texas (*Stipa leucotricha*),
  209
Wolf (*Canis lupus*), vii, 128, 216
Wolftail grass (*Lycurus phleoides*), 209

Yarrow, western (*Achillea millifolium*), 85
Yellowthroat (*Geothlypis trichas*), 70

Zooplankton, 122